Big Data Made Easy

A Working Guide to the Complete
Hadoop Toolset

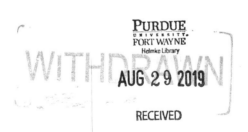

Michael Frampton

Apress®

Big Data Made Easy: A Working Guide to the Complete Hadoop Toolset

ISBN-13 (pbk): 978-1-4842-0095-7

ISBN-13 (electronic): 978-1-4842-0094-0

Managing Director: Welmoed Spahr
Acquisitions Editor: Jeff Olson
Developmental Editor: Linda Laflamme
Technical Reviewer: Andrzej Szymanski
Editorial Board: Steve Anglin, Mark Beckner, Gary Cornell, Louise Corrigan, James DeWolf, Jonathan Gennick, Robert Hutchinson, Michelle Lowman, James Markham, Matthew Moodie, Jeff Olson, Jeffrey Pepper, Douglas Pundick, Ben Renow-Clarke, Gwenan Spearing, Matt Wade, Steve Weiss
Coordinating Editor: Rita Fernando
Copy Editor: Carole Berglie
Compositor: SPi Global
Indexer: SPi Global

Distributed to the book trade worldwide by Springer Science+Business Media New York, 233 Spring Street, 6th Floor, New York, NY 10013. Phone 1-800-SPRINGER, fax (201) 348-4505, e-mail orders-ny@springer-sbm.com, or visit www.springeronline.com. Apress Media, LLC is a California LLC and the sole member (owner) is Springer Science + Business Media Finance Inc (SSBM Finance Inc). SSBM Finance Inc is a Delaware corporation.

For information on translations, please e-mail rights@apress.com, or visit www.apress.com.

Apress and friends of ED books may be purchased in bulk for academic, corporate, or promotional use. eBook versions and licenses are also available for most titles. For more information, reference our Special Bulk Sales–eBook Licensing web page at www.apress.com/bulk-sales.

Any source code or other supplementary materials referenced by the author in this text is available to readers at www.apress.com. For detailed information about how to locate your book's source code, go to www.apress.com/source-code/.

This book is dedicated to my family—to my wife, my son, and my parents.

Contents at a Glance

Contents

About the Author

Michael Frampton has been in the IT industry since 1990, working in a variety of roles (tester, developer, support, QA) and many sectors (telecoms, banking, energy, insurance). He has also worked for major corporations and banks as a contractor and a permanent member of staff, including Agilent, BT, IBM, HP, Reuters, and JPMorgan Chase. The owner of Semtech Solutions, an IT/Big Data consultancy, Mike Frampton currently lives by the beach in Paraparaumu, New Zealand, with his wife and son. Mike has a keen interest in new IT-based technologies and the way that technologies integrate. Being married to a Thai national, Mike divides his time between Paraparaumu or Wellington in New Zealand and their house in Roi Et, Thailand.

About the Technical Reviewer

 **Andrzej Szymanski** started his IT career in 1992, in the data mining, warehousing, and customer profiling industry, the very origins of what is big data today. His main focus has been data processing and analysis, as well as development, systems, and database administration across all main platforms, such as IBM Mainframe, Unix, and Windows, and all leading DBMSs, such as Sybase, Oracle, MS SQL, and MySQL. Szymanski's big data and DevOps adventure began in News International, in January 2011, where he was a key player in creating a fully scalable and distributable big data ecosystem, with an aim of sharing it with subsidiaries of News Corporation. This involved R&D, solution architecture, creating ETL workflows for big data, Continuous Integration Zero Touch deployment mechanisms, and system administration and knowledge transfer to sister companies, to name but few of the key areas. Szymanski was born in Poland, where he completed his primary and secondary education. He studied economics in Moscow, but his key passion has always been computers. He is currently based in Prague.

Acknowledgments

I would like to thank my wife and son for allowing me the time to write this book. Without your support, Teeruk, developing this book would not have been possible.

I would also like to thank all those who gladly answered my technical questions about the software covered in this book. I extend my gratitude to the Apache and Lucene organizations, without whom open-source-based projects like this one would not be possible. Also, specific thanks go to Deborah Wiltshire (Cloudera); Diya Soubra (ARM); Mary Starr (Nagios); Michael Armbrust (Spark); Rebecca G. Shomair, Daniel Bechtel, and Michael Mrstik (Pentaho); and Chris Taylor and Mark Balkenende (Talend).

Lastly, my thanks go to **Andrzej Szymanski**, who carried out a precise technical check, and to the editorial help afforded by Rita Fernando, Jeff Olson, and Linda Laflamme.

Introduction

If you would like to learn about the big data Hadoop-based toolset, then *Big Data Made Easy* is for you. It provides a wide overview of Hadoop and the tools you can use with it. I have based the Hadoop examples in this book on CentOS, the popular and easily accessible Linux version; each of its practical examples takes a step-by-step approach to installation and execution. Whether you have a pressing need to learn about Hadoop or are just curious, *Big Data Made Easy* will provide a starting point and offer a gentle learning curve through the functional layers of Hadoop-based big data.

Starting with a set of servers and with just CentOS installed, I lead you through the steps of downloading, installing, using, and error checking. The book covers following topics:

- Hadoop installation (V1 and V2)

- Web-based data collection (Nutch, Solr, Gora, HBase)

- Map Reduce programming (Java, Pig, Perl, Hive)

- Scheduling (Fair and Capacity schedulers, Oozie)

- Moving data (Hadoop commands, Sqoop, Flume, Storm)

- Monitoring (Hue, Nagios, Ganglia)

- Hadoop cluster management (Ambari, CDH)

- Analysis with SQL (Impala, Hive, Spark)

- ETL (Pentaho, Talend)

- Reporting (Splunk, Talend)

As you reach the end of each topic, having completed each example installation, you will be increasing your depth of knowledge and building a Hadoop-based big data system. No matter what your role in the IT world, appreciation of the potential in Hadoop-based tools is best gained by working along with these examples.

Having worked in development, support, and testing of systems based in data warehousing, I could see that many aspects of the data warehouse system translate well to big data systems. I have tried to keep this book practical and organized according to the topics listed above. It covers more than storage and processing; it also considers such topics as data collection and movement, scheduling and monitoring, analysis and management, and ETL and reporting.

This book is for anyone seeking a practical introduction to the world of Linux-based Hadoop big data tools. It does not assume knowledge of Hadoop, but it does require some knowledge of Linux and SQL. Each command use is explained at the point it is utilized.

Downloading the Code

The source code for this book is available in ZIP file format in the Downloads section of the Apress website, www.apress.com.

Contacting the Author

I hope that you find this book useful and that you enjoy the Hadoop system as much as I have. I am always interested in new challenges and understanding how people are using the technologies covered in this book. Tell me about what you're doing!

You can find me on LinkedIn at www.linkedin.com/profile/view?id=73219349.

In addition, you can contact me via my website at www.semtech-solutions.co.nz or by email at mike_frampton@hotmail.com.

■ ■ ■

## CHAPTER 1

The Problem with Data

The term "big data" refers to data sets so large and complex that traditional tools, like relational databases, are unable to process them in an acceptable time frame or within a reasonable cost range. Problems occur in sourcing, moving, searching, storing, and analyzing the data, but with the right tools these problems can be overcome, as you'll see in the following chapters. A rich set of big data processing tools (provided by the Apache Software Foundation, Lucene, and third-party suppliers) is available to assist you in meeting all your big data needs.

In this chapter, I present the concept of big data and describe my step-by-step approach for introducing each type of tool, from sourcing the software to installing and using it. Along the way, you'll learn how a big data system can be built, starting with the distributed file system and moving on to areas like data capture, Map Reduce programming, moving data, scheduling, and monitoring. In addition, this chapter offers a set of requirements for big data management that provide a standard by which you can measure the functionality of these tools and similar ones.

A Definition of "Big Data"

The term "big data" usually refers to data sets that exceed the ability of traditional tools to manipulate them—typically, those in the high terabyte range and beyond. Data volume numbers, however, aren't the only way to categorize big data. For example, in his now cornerstone 2001 article "3D Management: Controlling Data Volume, Velocity, and Variety," Gartner analyst Doug Laney described big data in terms of what is now known as the *3Vs*:

- **Volume**: The overall size of the data set

- **Velocity**: The rate at which the data arrives and also how fast it needs to be processed

- **Variety**: The wide range of data that the data set may contain—that is, web logs, audio, images, sensor or device data, and unstructured text, among many others types

Diya Soubra, a product marketing manager at ARM a company that designs and licenses microprocessors, visually elaborated on the 3Vs in his 2012 datasciencecentral.com article "The 3Vs that Define Big Data." He has kindly allowed me to reproduce his diagram from that article as Figure 1-1. As you can see, big data is expanding in multiple dimensions over time.

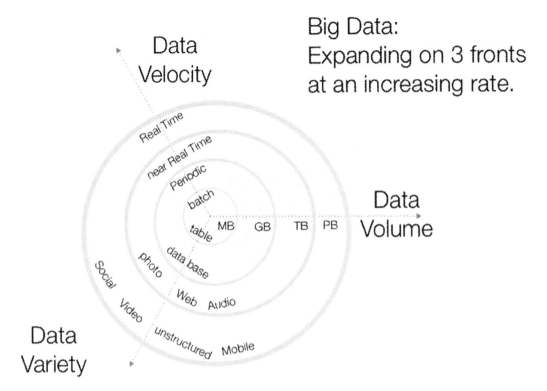

Figure 1-1. *Diya Soubra's multidimensional 3V diagram showing big data's expansion over time*

You can find real-world examples of current big data projects in a range of industries. In science, for example, a single genome file might contain 100 GB of data; the "1000 Genomes Project" has amassed 200 TB worth of information already. Or, consider the data output of the Large Hadron Collider, which produces 15 PB of detector data per year. Finally, eBay stores 40 PB of semistructured and relational data on its Singularity system.

The Potentials and Difficulties of Big Data

Big data needs to be considered in terms of how the data will be manipulated. The size of the data set will impact data capture, movement, storage, processing, presentation, analytics, reporting, and latency. Traditional tools quickly can become overwhelmed by the large volume of big data. Latency—the time it takes to access the data—is as an important a consideration as volume. Suppose you might need to run an ad hoc query against the large data set or a predefined report. A large data storage system is not a data warehouse, however, and it may not respond to queries in a few seconds. It is, rather, the organization-wide repository that stores all of its data and is the system that *feeds into* the data warehouses for management reporting.

One solution to the problems presented by very large data sets might be to discard parts of the data so as to reduce data volume, but this isn't always practical. Regulations might require that data be stored for a number of years, or competitive pressure could force you to save everything. Also, who knows what future benefits might be gleaned from historic business data? If parts of the data are discarded, then the detail is lost and so too is any potential future competitive advantage.

Instead, a parallel processing approach can do the trick—think divide and conquer. In this ideal solution, the data is divided into smaller sets and is processed in a parallel fashion. What would you need to implement such an environment? For a start, you need a robust storage platform that's able to scale to a very large degree (and at reasonable cost) as the data grows and one that will allow for system failure. Processing all this data may take thousands of servers, so the price of these systems must be affortable to keep the cost per unit of storage reasonable. In licensing terms, the software must also be affordable because it will need to be installed on thousands of servers. Further, the system must offer redundancy in terms of both data storage and hardware used. It must also operate on commodity hardware, such as generic, low-cost servers, which helps to keep costs down. It must additionally be able to scale to a very high degree because the data set will start large and will continue to grow. Finally, a system like this should take the processing to the data, rather than expect the data to come to the processing. If the latter were to be the case, networks would quickly run out of bandwidth.

Requirements for a Big Data System

This idea of a big data system requires a tool set that is rich in functionality. For example, it needs a unique kind of distributed storage platform that is able to move very large data volumes into the system without losing data. The tools must include some kind of configuration system to keep all of the system servers coordinated, as well as ways of finding data and streaming it into the system in some type of ETL-based stream. (ETL, or extract, transform, load, is a data warehouse processing sequence.) Software also needs to monitor the system and to provide downstream destination systems with data feeds so that management can view trends and issue reports based on the data. While this big data system may take hours to move an individual record, process it, and store it on a server, it also needs to monitor trends in real time.

In summary, to manipulate big data, a system requires the following:

- A method of collecting and categorizing data

- A method of moving data into the system safely and without data loss

- A storage system that

    - Is distributed across many servers

    - Is scalable to thousands of servers

    - Will offer data redundancy and backup

    - Will offer redundancy in case of hardware failure

    - Will be cost-effective

- A rich tool set and community support

- A method of distributed system configuration

- Parallel data processing

- System-monitoring tools

- Reporting tools

- ETL-like tools (preferably with a graphic interface) that can be used to build tasks that process the data and monitor their progress

- Scheduling tools to determine when tasks will run and show task status

- The ability to monitor data trends in real time

- Local processing where the data is stored to reduce network bandwidth usage

Later in this chapter I explain how this book is organized with these requirements in mind. But let's now consider which tools best meet the big data requirements listed above.

How Hadoop Tools Can Help

Hadoop tools are a good fit for your big data needs. When I refer to Hadoop tools, I mean the whole Apache (`www.apache.org`) tool set related to big data. A community-based, open-source approach to software development, the Apache Software Foundation (ASF) has had a huge impact on both software development for big data and the overall approach that has been taken in this field. It also fosters significant cross-pollination of both ideas and development by the parties involved—for example, Google, Facebook, and LinkedIn. Apache runs an incubator program in which projects are accepted and matured to ensure that they are robust and production worthy.

Hadoop was developed by Apache as a distributed parallel big data processing system. It was written in Java and released under an Apache license. It assumes that failures will occur, and so it is designed to offer both hardware and data redundancy automatically. The Hadoop platform offers a wide tool set for many of the big data functions that I have mentioned. The original Hadoop development was influenced by Google's MapReduce and the Google File System.

The following list is a sampling of tools available in the Hadoop ecosystem. Those marked in boldface are introduced in the chapters that follow:

- **Ambari**   Hadoop management and monitoring
- Avro   Data serialization system
- Chukwa   Data collection and monitoring
- **Hadoop**   Hadoop distributed storage platform
- Hama   BSP scientific computing framework
- **HBase**   Hadoop NoSQL non-relational database
- **Hive**   Hadoop data warehouse
- **Hue**   Hadoop web interface for analyzing data
- Mahout   Scalable machine learning platform
- **Map/Reduce**   Algorithm used by the Hadoop MR component
- **Nutch**   Web crawler
- **Oozie**   Workflow scheduler
- **Pentaho**   Open-source analytics tool set
- **Pig**   Data analysis high-level language
- **Solr**   Search platform
- **Sqoop**   Bulk data-transfer tool
- **Storm**   Distributed real-time computation system
- **Yarn**   Map/Reduce in Hadoop Version 2
- **ZooKeeper**   Hadoop centralized configuration system

When grouped together, the ASF, Lucene, and other provider tools, some of which are here, provide a rich functional set that will allow you to manipulate your data.

My Approach

My approach in this book is to build the various tools into one large system. Stage by stage, and starting with the Hadoop Distributed File System (HDFS), which is the big data file system, I do the following:

- Introduce the tool

- Show how to obtain the installation package

- Explain how to install it, with examples

- Employ examples to show how it can be used

Given that I have a lot of tools and functions to introduce, I take only a brief look at each one. Instead, I show you how each of these tools can be used as individual parts of a big data system. It is hoped that you will be able to investigate them further in your own time.

The Hadoop platform tool set is installed on CentOS Linux 6.2. I use Linux because it is free to download and has a small footprint on my servers. I use Centos rather than another free version of Linux because some of the Hadoop tools have been released for CentOS only. For instance, at the time of writing this, Ambari is not available for Ubuntu Linux.

Throughout the book, you will learn how you can build a big data system using low-cost, commodity hardware. I relate the use of these big data tools to various IT roles and follow a step-by-step approach to show how they are feasible for most IT professionals. Along the way, I point out some solutions to common problems you might encounter, as well as describe the benefits you can achieve with Hadoop tools. I use small volumes of data to demonstrate the systems, tools, and ideas; however, the tools scale to very large volumes of data.

Some level of knowledge of Linux, and to a certain extent Java, is assumed. Don't be put off by this; instead, think of it as an opportunity to learn a new area if you aren't familiar with the subject.

Overview of the Big Data System

While many organizations may not yet have the volumes of data that could be defined as big data, all need to consider their systems as a whole.A large organization might have a single big data repository. In any event, it is useful to investigate these technologies as preparation for meeting future needs.

Big Data Flow and Storage

Many of the principles governing business intelligence and data warehousing scale to big data proportions. For instance, Figure 1-2 depicts a data warehouse system in general terms.

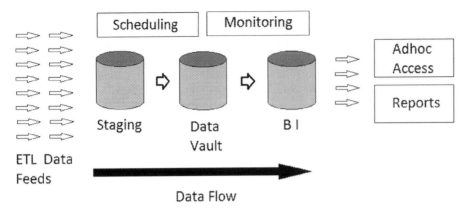

Figure 1-2. *A general data warehouse system*

As you can see in Figure 1-2, ETL (extraction, transformation, and loading of the data) feeds arrive at the staging schema of the warehouse and are loaded into their current raw format in staging area tables. The data is then transformed and moved to the data vault, which contains all the data in the repository. That data might be filtered, cleaned, enriched, and restructured. Lastly, the data is loaded into the BI, or Business Intelligence, schema of the warehouse, where the data could be linked to reference tables. It is at this point that the data is available for the business via reporting tools and adhoc reports. Figure 1-2 also illustrates the scheduling and monitoring tasks. *Scheduling* controls when feeds are run and the relationships between them, while *monitoring* determines whether the feeds have run and whether errors have occurred. Note also that scheduled feeds can be inputs to the system, as well as outputs.

■ **Note**   The data movement flows from extraction from raw sources, to loading, to staging and transformation, and to the data vault and the BI layer. The acronym for this process is ELT (extract, load, transfer), which better captures what is happening than the common term ETL.

Many features of this data warehouse system can scale up to and be useful in a big data system. Indeed, the big data system could feed data to data warehouses and datamarts. Such a big data system would need extraction, loading, and transform feeds, as well as scheduling, monitoring, and perhaps the data partitioning that a data warehouse uses, to separate the stages of data processing and access. By adding a big data repository to an IT architecture, you can extend future possibilities to mine data and produce useful reports. Whereas currently you might filter and aggregate data to make it fit a datamart, the new architecture allows you to store all of your raw data.

So where would a big data system fit in terms of other systems a large organization might have? Figure 1-3 represents its position in general terms, for there are many variations on this, depending on the type of company and its data feeds.

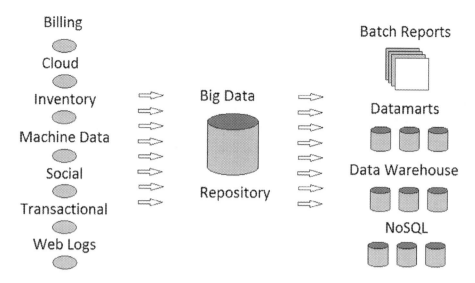

Figure 1-3. *A general big data environment*

Figure 1-3 does not include all types of feeds. Also, it does not have the feedback loops that probably would exist. For instance, data warehouse feeds might form inputs, have their data enriched, and feed outputs. Web log data might be inputs, then enriched with location and/or transaction data, and become enriched outputs. However, the idea here is that a single, central big data repository can exist to hold an organization's big data.

Benefits of Big Data Systems

Why investigate the use of big data and a parallel processing approach? First, if your data can no longer be processed by traditional relational database systems (RDBMS), that might mean your organization will have future data problems. You might have been forced to introduce NoSQL database technology so as to process very large data volumes in an acceptable time frame. Hadoop might not be the immediate solution to your processing problems, owing to its high latency, but it could provide a scalable big data storage platform.

Second, big data storage helps to establish a new skills base within the organization. Just as data warehousing brought with it the need for new skills to build, support, and analyze the warehouse, so big data leads to the same type of skills building. One of the biggest costs in building a big data system is the specialized staff needed to maintain it and use the data in it. By starting now, you can build a skills pool within your organization, rather than have to hire expensive consultants later. (Similarly, as an individual, accessing these technologies can help you launch a new and lucrative career in big data.)

Third, by adopting a platform that can scale to a massive degree, a company can extend the shelf life of its system and so save money, as the investment involved can be spread over a longer time. Limited to interim solutions, a company with a small cluster might reach capacity within a few years and require redevelopment.

Fourth, by getting involved in the big data field now, a company can future-proof itself and reduce risk by building a vastly scalable distributed platform. By introducing the technologies and ideas in a company now, there will be no shock felt in later years, when there is a need to adopt the technology.

In developing any big data system, your organization needs to keep its goals in mind. Why are you developing the system? What do you hope to achieve? How will the system be used? What will you store? You measure the system use over time against the goals that were established at its inception.

What's in This Book

This book is organized according to the particular features of a big data system, paralleling the general requirements of a big data system, as listed in the beginning of this chapter. This first chapter describes the features of big data and names the related tools that are introduced in the chapters that follow. My aim here is to describe as many big data tools as possible, using practical examples. (Keep in mind, however, that writing deadlines and software update schedules don't always mesh, so some tools or functions may have changed by the time you read this.)

All of the tools discussed in this book have been chosen because they are supported by a large user base, which fulfills big data's general requirements of a rich tool set and community support. Each Apache Hadoop-based tool has its own website and often its own help forum. The ETL and reporting tools introduced in Chapters 10 and 11, although non-Hadoop, are also supported by their own communities.

Storage: Chapter 2

Discussed in Chapter 2, storage represents the greatest number of big data requirements, as listed earlier:

- A storage system that

    - Is distributed across many servers

    - Is scalable to thousands of servers

    - Will offer data redundancy and backup

    - Will offer redundancy in case of hardware failure

    - Will be cost-effective

A distributed storage system that is highly scalable, Hadoop meets all of these requirements. It offers a high level of redundancy with data blocks being copied across the cluster. It is fault tolerant, having been designed with hardware failure in mind. It also offers a low cost per unit of storage. Hadoop versions 1.x and 2.x are installed and examined in Chapter 2, as well as a method of distributed system configuration. The Apache ZooKeeper system is used within the Hadoop ecosystem to provide a distributed configuration system for Apache Hadoop tools.

Data Collection: Chapter 3

Automated web crawling to collect data is a much-used technology, so we need a method of collecting and categorizing data. Chapter 3 describes two architectures using Nutch and Solr to search the web and store data. The first stores data directly to HDFS, while the second uses Apache HBase. The chapter provides examples of both.

Processing: Chapter 4

The following big data requirements relate to data processing:

- Parallel data processing

- Local processing where the data is stored to reduce network bandwidth usage

Chapter 4 introduces a variety of Map Reduce programming approaches, with examples. Map Reduce programs are developed in Java, Apache Pig, Perl, and Apache Hive.

Scheduling: Chapter 5

The big data requirement for scheduling encompasses the need to share resources and determine when tasks will run. For sharing Hadoop-based resources, Chapter 5 introduces the Capacity and Fair schedulers for Hadoop. It also introduces Apache Oozie, showing how simple ETL tasks can be created using Hadoop components like Apache Sqoop and Apache Pig. Finally, it demonstrates how to schedule Oozie tasks.

Data Movement: Chapter 6

Big data systems require tools to allow safe movement of a variety of data types, safely and without data loss. Chapter 6 introduces the Apache Sqoop tool for moving data into and out of relational databases. It also provides an example of how Apache Flume can be used to process log-based data. Apache Storm is introduced for data stream processing.

Monitoring: Chapter 7

The requirement for system monitoring tools for a big data system is discussed in Chapter 7. The chapter introduces the Hue tool as a single location to access a wide range of Apache Hadoop functionality. It also demonstrates the Ganglia and Nagios resource monitoring and alerting tools.

Cluster Management: Chapter 8

Cluster managers are introduced in Chapter 8 by using the Apache Ambari tool to install Horton Works HDP 2.1 and Cloudera's cluster manager to install Cloudera CDH5. A brief overview is then given of their functionality.

Analysis: Chapter 9

Big data requires the ability to monitor data trends in real time. To that end, Chapter 9 introduces the Apache Spark real-time, in-memory distributed processing system. It also shows how Spark SQL can be used, via an example. It also includes a practical demonstration of the features of the Apache Hive and Cloudera Impala query languages.

ETL: Chapter 10

Although ETL was briefly introduced in Chapter 5, this chapter discusses the need for graphic tools for ETL chain building and management. ETL-like tools (preferably with a graphic interface) can be used to build tasks to process the data and monitor their progress. Thus, Chapter 10 introduces the Pentaho and Talend graphical ETL tools for big data. This chapter investigates their visual object based approach to big data ETL task creation. It also shows that these tools offer an easier path into the work of Map Reduce development.

Reports: Chapter 11

Big data systems need reporting tools. In Chapter 11, some reporting tools are discussed and a typical dashboard is built using the Splunk/Hunk tool. Also, the evaluative data-quality capabilities of Talend are investigated by using the profiling function.

Summary

While introducing the challenges and benefits of big data, this chapter also presents a set of requirements for big data systems and explains how they can be met by utilizing the tools discussed in the remaining chapters of this book.

The aim of this book has been to explain the building of a big data processing system by using the Hadoop tool set. Examples are used to explain the functionality provided by each Hadoop tool. Starting with HDFS for storage, followed by Nutch and Solr for data capture, each chapter covers a new area of functionality, providing a simple overview of storage, processing, and scheduling. With these examples and the step-by-step approach, you can build your knowledge of big data possibilities and grow your familiarity with these tools. By the end of Chapter 11, you will have learned about most of the major functional areas of a big data system.

As you read through this book, you should consider how to use the individual Hadoop components in your own systems. You will also notice a trend toward easier methods of system management and development. For instance, Chapter 2 starts with a manual installation of Hadoop, while Chapter 8 uses cluster managers. Chapter 4 shows handcrafted code for Map Reduce programming, but Chapter 10 introduces visual object based Map Reduce task development using Talend and Pentaho.

Now it's time to start, and we begin by looking at Hadoop itself. The next chapter introduces the Hadoop application and its uses, and shows how to configure and use it.

CHAPTER 2

■ ■ ■

Storing and Configuring Data with Hadoop, YARN, and ZooKeeper

This chapter introduces Hadoop versions V1 and V2, laying the groundwork for the chapters that follow. Specifically, you first will source the V1 software, install it, and then configure it. You will test your installation by running a simple word-count Map Reduce task. As a comparison, you will then do the same for V2, as well as install a ZooKeeper quorum. You will then learn how to access ZooKeeper via its commands and client to examine the data that it stores. Lastly, you will learn about the Hadoop command set in terms of shell, user, and administration commands. The Hadoop installation that you create here will be used for storage and processing in subsequent chapters, when you will work with Apache tools like Nutch and Pig.

An Overview of Hadoop

Apache Hadoop is available as three download types via the hadoop.apache.org website. The releases are named as follows:

- Hadoop-1.2.1
- Hadoop-0.23.10
- Hadoop-2.3.0

The first release relates to Hadoop V1, while the second two relate to Hadoop V2. There are two different release types for V2 because the version that is numbered 0.xx is missing extra components like NN and HA. (NN is "name node" and HA is "high availability.") Because they have different architectures and are installed differently, I first examine both Hadoop V1 and then Hadoop V2 (YARN). In the next section, I will give an overview of each version and then move on to the interesting stuff, such as how to source and install both.

Because I have only a single small cluster available for the development of this book, I install the different versions of Hadoop and its tools on the same cluster nodes. If any action is carried out for the sake of demonstration, which would otherwise be dangerous from a production point of view, I will flag it. This is important because, in a production system, when you are upgrading, you want to be sure that you retain all of your data. However, for demonstration purposes, I will be upgrading and downgrading periodically.

So, in general terms, what is Hadoop? Here are some of its characteristics:

- It is an open-source system developed by Apache in Java.
- It is designed to handle very large data sets.
- It is designed to scale to very large clusters.
- It is designed to run on commodity hardware.

- It offers resilience via data replication.

- It offers automatic failover in the event of a crash.

- It automatically fragments storage over the cluster.

- It brings processing to the data.

- Its supports large volumes of files—into the millions.

The third point comes with a caveat: Hadoop V1 has problems with very large scaling. At the time of writing, it is limited to a cluster size of around 4,000 nodes and 40,000 concurrent tasks. Hadoop V2 was developed in part to offer better resource usage and much higher scaling.

Using Hadoop V2 as an example, you see that there are four main component parts to Hadoop. *Hadoop Common* is a set of utilities that support Hadoop as a whole. *Hadoop Map Reduce* is the parallel processing system used by Hadoop. It involves the steps Map, Shuffle, and Reduce. A big volume of data (the text of this book, for example) is *mapped* into smaller elements (the individual words), then an operation (say, a word count) is carried out locally on the small elements of data. These results are then *shuffled* into a whole, and *reduced* to a single list of words and their counts. *Hadoop YARN* handles scheduling and resource management. Finally, *Hadoop Distributed File System (HDFS)* is the distributed file system that works on a master/slave principle whereby a name node manages a cluster of slave data nodes.

The Hadoop V1 Architecture

In the V1 architecture, a master Job Tracker is used to manage Task Trackers on slave nodes (Figure 2-1). Hadoop's data node and Task Trackers co-exist on the same slave nodes.

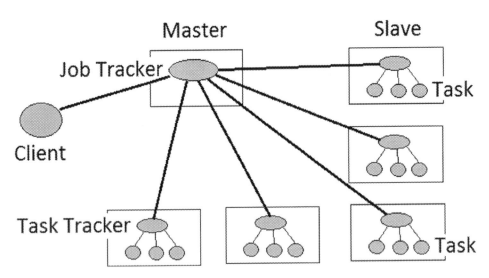

Figure 2-1. *Hadoop V1 architecture*

The cluster-level Job Tracker handles client requests via a Map Reduce (MR) API. The clients need only process via the MR API, as the Map Reduce framework and system handle the scheduling, resources, and failover in the event of a crash. Job Tracker handles jobs via data node–based Task Trackers that manage the actual tasks or processes. Job Tracker manages the whole client-requested job, passing subtasks to individual slave nodes and monitoring their availability and the tasks' completion.

Hadoop V1 only scales to clusters of around 4,000 to 5,000 nodes, and there are also limitations on the number of concurrent processes that can run. It has only a single processing type, Map Reduce, which although powerful does not allow for requirements like graph or real-time processing.

The Differences in Hadoop V2

With YARN, Hadoop V2's Job Tracker has been split into a master Resource Manager and slave-based Application Master processes. It separates the major tasks of the Job Tracker: resource management and monitoring/scheduling. The Job History server now has the function of providing information about completed jobs. The Task Tracker has been replaced by a slave-based Node Manager, which handles slave node–based resources and manages tasks on the node. The actual tasks reside within containers launched by the Node Manager. The Map Reduce function is controlled by the Application Master process, while the tasks themselves may be either Map or Reduce tasks.

Hadoop V2 also offers the ability to use non-Map Reduce processing, like Apache Giraph for graph processing, or Impala for data query. Resources on YARN can be shared among all three processing systems.

Figure 2-2 shows client task requests being sent to the global Resource Manager and the slave-based Node Managers launching containers, which have the actual tasks. It also monitors their resource usage. The Application Master requests containers from the scheduler and receives status updates from the container-based Map Reduce tasks.

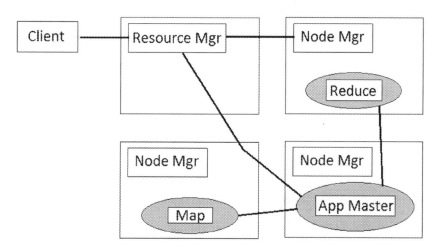

Figure 2-2. *Hadoop V2 architecture*

This architecture enables Hadoop V2 to scale to much larger clusters and provides the ability to have a higher number of concurrent processes. It also now offers the ability, as mentioned earlier, to run different types of processes concurrently, not just Map Reduce.

This is an introduction to the Hadoop V1 and V2 architectures. You might have the opportunity to work with both versions, so I give examples for installation and use of both. The architectures are obviously different, as seen in Figures 2-1 and 2-2, and so the actual installation/build and usage differ as well. For example, for V1 you will carry out a manual install of the software while for V2 you will use the Cloudera software stack, which is described next.

The Hadoop Stack

Before we get started with the Hadoop V1 and V2 installations, it is worth discussing the work of companies like Cloudera and Hortonworks. They have built stacks of Hadoop-related tools that have been tested for interoperability. Although I describe how to carry out a manual installation of software components for V1, I show how to use one of the software stacks for the V2 install.

When you're trying to use multiple Hadoop platform tools together in a single stack, it is important to know what versions will work together without error. If, for instance, you are using ten tools, then the task of tracking compatible version numbers quickly becomes complex. Luckily there are a number of Hadoop stacks available. Suppliers can provide a single tested package that you can download. Two of the major companies in this field are Cloudera and Hortonworks. Apache Bigtop, a testing suite that I will demonstrate in Chapter 8, is also used as the base for the Cloudera Hadoop stack.

Table 2-1 shows the current stacks from these companies, listing components and versions of tools that are compatible at the time of this writing.

Table 2-1. *Hadoop Stack Tool Version Details*

	Cloudera CDH 4.6.0	Hortonworks Data Platform 2.0
Ambari		1.4.4
DataFu	0.0.4	
Flume	1.4.0	1.4.0
Hadoop	2.0.0	2.2.0
HCatalog	0.5.0	0.12.0
HBase	0.94	0.96.1
Hive	0.10.0	0.12.0
Hue	2.5.0	2.3.0
Mahout	0.7	0.8.0
Oozie	3.3.2	4.0.0
Parquet	1.2.5	
Pig	0.11	0.12.0
Sentry	1.1.0	
Sqoop	1.4.3	1.4.4
Sqoop2	1.99.2	
Whirr	0.8.2	
ZooKeeper	3.4.5	3.4.5

While I use a Hadoop stack in the rest of the book, here I will show the process of downloading, installing, configuring, and running Hadoop V1 so that you will be able to compare the use of V1 and V2.

Environment Management

Before I move into the Hadoop V1 and V2 installations, I want to point out that I am installing both Hadoop V1 and V2 on the same set of servers. Hadoop V1 is installed under /usr/local while Hadoop V2 is installed as a Cloudera CDH release and so will have a defined set of directories:

- Logging under /var/log; that is, /var/log/hadoop-hdfs/
- Configuration under /etc/hadoop/conf/
- Executables defined as servers under /etc/init.d/; that is, hadoop-hdfs-namenode

I have also created two sets of .bashrc environment configuration files for the Linux Hadoop user account:

```
[hadoop@hc1nn ~]$ pwd
/home/hadoop

[hadoop@hc1nn ~]$ ls -l .bashrc*
lrwxrwxrwx. 1 hadoop hadoop   16 Jun 30 17:59 .bashrc -> .bashrc_hadoopv2
-rw-r--r--. 1 hadoop hadoop 1586 Jun 18 17:08 .bashrc_hadoopv1
-rw-r--r--. 1 hadoop hadoop 1588 Jul 27 11:33 .bashrc_hadoopv2
```

By switching the .bashrc symbolic link between the Hadoop V1 (.bashrc_hadoopv1) and V2 (.bashrc_hadoopv2) files, I can quickly navigate between the two environments. Each installation has a completely separate set of resources. This approach enables me to switch between Hadoop versions on my single set of testing servers while writing this guide. From a production viewpoint, however, you would install only one version of Hadoop at a time.

Hadoop V1 Installation

Before you attempt to install Hadoop, you must ensure that Java 1.6.x is installed and that SSH (secure shell) is installed and running. The master name node must be able to create an SSH session to reach each of its data nodes without using a password in order to manage them. On CentOS, you can install SSH via the root account as follows:

```
yum  install openssh-server
```

This will install the secure shell daemon process. Repeat this installation on all of your servers, then start the service (as root):

```
service sshd restart
```

Now, in order to make the SSH sessions from the name node to the data nodes operate without a password, you must create an SSH key on the name node and copy the key to each of the data nodes. You create the key with the keygen command as the hadoop user (I created the hadoop user account during the installation of the CentOS operating system on each server), as follows:

```
ssh-keygen
```

A key is created automatically as $HOME/.ssh/id_rsa.pub. You now need to copy this key to the data nodes. You run the following command to do that:

```
ssh-copy-id hadoop@hc1r1m1
```

This copies the new SSH key to the data node hc1r1m1 as user hadoop; you change the server name to copy the key to the other data node servers.

The remote passwordless secure shell access can now be tested with this:

```
ssh hadoop@hc1r1m1
```

A secure shell session should now be created on the host hc1r1m1 without need to prompt a password.

As Hadoop has been developed using Java, you must also ensure that you have a suitable version of Java installed on each machine. I will be using four machines in a mini cluster for this test:

- hc1nn - A Linux CentOS 6 server for a name node

- hc1r1m1 - A Linux CentOS 6 server for a data node

- hc1r1m2 - A Linux CentOS 6 server for a data node

- hc1r1m3 - A Linux CentOS 6 server for a data node

Can the name node access all of the data nodes via SSH (secure shell) without being prompted for a password? And is a suitable Java version installed? I have a user account called hadoop on each of these servers that I use for this installation. For instance, the following command line shows hadoop@hc1nn, which means that we are logged into the server hc1nn as the Linux user hadoop:

```
[hadoop@hc1nn ~]$ java -version
java version "1.6.0_30"
OpenJDK Runtime Environment (IcedTea6 1.13.1) (rhel-3.1.13.1.el6_5-i386)
OpenJDK Client VM (build 23.25-b01, mixed mode)
```

This command, java -version, shows that we have OpenJDK java version 1.6.0_30 installed. The following commands create an SSH session on each of the data nodes and checks the Java version on each:

```
[hadoop@hc1nn ~]$ ssh hadoop@hc1r1m3
Last login: Thu Mar 13 19:41:12 2014 from hc1nn
[hadoop@hc1r1m3 ~]$
[hadoop@hc1r1m3 ~]$ java -version
java version "1.6.0_30"
OpenJDK Runtime Environment (IcedTea6 1.13.1) (rhel-3.1.13.1.el6_5-i386)
OpenJDK Server VM (build 23.25-b01, mixed mode)
[hadoop@hc1r1m3 ~]$ exit
logout
Connection to hc1r1m3 closed.

[hadoop@hc1nn ~]$ ssh hadoop@hc1r1m2
Last login: Thu Mar 13 19:40:45 2014 from hc1nn
[hadoop@hc1r1m2 ~]$ java -version
java version "1.6.0_30"
OpenJDK Runtime Environment (IcedTea6 1.13.1) (rhel-3.1.13.1.el6_5-i386)
OpenJDK Server VM (build 23.25-b01, mixed mode)
```

```
[hadoop@hc1r1m2 ~]$ exit
logout
Connection to hc1r1m2 closed.

[hadoop@hc1nn ~]$ ssh hadoop@hc1r1m1
Last login: Thu Mar 13 19:40:22 2014 from hc1r1m3
[hadoop@hc1r1m1 ~]$ java -version
java version "1.6.0_30"
OpenJDK Runtime Environment (IcedTea6 1.13.1) (rhel-3.1.13.1.el6_5-x86_64)
OpenJDK 64-Bit Server VM (build 23.25-b01, mixed mode)
[hadoop@hc1r1m1 ~]$ exit
logout
Connection to hc1r1m1 closed.
```

These three SSH statements show that a secure shell session can be created from the name node, hc1nn, to each of the data nodes.

Notice that I am using the Java OpenJDK (http://openjdk.java.net/) here. Generally it's advised that you use the Oracle Sun JDK. However, Hadoop has been tested against the OpenJDK, and I am familiar with its use. I don't need to register to use OpenJDK, and I can install it on Centos using a simple yum command. Additionally, the Sun JDK install is more complicated.

Now let's download and install a version of Hadoop V1. In order to find the release of Apache Hadoop to download, start here: http://hadoop.apache.org.

Next, choose Download Hadoop, click the release option, then choose Download, followed by Download a Release Now! This will bring you to this page: http://www.apache.org/dyn/closer.cgi/hadoop/common/. It suggests a local mirror site that you can use to download the software. It's a confusing path to follow; I'm sure that this website could be simplified a little. The suggested link for me is http://apache.insync.za.net/hadoop/common. You may be offered a different link.

On selecting that site, I'm offered a series of releases. I choose 1.2.1, and then I download the file: Hadoop-1.2.1.tar.gz. Why choose this particular format over the others? From past experience, I know how to unpack it and use it; feel free to choose the format with which you're most comfortable.

Download the file to /home/hadoop/Downloads. (This download and installation must be carried out on each server.) You are now ready to begin the Hadoop single-node installation for Hadoop 1.2.1.

The approach from this point on will be to install Hadoop onto each server separately as a single-node installation, configure it, and try to start the servers. This will prove that each node is correctly configured individually. After that, the nodes will be grouped into a Hadoop master/slave cluster. The next section describes the single-node installation and test, which should be carried out on all nodes. This will involve unpacking the software, configuring the environment files, formatting the file system, and starting the servers. This is a manual process; if you have a very large production cluster, you would need to devise a method of automating the process.

Hadoop 1.2.1 Single-Node Installation

From this point on, you will be carrying out a single-node Hadoop installation (until you format the Hadoop file system on this node). First, you ftp the file hadoop-1.2.1.tar.gz to all of your nodes and carry out the steps in this section on all nodes.

So, given that you are logged in as the user hadoop, you see the following file in the $HOME/Downloads directory:

```
[hadoop@hc1nn Downloads]$ ls -l
total 62356
-rw-rw-r--. 1 hadoop hadoop 63851630 Mar 15 15:01 hadoop-1.2.1.tar.gz
```

This is a gzipped tar file containing the Hadoop 1.2.1 software that you are interested in. Use the Linux gunzip tool to unpack the gzipped archive:

```
[hadoop@hc1nn Downloads]$ gunzip hadoop-1.2.1.tar.gz
[hadoop@hc1nn Downloads]$ ls -l
total 202992
-rw-rw-r--. 1 hadoop hadoop 207861760 Mar 15 15:01 hadoop-1.2.1.tar
```

Then, unpack the tar file:

```
[hadoop@hc1nn Downloads]$ tar xvf hadoop-1.2.1.tar
[hadoop@hc1nn Downloads]$ ls -l
total 202996
drwxr-xr-x. 15 hadoop hadoop      4096 Jul 23  2013 hadoop-1.2.1
-rw-rw-r--.  1 hadoop hadoop 207861760 Mar 15 15:01 hadoop-1.2.1.tar
```

Now that the software is unpacked to the local directory hadoop-1.2.1, you move it into a better location. To do this, you will need to be logged in as root:

```
[hadoop@hc1nn Downloads]$ su -
Password:
[root@hc1nn ~]# cd /home/hadoop/Downloads
[root@hc1nn Downloads]# mv hadoop-1.2.1 /usr/local
[root@hc1nn Downloads]# cd /usr/local
```

You have now moved the installation to /usr/local, but make sure that the hadoop user owns the installation. Use the Linux chown command to recursively change the ownership and group membership for files and directories within the installation:

```
[root@hc1nn local]# chown -R hadoop:hadoop hadoop-1.2.1
[root@hc1nn local]# ls -l
total 40
drwxr-xr-x. 15 hadoop hadoop 4096 Jul 23  2013 hadoop-1.2.1
```

You can see from the last line in the output above that the directory is now owned by hadoop and is a member of the hadoop group.

You also create a symbolic link to refer to your installation so that you can have multiple installations on the same host for testing purposes:

```
[root@hc1nn local]# ln -s hadoop-1.2.1 hadoop
[root@hc1nn local]# ls -l
lrwxrwxrwx.  1 root   root     12 Mar 15 15:11 hadoop -> hadoop-1.2.1
drwxr-xr-x. 15 hadoop hadoop 4096 Jul 23  2013 hadoop-1.2.1
```

The last two lines show that there is a symbolic link called hadoop under the directory /usr/local that points to our hadoop-1.2.1 installation directory at the same level. If you later upgrade and install a new version of the Hadoop V1 software, you can just change this link to point to it. Your environment and scripts can then remain static and always use the path /usr/local/hadoop.

Now, you follow these steps to proceed with installation.

1. Set up Bash shell file for hadoop $HOME/.bashrc

When logged in as hadoop, you add the following text to the end of the file $HOME/.bashrc. When you create this Bash shell, environmental variables like JAVA_HOME and HADOOP_PREFIX are set. The next time a Bash shell is created by the hadoop user account, these variables will be pre-defined.

```
#######################################################
# Set Hadoop related env variables

export HADOOP_PREFIX=/usr/local/hadoop


# set JAVA_HOME (we will also set a hadoop specific value later)
export JAVA_HOME=/usr/lib/jvm/jre-1.6.0-openjdk

# some handy aliases and functions
unalias fs 2>/dev/null
alias fs="hadoop fs"
unalias hls 2>/dev/null
alias hls="fs -l"

# add hadoop to the path
export PATH=$PATH:$HADOOP_PREFIX
export PATH=$PATH:$HADOOP_PREFIX/bin
export PATH=$PATH:$HADOOP_PREFIX/sbin
```

Note that you are not using the $HADOOP_HOME variable, because with this release it has been superseded. If you use it instead of $HADOOP_PREFIX, you will receive warnings.

2. Set up conf/hadoop-env.sh

You now modify the configuration file hadoop-env.sh to specify the location of the Java installation by setting the JAVA_HOME variable. In the file conf/hadoop-env.sh, you change:

```
# export JAVA_HOME=/usr/lib/j2sdk1.5-sun
```

to

```
export JAVA_HOME=/usr/lib/jvm/jre-1.6.0-openjdk
```

Note: When referring to the Hadoop installation configuration directory in this section, and all subsequent sections for the V1 installation, I mean the /usr/local/hadoop/conf directory.

3. Create Hadoop temporary directory

On the Linux file system, you create a Hadoop temporary directory, as shown below. This will give Hadoop a working area. Set the ownership to the hadoop user and also set the directory permissions:

```
[root@hc1nn local]# mkdir -p /app/hadoop/tmp
[root@hc1nn local]# chown -R hadoop:hadoop /app/hadoop
[root@hc1nn local]# chmod 750 /app/hadoop/tmp
```

4. Set up conf/core-site.xml

You set up the configuration for the Hadoop core component. This file configuration is based on XML; it defines the Hadoop temporary directory and default file system access. There are many more options that can be specified; see the Hadoop site (hadoop.apache.org) for details.

Add the following text to the file between the configuration tags:

```
<property>
  <name>hadoop.tmp.dir</name>
  <value>/app/hadoop/tmp</value>
  <description>A base for other temporary directories.</description>
</property>

<property>
  <name>fs.default.name</name>
  <value>hdfs://localhost:54310</value>
  <description>The name of the default file system.</description>
</property>
```

5. Set up conf/mapred-site.xml

Next, you set up the basic configuration for the Map Reduce component, adding the following between the configuration tags. This defines the host and port name for each Job Tracker server.

```
<property>
  <name>mapred.job.tracker</name>
  <value>localhost:54311</value>
  <description>The host and port for the Map Reduce job tracker
    </description>
</property>

<property>
  <name>mapred.job.tracker.http.address</name>
  <value>localhost:50030</value>
</property>

<property>
  <name>mapred.task.tracker.http.address</name>
  <value>localhost:50060</value>
</property>
```

The example configuration file here is for the server hc1r1m1. When the configuraton is changed to a cluster, these Job Tracker entries will refer to Name Node machine hc1nn.

6. Set up file conf/hdfs-site.xml

Set up the basic configuration for the HDFS, adding the following between the configuration tags. This defines the replication level for the HDFS; it shows that a single block will be copied twice. It also specifies the address of the Name Node web user interface as dfs.http.address:

```
<property>
  <name>dfs.replication</name>
  <value>3</value>
  <description>The replication level</description>
</property>
<property>
  <name>dfs.http.address</name>
  <value>http://localhost:50070/</value>
</property>
```

7. Format the file system

Run the following command as the Hadoop user to format the file system:
```
hadoop namenode -format
```

▓ **Warning**   Do not execute this command on a running HDFS or you will lose your data!

The output should look like this:

```
14/03/15 16:08:19 INFO namenode.NameNode: STARTUP_MSG:
/************************************************************
STARTUP_MSG: Starting NameNode
STARTUP_MSG:   host = hc1nn/192.168.1.107
STARTUP_MSG:   args = [-format]
STARTUP_MSG:   version = 1.2.1
STARTUP_MSG:   build = https://svn.apache.org/repos/asf/hadoop/common/branches/branch-1.2 -r
1503152; compiled by 'mattf' on Mon Jul 22 15:23:09 PDT 2013
STARTUP_MSG:   java = 1.6.0_30
************************************************************/
14/03/15 16:08:20 INFO util.GSet: Computing capacity for map BlocksMap
14/03/15 16:08:20 INFO util.GSet: VM type       = 32-bit
14/03/15 16:08:20 INFO util.GSet: 2.0% max memory = 1013645312
14/03/15 16:08:20 INFO util.GSet: capacity      = 2^22 = 4194304 entries
14/03/15 16:08:20 INFO util.GSet: recommended=4194304, actual=4194304
14/03/15 16:08:20 INFO namenode.FSNamesystem: fsOwner=hadoop
14/03/15 16:08:20 INFO namenode.FSNamesystem: supergroup=supergroup
14/03/15 16:08:20 INFO namenode.FSNamesystem: isPermissionEnabled=true
14/03/15 16:08:20 INFO namenode.FSNamesystem: dfs.block.invalidate.limit=100
```

```
14/03/15 16:08:20 INFO namenode.FSNamesystem: isAccessTokenEnabled=false accessKeyUpdateInterval=0
min(s), accessTokenLifetime=0 min(s)
14/03/15 16:08:20 INFO namenode.FSEditLog: dfs.namenode.edits.toleration.length = 0
14/03/15 16:08:20 INFO namenode.NameNode: Caching file names occuring more than 10 times
14/03/15 16:08:20 INFO common.Storage: Image file /app/hadoop/tmp/dfs/name/current/fsimage of size
112 bytes saved in 0 seconds.
14/03/15 16:08:20 INFO namenode.FSEditLog: closing edit log: position=4, editlog=/app/hadoop/tmp/
dfs/name/current/edits
14/03/15 16:08:20 INFO namenode.FSEditLog: close success: truncate to 4, editlog=/app/hadoop/tmp/
dfs/name/current/edits
14/03/15 16:08:21 INFO common.Storage: Storage directory /app/hadoop/tmp/dfs/name has been
successfully formatted.
14/03/15 16:08:21 INFO namenode.NameNode: SHUTDOWN_MSG:
/************************************************************
SHUTDOWN_MSG: Shutting down NameNode at hc1nn/192.168.1.107
************************************************************/
```

Now you test that you can start, check, and stop the Hadoop servers on a standalone node without errors. Start the servers by using:

```
start-all.sh
```

You will see this:

```
starting namenode, logging to
/usr/local/hadoop-1.2.1/libexec/../logs/hadoop-hadoop-namenode-hc1nn.out
localhost: starting datanode, logging to
/usr/local/hadoop-1.2.1/libexec/../logs/hadoop-hadoop-datanode-hc1nn.out
localhost: starting secondarynamenode, logging to
/usr/local/hadoop-1.2.1/libexec/../logs/hadoop-hadoop-secondarynamenode-hc1nn.out
starting jobtracker, logging to
/usr/local/hadoop-1.2.1/libexec/../logs/hadoop-hadoop-jobtracker-hc1nn.out
localhost: starting tasktracker, logging to
/usr/local/hadoop-1.2.1/libexec/../logs/hadoop-hadoop-tasktracker-hc1nn.out
```

Now, check that the servers are running. Note that you should expect to see the following:

- Name node
- Secondary name node
- Job Tracker
- Task Tracker
- Data node

Running on the master server hc1nn, use the jps command to list the servers that are running:

```
[hadoop@hc1nn ~]$ jps
2116 SecondaryNameNode
2541 Jps
2331 TaskTracker
2194 JobTracker
```

```
1998 DataNode
1878 NameNode
```

If you find that the jps command is not available, check that it exists as $JAVA_HOME/bin/jps. Ensure that you installed the Java JDK in the previous step. If that does not work, then try installing the Java OpenJDK development package as root:

```
[root@hc1nn ~]$ yum install java-1.6.0-openjdk-devel
```

Your result shows that the servers are running. If you need to stop them, use the stop-all.sh command, as follows:

```
[hadoop@hc1nn ~]$ stop-all.sh
stopping jobtracker
localhost: stopping tasktracker
stopping namenode
localhost: stopping datanode
localhost: stopping secondarynamenode
```

You have now completed a single-node Hadoop installation. Next, you repeat the steps for the Hadoop V1 installation on all of the nodes that you plan to use in your Hadoop cluster. When that is done, you can move to the next section, "Setting up the Cluster," where you'll combine all of the single-node machines into a Hadoop cluster that's run from the Name Node machine.

Setting up the Cluster

Now you are ready to set up the Hadoop cluster. Make sure that all servers are stopped on all nodes by using the stop-all.sh script.

First, you must tell the name node where all of its slaves are. To do so, you add the following lines to the master and slaves files. (You only do this on the Name Node server [hc1nn], which is the master. It then knows that it is the master and can identify its slave data nodes.) You add the following line to the file $HADOOP_PREFIX/conf/masters to identify it as the master:

```
hc1nn
```

Then, you add the following lines to the file $HADOOP_PREFIX/conf/slaves to identify those servers as slaves:

```
hc1nn
hc1r1m1
hc1r1m2
hc1r1m3
```

These are all of the machines in my cluster. Your machine names may be different, so you would insert your own machine names. Note also that I am using the Name Node machine (hc1nn) as a master and a slave. In a production cluster you would have name nodes and data nodes on separate servers.

On all nodes, you change the value of fs.default.name in the file $HADOOP_PREFIX/conf/core-site.xml to be:

```
hdfs://hc1nn:54310
```

This configures all nodes for the core Hadoop component to access the HDFS using the same address.

On all nodes, you change the value of mapred.job.tracker in the file $HADOOP_PREFIX/conf/mapred-site.xml to be:

```
hc1nn:54311
```

This defines the host and port names on all servers for the Map Reduce Job Tracker server to point to the Name Node machine.

On all nodes, check that the value of dfs.replication in the file $HADOOP_PREFIX/conf/hdfs-site.xml is set to 3. This means that three copies of each block of data will automatically be kept by HDFS.

In the same file, ensure that the line http://localhost:50070/ for the variable dfs.http.address is changed to:

```
http://hc1nn:50070/
```

This sets the HDFS web/http address to point to the Name Node master machine hc1nn. With none of the Hadoop servers running, you format the cluster from the Name Node server—in this instance, hc1nn:

```
hadoop namenode -format
```

At this point, a common problem can occur with Hadoop file system versioning between the name node and data nodes. Within HDFS, there are files named VERSION that contain version numbering information that is regenerated each time the file system is formatted, such as:

```
[hadoop@hc1nn dfs]$ pwd
/app/hadoop/tmp/dfs

[hadoop@hc1nn dfs]$ find . -type f -name VERSION -exec grep -H namespaceID {} \;
./data/current/VERSION:namespaceID=1244166645
./name/current/VERSION:namespaceID=1244166645
./name/previous.checkpoint/VERSION:namespaceID=1244166645
./namesecondary/current/VERSION:namespaceID=1244166645
```

The Linux command shown here is executed as the hadoop user searches for the VERSION files under /app/hadoop/tmp/dfs and strips the namespace ID information out of them. If this command was executed on the Name Node server and the Data Node servers, you would expect to see the same value 1244166645. When this versioning gets out of step on the data nodes, an error occurs, such as follows:

```
ERROR org.apache.hadoop.hdfs.server.datanode.DataNode: java.io.IOException: Incompatible
namespaceIDs
```

While this problem seems to have two solutions, only one is viable. Although you could delete the data directory /app/hadoop/tmp/dfs/data on the offending data node, reformat the file system, and then start the servers, this approach will cause data loss. The second, more effective method involves editing the VERSION files on the data nodes so that the namespace ID values match those found on the Name Node machine.

You need to ensure that your firewall will enable port access for Hadoop to communicate. When you attempt to start the Hadoop servers, check the logs in the log directory (/usr/local/hadoop/logs).

Now, start the cluster from the name node; this time, you will start the HDFS servers using the script start-dfs.sh:

```
[hadoop@hc1nn logs]$ start-dfs.sh
starting namenode, logging to /usr/local/hadoop-1.2.1/libexec/../logs/hadoop-hadoop-namenode-
hc1nn.out
hc1r1m2: starting datanode, logging to /usr/local/hadoop-1.2.1/libexec/../logs/hadoop-hadoop-
datanode-hc1r1m2.out
```

```
hc1r1m1: starting datanode, logging to /usr/local/hadoop-1.2.1/libexec/../logs/hadoop-hadoop-
datanode-hc1r1m1.out
hc1r1m3: starting datanode, logging to /usr/local/hadoop-1.2.1/libexec/../logs/hadoop-hadoop-
datanode-hc1r1m3.out
hc1nn: starting datanode, logging to /usr/local/hadoop-1.2.1/libexec/../logs/hadoop-hadoop-
datanode-hc1nn.out
hc1nn: starting secondarynamenode, logging to /usr/local/hadoop-1.2.1/libexec/../logs/hadoop-
hadoop-secondarynamenode-hc1nn.out
```

As mentioned, check the logs for errors under $HADOOP_PREFIX/logs on each server. If you get errors like "No Route to Host," it is a good indication that your firewall is blocking a port. It will save a great deal of time and effort if you ensure that the firewall port access is open. (If you are unsure how to do this, then approach your systems administrator.)

You can now check that the servers are running on the name node by using the jps command:

```
[hadoop@hc1nn ~]$ jps
2116 SecondaryNameNode
2541 Jps
1998 DataNode
1878 NameNode
```

If you need to stop the HDFS servers, you can use the stop-dfs.sh script. Don't do it yet, however, as you will start the Map Reduce servers next.

With the HDFS servers running, it is now time to start the Map Reduce servers. The HDFS servers should always be started first and stopped last. Use the start-mapred.sh script to start the Map Reduce servers, as follows:

```
[hadoop@hc1nn logs]$ start-mapred.sh
starting jobtracker, logging to
/usr/local/hadoop-1.2.1/libexec/../logs/hadoop-hadoop-jobtracker-hc1nn.out
hc1r1m2: starting tasktracker, logging to
/usr/local/hadoop-1.2.1/libexec/../logs/hadoop-hadoop-tasktracker-hc1r1m2.out
hc1r1m3: starting tasktracker, logging to
/usr/local/hadoop-1.2.1/libexec/../logs/hadoop-hadoop-tasktracker-hc1r1m3.out
hc1r1m1: starting tasktracker, logging to
/usr/local/hadoop-1.2.1/libexec/../logs/hadoop-hadoop-tasktracker-hc1r1m1.out
hc1nn: starting tasktracker, logging to
/usr/local/hadoop-1.2.1/libexec/../logs/hadoop-hadoop-tasktracker-hc1nn.out
```

Note that the Job Tracker has been started on the name node and a Task Tracker on each of the data nodes. Again, check all of the logs for errors.

Running a Map Reduce Job Check

When your Hadoop V1 system has all servers up and there are no errors in the logs, you're ready to run a sample Map Reduce job to check that you can run tasks. For example, try using some data based on works by Edgar Allan Poe. I have downloaded this data from the Internet and have stored it on the Linux file system under /tmp/edgar. You could use any text-based data, however, as you just want to run a test to count some words using Map Reduce. It is not the data that is important but, rather, the correct functioning of Hadoop. To begin, go to the edgar directory, as follows:

```
cd /tmp/edgar
```

```
[hadoop@hc1nn edgar]$ ls -l
total 3868
-rw-rw-r--. 1 hadoop hadoop 632294 Feb  5  2004 10947-8.txt
-rw-r--r--. 1 hadoop hadoop 559342 Feb 23  2005 15143-8.txt
-rw-rw-r--. 1 hadoop hadoop  66409 Oct 27  2010 17192-8.txt
-rw-rw-r--. 1 hadoop hadoop 550284 Mar 16  2013 2147-8.txt
-rw-rw-r--. 1 hadoop hadoop 579834 Dec 31  2012 2148-8.txt
-rw-rw-r--. 1 hadoop hadoop 596745 Feb 17  2011 2149-8.txt
-rw-rw-r--. 1 hadoop hadoop 487087 Mar 27  2013 2150-8.txt
-rw-rw-r--. 1 hadoop hadoop 474746 Jul  1  2013 2151-8.txt
```

There are eight Linux text files in this directory that contain the test data. First, you copy this data from the Linux file system into the HDFS directory /user/hadoop/edgar using the Hadoop file system copyFromLocal command:

```
[hadoop@hc1nn edgar]$ hadoop fs -copyFromLocal /tmp/edgar /user/hadoop/edgar
```

Now, you check the files that have been loaded to HDFS:

```
[hadoop@hc1nn edgar]$ hadoop dfs -ls  /user/hadoop/edgar
```

```
Found 1 items
drwxr-xr-x   - hadoop hadoop          0 2014-09-05 20:25 /user/hadoop/edgar/edgar
```

```
[hadoop@hc1nn edgar]$ hadoop dfs -ls  /user/hadoop/edgar/edgar
```

```
Found 8 items
-rw-r--r--   2 hadoop hadoop    632294 2014-03-16 13:50 /user/hadoop/edgar/edgar/10947-8.txt
-rw-r--r--   2 hadoop hadoop    559342 2014-03-16 13:50 /user/hadoop/edgar/edgar/15143-8.txt
-rw-r--r--   2 hadoop hadoop    66409  2014-03-16 13:50 /user/hadoop/edgar/edgar/17192-8.txt
-rw-r--r--   2 hadoop hadoop    550284 2014-03-16 13:50 /user/hadoop/edgar/edgar/2147-8.txt
-rw-r--r--   2 hadoop hadoop    579834 2014-03-16 13:50 /user/hadoop/edgar/edgar/2148-8.txt
-rw-r--r--   2 hadoop hadoop    596745 2014-03-16 13:50 /user/hadoop/edgar/edgar/2149-8.txt
-rw-r--r--   2 hadoop hadoop    487087 2014-03-16 13:50 /user/hadoop/edgar/edgar/2150-8.txt
-rw-r--r--   2 hadoop hadoop    474746 2014-03-16 13:50 /user/hadoop/edgar/edgar/2151-8.txt
```

Next, you run the Map Reduce job, using the Hadoop jar command to pick up the word count from an examples jar file. This will run a word count on the Edgar Allan Poe data:

```
[hadoop@hc1nn edgar]$ cd $HADOOP_PREFIX
```

```
[hadoop@hc1nn hadoop-1.2.1]$ hadoop jar ./hadoop-examples-1.2.1.jar  wordcount
/user/hadoop/edgar  /user/hadoop/edgar-results
```

This job executes the word-count task in the jar file hadoop-examples-1.2.1.jar. It takes data from HDFS under /user/hadoop/edgar and outputs the results to /user/hadoop/edgar-results. The output of this command is as follows:

```
14/03/16 14:08:07 INFO input.FileInputFormat: Total input paths to process : 8
14/03/16 14:08:07 INFO util.NativeCodeLoader: Loaded the native-hadoop library
14/03/16 14:08:07 INFO mapred.JobClient: Running job: job_201403161357_0002
14/03/16 14:08:08 INFO mapred.JobClient:  map 0% reduce 0%
14/03/16 14:08:18 INFO mapred.JobClient:  map 12% reduce 0%
```

```
14/03/16 14:08:19 INFO mapred.JobClient:  map 50% reduce 0%
14/03/16 14:08:23 INFO mapred.JobClient:  map 75% reduce 0%
14/03/16 14:08:26 INFO mapred.JobClient:  map 75% reduce 25%
14/03/16 14:08:28 INFO mapred.JobClient:  map 87% reduce 25%
14/03/16 14:08:29 INFO mapred.JobClient:  map 100% reduce 25%
14/03/16 14:08:33 INFO mapred.JobClient:  map 100% reduce 100%
14/03/16 14:08:34 INFO mapred.JobClient: Job complete: job_201403161357_0002
14/03/16 14:08:34 INFO mapred.JobClient: Counters: 29
14/03/16 14:08:34 INFO mapred.JobClient:   Job Counters
14/03/16 14:08:34 INFO mapred.JobClient:     Launched reduce tasks=1
14/03/16 14:08:34 INFO mapred.JobClient:     SLOTS_MILLIS_MAPS=77595
14/03/16 14:08:34 INFO mapred.JobClient:     Total time spent by all reduces
waiting after reserving slots (ms)=0
14/03/16 14:08:34 INFO mapred.JobClient:     Total time spent by all maps
waiting after reserving slots (ms)=0
14/03/16 14:08:34 INFO mapred.JobClient:     Launched map tasks=8
14/03/16 14:08:34 INFO mapred.JobClient:     Data-local map tasks=8
14/03/16 14:08:34 INFO mapred.JobClient:     SLOTS_MILLIS_REDUCES=15037
14/03/16 14:08:34 INFO mapred.JobClient:   File Output Format Counters
14/03/16 14:08:34 INFO mapred.JobClient:     Bytes Written=769870
14/03/16 14:08:34 INFO mapred.JobClient:   FileSystemCounters
14/03/16 14:08:34 INFO mapred.JobClient:     FILE_BYTES_READ=1878599
14/03/16 14:08:34 INFO mapred.JobClient:     HDFS_BYTES_READ=3947632
14/03/16 14:08:34 INFO mapred.JobClient:     FILE_BYTES_WRITTEN=4251698
14/03/16 14:08:34 INFO mapred.JobClient:     HDFS_BYTES_WRITTEN=769870
14/03/16 14:08:34 INFO mapred.JobClient:   File Input Format Counters
14/03/16 14:08:34 INFO mapred.JobClient:     Bytes Read=3946741
14/03/16 14:08:34 INFO mapred.JobClient:   Map-Reduce Framework
14/03/16 14:08:34 INFO mapred.JobClient:     Map output materialized
bytes=1878641
14/03/16 14:08:34 INFO mapred.JobClient:     Map input records=72369
14/03/16 14:08:34 INFO mapred.JobClient:     Reduce shuffle bytes=1878641
14/03/16 14:08:34 INFO mapred.JobClient:     Spilled Records=256702
14/03/16 14:08:34 INFO mapred.JobClient:     Map output bytes=6493886
14/03/16 14:08:34 INFO mapred.JobClient:     CPU time spent (ms)=25930
14/03/16 14:08:34 INFO mapred.JobClient:     Total committed heap usage
(bytes)=1277771776
14/03/16 14:08:34 INFO mapred.JobClient:     Combine input records=667092
14/03/16 14:08:34 INFO mapred.JobClient:     SPLIT_RAW_BYTES=891
14/03/16 14:08:34 INFO mapred.JobClient:     Reduce input records=128351
14/03/16 14:08:34 INFO mapred.JobClient:     Reduce input groups=67721
14/03/16 14:08:34 INFO mapred.JobClient:     Combine output records=128351
14/03/16 14:08:34 INFO mapred.JobClient:     Physical memory (bytes)
snapshot=1508696064
14/03/16 14:08:34 INFO mapred.JobClient:     Reduce output records=67721
14/03/16 14:08:34 INFO mapred.JobClient:     Virtual memory (bytes)
snapshot=4710014976
14/03/16 14:08:34 INFO mapred.JobClient:     Map output records=667092
```

To take a look at the results (found in the HDFS directory /user/hadoop/edgar-results), use the Hadoop file system ls command:

```
[hadoop@hc1nn hadoop-1.2.1]$ hadoop fs -ls /user/hadoop/edgar-results
Found 3 items
-rw-r--r--   1 hadoop supergroup             0 2014-03-16 14:08
/user/hadoop/edgar-results/_SUCCESS
drwxr-xr-x   - hadoop supergroup             0 2014-03-16 14:08
/user/hadoop/edgar-results/_logs
-rw-r--r--   1 hadoop supergroup        769870 2014-03-16 14:08
/user/hadoop/edgar-results/part-r-00000
```

This shows that the the word-count job has created a file called _SUCCESS to indicate a positive outcome. It has created a log directory called _logs and a data file called part-r-00000. The last file in the list, the part file, is of the most interest. You can extract it from HDFS and look at the contents by using the Hadoop file system cat command:

```
doop@hc1nn hadoop-1.2.1]$ mkdir -p /tmp/hadoop/
doop@hc1nn hadoop-1.2.1]$ hadoop fs -cat
/user/hadoop/edgar-results/part-r-00000 > /tmp/hadoop/part-r-00000
```

The results reveal that the test job produced a results file containing 67,721 records. You can show this by using the Linux command wc -l to produce a line count of the results file:

```
[hadoop@hc1nn hadoop-1.2.1]$ wc -l /tmp/hadoop/part-r-00000
67721 /tmp/hadoop/part-r-00000
```

By using the Linux head command with a -20 option, you can look at the first 20 lines of the output part file on the Linux file system:

```
[hadoop@hc1nn hadoop-1.2.1]$ head -20 /tmp/hadoop/part-r-00000
!             1
"             22
"''T          1
"'--          1
"'A           1
"'After       1
"'Although    1
"'Among       2
"'And         2
"'Another     1
"'As          2
"'At          1
"'Aussi       1
"'Be          2
"'Being       1
"'But         1
"'But,'       1
"'But--still--monsieur----' 1
"'Catherine, 1
"'Comb        1
```

Clearly, the Hadoop V1 installation is working and can run a Map Reduce task. (The word-count algorithm does not seem to strip out characters like quotation marks; this is not an issue for our purposes here, but might be if you wanted a truly accurate count of the number of times Poe used particular words.)

Hadoop User Interfaces

Up to this point you have installed the release, configured it, and run a simple Map Reduce task to prove that it is working. But how can you visually examine the Hadoop distributed file system or determine the state of the Hadoop servers? Well, Hadoop provides a set of built-in user interfaces for this purpose. They are quite basic, but it is worthwhile knowing about them. (In a large production system, of course, you would use one of the more functional systems like Ambari for monitoring.)

In this example configuration, you can find the name node UI on port 50070 with a URL of http://hc1nn:50070/ (on the name node hc1nn). This port was defined via the value of the dfs.http.address in the configuration file hdfs-site.xml. The name node UI shows storage information and node basics, as illustrated in Figure 2-3. It is also possible to browse the Hadoop file system and logs to determine the state of the nodes. The levels of HDFS storage used and free can also be determined.

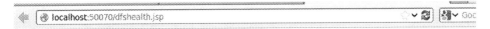

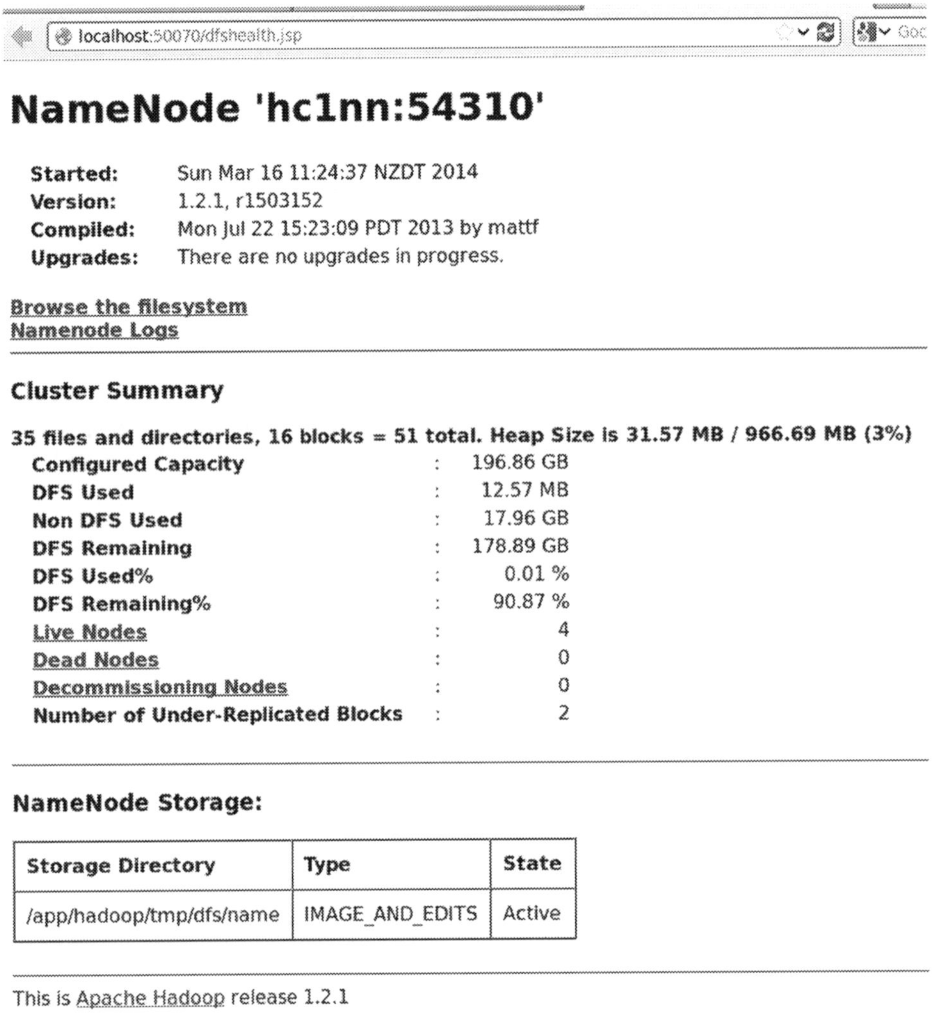

localhost:50070/dfshealth.jsp
NameNode 'hc1nn:54310'
Started: Sun Mar 16 11:24:37 NZDT 2014
Version: 1.2.1, r1503152
Compiled: Mon Jul 22 15:23:09 PDT 2013 by mattf
Upgrades: There are no upgrades in progress.
Browse the filesystem
Namenode Logs
Cluster Summary
35 files and directories, 16 blocks = 51 total. Heap Size is 31.57 MB / 966.69 MB (3%)
Configured Capacity : 196.86 GB
DFS Used : 12.57 MB
Non DFS Used : 17.96 GB
DFS Remaining : 178.89 GB
DFS Used% : 0.01 %
DFS Remaining% : 90.87 %
Live Nodes : 4
Dead Nodes : 0
Decommissioning Nodes : 0
Number of Under-Replicated Blocks : 2
NameNode Storage:
Storage Directory | Type | State
/app/hadoop/tmp/dfs/name | IMAGE_AND_EDITS | Active
This is Apache Hadoop release 1.2.1

Figure 2-3. *The name node user interface*

To see the administration information for Map Reduce, go to port 50030 by using the URL http://hc1nn:50030/, shown in Figure 2-4. This port number refers to the value of the variable mapred.job.tracker.http.address already defined in the mapred-site.xml configuration file. Figure 2-4 shows the state of the jobs, as well as the capacity per node in terms of Map and Reduce. It also shows an example word-count job that's running currently, which that has completed the Map stage and is 8.33 percent into its Reduce phase.

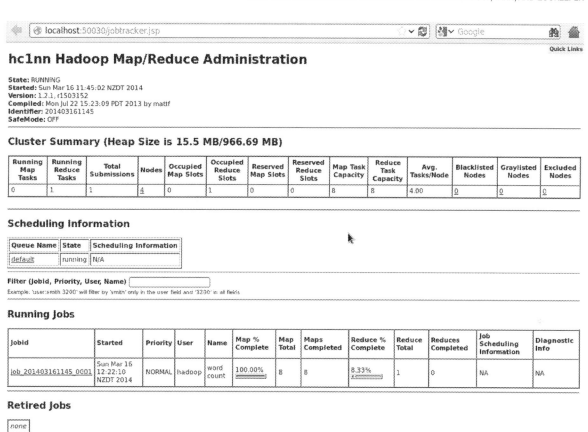

Figure 2-4. *Hadoop Job Tracker user interface*

The Task Tracker UI is on port 50060; this is the value defined by the variable mapred.task.tracker.http.address in the configuration file mapred-site.xml. Use the URL http://hc1nn:50060/ (on the name node hc1nn) to access it and check the status of current tasks. Figure 2-5 shows running and non-running tasks, as well as providing a link to the log files. It also offers a basic list of task statuses and their progress.

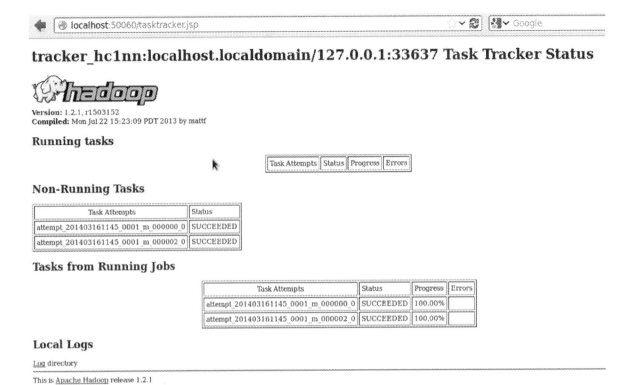

Figure 2-5. *The Task Tracker user interface*

Now that you have tasted the flavor of Hadoop V1, shut it down and get ready to install Hadoop V2.

Hadoop V2 Installation

In moving on to Hadoop V2, you will this time download and use the Cloudera stack. Specifically, you will install CDH 4 because it is available for both 32-bit and 64-bit machines and it supports YARN. I have chosen to install the latest manual CDH release available at the time of this writing.

In this section, you will not only learn how to obtain and install the Cloudera Hadoop packages; you'll also find out how to install, run, and use ZooKeeper, as well as how to configure Hadoop V2. You'll tour the necessary directories on the Linux file system and HDFS; lastly, you'll find out how to control the servers as Linux services.

To begin, you need to add a CDH repository file on all machines as root under /etc/yum.repos.d/. You create a file named cloudera-cdh4.repo on each server, with the following contents:

```
root@hc1r1m1 yum.repos.d]# cat cloudera-cdh4.repo
[cloudera-cdh4]
name=Cloudera's Distribution for Hadoop, Version 4
baseurl=http://archive.cloudera.com/cdh4/redhat/6/x86_64/cdh/4/
gpgkey = http://archive.cloudera.com/cdh4/redhat/6/x86_64/cdh/RPM-GPG-KEY-cloudera
gpgcheck = 1
```

The Linux cat command shows the contents of the cloudera-cdh4.repo file. The ls -l command shows that it is owned by the root Linux user:

```
[root@hc1r1m1 yum.repos.d]# ls -l cloudera-cdh4.repo
-rw-r--r-- 1 root root 229 Sep  6 09:24 cloudera-cdh4.repo
```

This repository configuration file tells the yum command where to source Cloudera cdh4 software. The above file is configured for a 64-bit machine (x86_64); a 32-bit machine would have the following lines:

```
baseurl=http://archive.cloudera.com/cdh4/redhat/6/i386/cdh/4/
gpgkey = http://archive.cloudera.com/cdh4/redhat/6/i386/cdh/RPM-GPG-KEY-cloudera
```

At this point, Cloudera advises you to install ZooKeeper so that you have a ZooKeeper cluster; you'll do so in the next section. Tools like HBase (the distributed database that will be introduced in Chapter 3) rely on it.

ZooKeeper Installation

ZooKeeper is a centralized service for maintaining configuration information in distributed applications. Many of the tools in the Hadoop ecosystem use it, so it will be helpful to install it now because you will need it later. You install the ZooKeeper base package as follows:

```
yum install zookeeper
```

When you install ZooKeeper, remember that it needs to be installed on an odd number of servers—for example, three machines. (When voting on an outcome, the odd number of servers makes it possible for ZooKeper to reach a majority decision.) Next, you install the ZooKeeper server on each node:

```
yum install zookeeper-server
```

After installation, the ZooKeeper configuration needs to be altered for your environment. By default, the configuration file is stored under /etc/zookeeper/conf/zoo.cfg. Its initial contents are the following:

```
maxClientCnxns=50
# The number of milliseconds of each tick
tickTime=2000
# The number of ticks that the initial
# synchronization phase can take
initLimit=10
# The number of ticks that can pass between
# sending a request and getting an acknowledgement
syncLimit=5
# the directory where the snapshot is stored.
dataDir=/var/lib/zookeeper
# the port at which the clients will connect
clientPort=2181
```

You need to add to these a section at the bottom of the file to define the port ranges used by ZooKeeper clients on each of the servers. For instance, on server hc1r1m1, the port range for ZooKeeper server 1 is 60050 to 61050. That allows for 1,000 client connections per ZooKeeper server.

```
server.1=hc1r1m1:60050:61050
server.2=hc1r1m2:60050:61050
server.3=hc1r1m3:60050:61050
#server.4=hc1nn:60050:61050
```

Note also that you have three servers defined and a fourth set up, but commented out to ensure that there will be an odd number of ZooKeeper instances to form a quorum. Now, you initialize the Hadoop V2 installation:ZooKeeper installation:

```
service zookeeper-server init
```

Edit the files /var/lib/zookeeper/myid on each server, entering integer numbers to match the configuration file. For instance, for the setup on hc1r1m1, you add a "1" to the file, and so on. This allows each ZooKeeper server to determine its ID number and so recognize its port range from the configuration file.

You now start ZooKeeper on hc1r1m1, hc1r1m2, and hc1r1m3, using the service command:

```
service zookeeper-server start

JMX enabled by default
Using config: /etc/zookeeper/conf/zoo.cfg
Starting zookeeper ... STARTED
```

Under /var/log/zookeeper/, you check the logs to ensure everything is running correctly:

```
-rw-r--r--. 1 zookeeper zookeeper 21450 Mar 20 18:54 zookeeper.log
-rw-r--r--. 1 zookeeper zookeeper     0 Mar 20 18:50 zookeeper.out
```

You'll likely see errors indicating that the servers can't reach each other, meaning that the firewall is interfering again. You need to open the ports that ZooKeeper uses and then restart both Iptables and the ZooKeeper server for the changes to be picked up. If you are unsure how to configure your firewall, approach your systems administrator.

```
[root@hc1r1m3 conf]# service zookeeper-server restart

Restarting ZooKeeper daemon: JMX enabled by default
Using config: /etc/zookeeper/conf/zoo.cfg
Stopping zookeeper ... STOPPED
JMX enabled by default
Using config: /etc/zookeeper/conf/zoo.cfg
Starting zookeeper ... STARTED
```

After you restart all of the ZooKeeper servers, they now will run as desired and will contain no errors in their log files, as shown:

```
2014-03-20 19:09:34,011 [myid:3] - INFO
[QuorumPeer[myid=3]/0.0.0.0:2181:Environment@100] - Server
environment:os.arch=i386
2014-03-20 19:09:34,012 [myid:3] - INFO
[QuorumPeer[myid=3]/0.0.0.0:2181:Environment@100] - Server
```

```
environment:os.version=2.6.32-220.el6.i686
2014-03-20 19:09:34,012 [myid:3] - INFO
[QuorumPeer[myid=3]/0.0.0.0:2181:Environment@100] - Server
environment:user.name=zookeeper
2014-03-20 19:09:34,013 [myid:3] - INFO
[QuorumPeer[myid=3]/0.0.0.0:2181:Environment@100] - Server
environment:user.home=/var/run/zookeeper
2014-03-20 19:09:34,014 [myid:3] - INFO
[QuorumPeer[myid=3]/0.0.0.0:2181:Environment@100] - Server
environment:user.dir=/
2014-03-20 19:09:34,028 [myid:3] - INFO
[QuorumPeer[myid=3]/0.0.0.0:2181:ZooKeeperServer@162] - Created server with
tickTime 2000 minSessionTimeout 4000 maxSessionTimeout 40000 datadir
/var/lib/zookeeper/version-2 snapdir /var/lib/zookeeper/version-2
2014-03-20 19:09:34,030 [myid:3] - INFO
[QuorumPeer[myid=3]/0.0.0.0:2181:Follower@63] - FOLLOWING - LEADER ELECTION
TOOK - 83
2014-03-20 19:09:34,145 [myid:3] - INFO
[QuorumPeer[myid=3]/0.0.0.0:2181:Learner@325] - Getting a snapshot from leader
2014-03-20 19:09:34,154 [myid:3] - INFO
[QuorumPeer[myid=3]/0.0.0.0:2181:FileTxnSnapLog@273] - Snapshotting:
0x100000000 to /var/lib/zookeeper/version-2/snapshot.100000000
```

Manually Accessing the ZooKeeper Servers

Using the server name, main port number, and some four-letter words, you can manually access the ZooKeeper servers. Specifically, you can use the nc command to issue additional four-letter commands. This type of access to ZooKeeper might be useful when you're investigating problems with the servers or just checking that all is okay.

For this setup, the configuration file lists the main port on each server as 2181. To access the configuration details for server hc1r1m2, therefore, you use the nc command to issue a conf command. Press Enter after both the nc command line and the conf command on the following line:

```
[hadoop@hc1r1m2 ~]$ nc  hc1r1m2 2181
conf

clientPort=2181
dataDir=/var/lib/zookeeper/version-2
dataLogDir=/var/lib/zookeeper/version-2
tickTime=2000
maxClientCnxns=50
minSessionTimeout=4000
maxSessionTimeout=40000
serverId=2
initLimit=10
syncLimit=5
electionAlg=3
electionPort=61050
quorumPort=60050
peerType=0
```

This has outputted the configuration of the ZooKeeper server on hc1r1m2. While it's on server hc1r1m2, you can check that the ZooKeeper server on hc1r1m1 is still running by using the ruok (running okay?) command:

```
[hadoop@hc1r1m2 ~]$ nc  hc1r1m1 2181
ruok

imok
```

The returned value of imok (I'm OK!) tells you the contacted server is running. You can then get the connection details for server hc1r1m1 by using the cons command:

```
[hadoop@hc1r1m2 ~]$ nc  hc1r1m1 2181
Cons

 /192.168.1.106:42731[0](queued=0,recved=1,sent=0)
```

The response says there is just a single connection to that server. To supplement this with some basic server details, you use the stat command:

```
[hadoop@hc1r1m2 ~]$ nc  hc1r1m1 2181
stat

Zookeeper version: 3.4.5-cdh4.6.0--1, built on 02/26/2014 09:19 GMT
Clients:
 /192.168.1.106:42732[0](queued=0,recved=1,sent=0)
Latency min/avg/max: 0/11/93
Received: 18
Sent: 17
Connections: 1
Outstanding: 0
Zxid: 0x300000006
Mode: follower
Node count: 4
```

This result provides status information for the ZooKeeper server on hc1r1m1. It has listed information like the installed version and the volume of messaging. (You can find a full list of ZooKeeper commands on the Cloudera site at archive.cloudera.com/cdh4/4/zookeeper.)

The ZooKeeper Client

An alternative to the nc command method is to use the built-in ZooKeeper client to access your servers. You can find it with the type command, as follows:

```
[hadoop@hc1r1m3 ~]$ type zookeeper-client
zookeeper-client is /usr/bin/zookeeper-client
```

By default, the client connects to ZooKeeper on the local server:

```
[hadoop@hc1r1m3 ~]$ zookeeper-client
Connecting to localhost:2181
```

You can also get a list of possible commands by entering any unrecognized command, such as help:

```
[zk: localhost:2181(CONNECTED) 1] help
ZooKeeper -server host:port cmd args
  connect host:port
  get path [watch]
  ls path [watch]
  set path data [version]
  rmr path
  delquota [-n|-b] path
  quit
  printwatches on|off
  create [-s] [-e] path data acl
  stat path [watch]
  close
  ls2 path [watch]
  history
  listquota path
  setAcl path acl
  getAcl path
  sync path
  redo cmdno
  addauth scheme auth
  delete path [version]
  setquota -n|-b val path
```

To connect to one of the other ZooKeeper servers in the quorum, you would use the connect command, specifying the server and its connection port. If, for example, you're currently on hc1r1m3, but want to connect to hc1r1m1 (remember, all servers are connected via port 2181), you use the following:

```
[zk: localhost:2181(CONNECTED) 0] connect hc1r1m1:2181

[zk: hc1r1m1:2181(CONNECTING) 1] 2014-03-22 13:28:05,898 [myid:] - INFO  [main-SendThread(hc1r1m1
:2181):ClientCnxn$SendThread@852] - Socket connection established to hc1r1m1/192.168.1.104:2181,
initiating session
2014-03-22 13:28:05,913 [myid:] - INFO  [main-SendThread(hc1r1m1:2181):ClientCnxn$SendThread@1214] -
Session establishment complete on server hc1r1m1/192.168.1.104:2181, sessionid = 0x144e6df88b70004,
negotiated timeout = 30000

[zk: hc1r1m1:2181(CONNECTED) 1]
```

The result tells you that you are now connected to a ZooKeeper session on client node hc1r1m1 from hc1r1m3.

So far, you have ZooKeeper installed and have learned how to access it manually. In a large cluster, however, a distributed application would connect to ZooKeeper automatically through one of its language APIs. It would use the hierarchy of ZNodes to store configuration information. So, at this point you should examine the ZooKeeper architecture in terms of those ZNodes to understand how they might be used for cluster configuration and monitoring.

ZooKeeper stores its data in a hierarchy of nodes called ZNodes, each designed to contain a small amount of data. When you log into the client, you can think of your session as similar to a Unix shell. Just as you can create directories and files in a shell, so you can create ZNodes and data in the client.

Try creating an empty topmost node named "zk-top," using this syntax:

```
[zk: localhost:2181(CONNECTED) 4] create /zk-top  ''
Created /zk-top
```

You can create a subnode, node1, of zk-top as well; you can add the contents cfg1 at the same time:

```
[zk: localhost:2181(CONNECTED) 5] create /zk-top/node1 'cfg1'
Created /zk-top/node1
```

To check the contents of the subnode (or any node), you use the get command:

```
 [zk: localhost:2181(CONNECTED) 6] get /zk-top/node1
'cfg1'
```

The delete command, not surprisingly, deletes a node:

```
[zk: localhost:2181(CONNECTED) 8] delete /zk-top/node2
```

The set command changes the context of a node. This command sequence first checks the contents of node1 with get, changes it with set, and then displays the new contents:

```
[zk: localhost:2181(CONNECTED) 9] get /zk-top/node1
'cfg1'
```

```
[zk: localhost:2181(CONNECTED) 10] set /zk-top/node1 'cfg2'
```

```
[zk: localhost:2181(CONNECTED) 11] get /zk-top/node1
'cfg2'
```

The contents of node1 changed from cfg1 to cfg2. Although this is a simple example, it explains the principle.

You can also list the subnodes of a node with the ls command. For example, node1 has no subnodes, but zk-top contains node1:

```
[zk: localhost:2181(CONNECTED) 0] ls /zk-top/node1
[]
[zk: localhost:2181(CONNECTED) 1] ls /zk-top
[node1]
```

You can place watches on the nodes to check whether they change. Watches are one-time events. If the contents change, then the watch fires and you will need to reset it. To demonstrate, I create a subnode node2 that contains data2:

```
[zk: localhost:2181(CONNECTED) 9] create /zk-top/node2  'data2'
```

```
[zk: localhost:2181(CONNECTED) 10] get /zk-top/node2
'data2'
```

Now, I use get to set a watcher on that node. The first command sets a watcher on the node2 data item "data2." When I change the data to "data3" with the next set command, the watcher notices the data change and fires, as shown:

```
[zk: localhost:2181(CONNECTED) 11] get /zk-top/node2 true
'data2'

[zk: localhost:2181(CONNECTED) 12] set /zk-top/node2 'data3'

WATCHER::

WatchedEvent state:SyncConnected type:NodeDataChanged path:/zk-top/node2
```

In addition to the basic nodes you've been working with, you can create sequential and ephemeral nodes with the create command. *Ephemeral* nodes exist only for the lifetime of the current session, while *sequential* nodes have a sequence number applied to the node name and will persist. To create a sequential node, use the –s option, as shown:

```
[zk: localhost:2181(CONNECTED) 13] ls /zk-top
[node2, node1]
[zk: localhost:2181(CONNECTED) 14] create -s /zk-top/node3 'data3'
Created /zk-top/node30000000005
[zk: localhost:2181(CONNECTED) 15] ls /zk-top
[node2, node1, node30000000005]
```

To create an ephemeral node, use the –e option:

```
[zk: localhost:2181(CONNECTED) 16] create -e /zk-top/node4 'data4'
Created /zk-top/node4
[zk: localhost:2181(CONNECTED) 17] ls /zk-top
[node4, node2, node1, node30000000005]
```

So, node 4 exists in this session under zk-top. It will disappear when you log out of the session (with quit) and start a new one. Notice that the ls command at the end of this sequence no longer lists node4:

```
[zk: localhost:2181(CONNECTED) 18] quit
Quitting...
2014-03-22 14:02:56,572 [myid:] - INFO  [main:ZooKeeper@684] - Session: 0x344e6df92ab0005 closed
2014-03-22 14:02:56,572 [myid:] - INFO  [main-EventThread:ClientCnxn$EventThread@512] - EventThread
shut down
[hadoop@hc1r1m3 ~]$
[hadoop@hc1r1m3 ~]$ zookeeper-client

[zk: localhost:2181(CONNECTED) 0] ls /zk-top
[node2, node1, node30000000005]
```

As you can see, node 4 is gone; it only existed for the session in which it was created. Note: you can also make sequential nodes ephemeral with a command like this:

```
[zk: localhost:2181(CONNECTED) 16] create -s -e /zk-top/node4 'data4'
```

ZooKeeper ZNodes and their commands are building blocks for meeting your needs. Ephemeral nodes are especially useful in a distributed clustered environment, for example. In each session, you could create a node to see which application nodes were connected. You could also store all your configuration information in a series of ZooKeeper ZNodes and have the applications use that configuration information on each node. You would then be able to ensure that the nodes were using the same configuration information.

This has been a basic introduction to ZooKeeper. For further reading, have a look at Cloudera's site or perhaps have a go at building your own distributed application.

Hadoop MRv2 and YARN

With ZooKeeper in place, you can continue installing the Cloudera CDH 4 release. The components will be installed using yum commands as root to install Cloudera packages. I chose to install a Cloudera stack because the installation has been professionally tested and packaged. The components are guaranteed to work together and with a range of Hadoop client applications. The instructions that follow describe the installation of the Name Node, Data Node, Resource Manager, Node Manager, Job History, and Proxy servers.

In comparison to the V1 installation, you do not have to choose the location for your installation; that is done automatically and the different parts of the installation are placed in meaningful locations. Configuration is placed under /etc/hadoop, logs are placed under /var/log, and executables are created as Linux servers under /etc/init.d. Here's the process:

1. Install the HDFS Name Node component on the master server hc1nn:

    ```
    [root@hc1nn ~]# yum install hadoop-hdfs-namenode
    ```

2. Install the HDFS Data Node component on the slave servers hc1r1m1 through 3:

    ```
    [root@hc1r1m1 ~]# yum install   hadoop-hdfs-datanode
    ```

3. Install the Resource Manager component on the Name Node machine hc1nn:

    ```
    [root@hc1nn ~]# yum install hadoop-yarn-resourcemanager
    ```

4. Install the Node Manager and Map Reduce on all of the Data Node slave servers hc1r1m1 through 3:

    ```
    [root@hc1r1m1 ~]# yum install hadoop-yarn-nodemanager   hadoop-mapreduce
    ```

5. Install the Job History and Proxy servers on a single node:

    ```
    yum install hadoop-mapreduce-historyserver hadoop-yarn-proxyserver
    ```

That concludes the component package installations.

Now that the software is installed, you need to set up the configuration files that they depend upon. You can find these configuration files under the directory /etc/hadoop/conf. They all have names like <component>-site.xml, where <component> is replaced by yarn, hdfs, mapred, or core.

The HDFS term you have come across already; it is the Hadoop distributed file system. YARN stands for "yet another resource negotiator." The MAPRED component is short for "Map Reduce," and CORE is the configuration for the Hadoop common utilities that support other Hadoop functions.

Use the ls command to view the configuration files that need to be altered:

```
[root@hc1nn conf]# cd /etc/hadoop/conf
[root@hc1nn conf]# ls -l *-site.xml
-rw-r--r--. 1 root root  904 Feb 26 23:17 core-site.xml
-rw-r--r--. 1 root root 1023 Feb 26 23:17 hdfs-site.xml
-rw-r--r--. 1 root root  904 Feb 26 23:17 mapred-site.xml
-rw-r--r--. 1 root root 2262 Feb 26 23:17 yarn-site.xml
```

On each node, set up the core-site.xml as follows:

```
<configuration>

  <property>
    <name>fs.defaultFS</name>
    <value>hdfs://hc1nn/</value>
  </property>

</configuration>
```

You need hdfs-site.xml on each node, as well. To set it up, use the form:

```
<configuration>

  <property>
     <name>dfs.namenode.name.dir</name>
     <value>/var/lib/hadoop-hdfs/cache/hdfs/dfs/name</value>
     <description> Can be a comma separated list of values </description>
  </property>

  <property>
    <name>dfs.permissions.superusergroup</name>
    <value>hadoop</value>
  </property>

  <property>
    <name>dfs.replication</name>
    <value>2</value>
  </property>

</configuration>
```

Remember, however, to define dfs.namenode.name.dir on the name node and dfs.datanode.data.dir on the data node. So, the entry for the data nodes would be as follows:

```
  <property>
     <name>dfs.datanode.name.dir</name>
     <value>/var/lib/hadoop-hdfs/cache/hdfs/dfs/name</value>
     <description> Can be a comma separated list of values </description>
  </property>
```

On each node, make sure that the directories used in the configuration files exist:

```
[root@hc1nn conf]# mkdir -p /var/lib/hadoop-hdfs/cache/hdfs/dfs/name
```

Next, set the ownership of these directories:

```
[root@hc1nn conf]# chown -R hdfs:hdfs /var/lib/hadoop-hdfs/cache/hdfs/dfs/name
```

```
[root@hc1nn conf]# chmod 700 /var/lib/hadoop-hdfs/cache/hdfs/dfs/name
```

The preceding commands create the name directory, change the ownership to the hdfs user, and set its permissions. Before starting the name node, though, you have to format the file system (as the hdfs user):

```
 [root@hc1nn conf]# su - hdfs
-bash-4.1$ hdfs namenode -format
```

Now, you set up the file mapred-site.xml, setting the framework to be yarn:

```
<configuration>

  <property>
    <name>mapreduce.framework.name</name>
    <value>yarn</value>
  </property>

</configuration>
```

Next, you set up the file yarn-site.xml. There is a lot of configuration information here, including port addresses, file system paths, and class path information. (For a full list of available configuration file options, check the Hadoop site at hadoop.apache.org.)

```
<configuration>

  <property>
    <name>yarn.nodemanager.aux-services</name>
    <value>mapreduce.shuffle</value>
  </property>

  <property>
    <name>yarn.nodemanager.aux-services.mapreduce.shuffle.class</name>
    <value>org.apache.hadoop.mapred.ShuffleHandler</value>
  </property>

  <property>
    <name>yarn.resourcemanager.address</name>
    <value>hc1nn:8032</value>
  </property>

  <property>
    <name>yarn.resourcemanager.scheduler.address</name>
    <value>hc1nn:8030</value>
  </property>
```

```
<property>
  <name> yarn.resourcemanager.resource-tracker.address</name>
  <value>hc1nn:8031</value>
</property>

<property>
  <name>yarn.resourcemanager.admin.address</name>
  <value>hc1nn:8033</value>
</property>

<property>
  <name>yarn.resourcemanager.webapp.address</name>
  <value>hc1nn:8088</value>
</property>

<property>
  <name>yarn.log-aggregation-enable</name>
  <value>true</value>
</property>

<property>
  <description>List of directories to store localized files in.</description>
  <name>yarn.nodemanager.local-dirs</name>
  <value>/var/lib/hadoop-yarn/cache/${user.name}/nm-local-dir</value>
</property>

<property>
  <description>Where to store container logs.</description>
  <name>yarn.nodemanager.log-dirs</name>
  <value>/var/log/hadoop-yarn/containers</value>
</property>

<property>
  <description>Where to aggregate logs to.</description>
  <name>yarn.nodemanager.remote-app-log-dir</name>
  <value>/var/log/hadoop-yarn/apps</value>
</property>

<property>
  <description>Classpath for typical applications.</description>
   <name>yarn.application.classpath</name>
   <value>
      $HADOOP_CONF_DIR,
      $HADOOP_COMMON_HOME/*,$HADOOP_COMMON_HOME/lib/*,
      $HADOOP_HDFS_HOME/*,$HADOOP_HDFS_HOME/lib/*,
      $HADOOP_MAPRED_HOME/*,$HADOOP_MAPRED_HOME/lib/*,
      $YARN_HOME/*,$YARN_HOME/lib/*
   </value>
</property>

</configuration>
```

Now, the directories used by YARN, as shown in the configuration above, will need to be created and ownership and group membership will have to be set to the YARN Linux user and group, as follows:

```
[root@hc1nn conf]# mkdir -p /var/log/hadoop-yarn/containers
[root@hc1nn conf]# mkdir -p /var/log/hadoop-yarn/apps
[root@hc1nn conf]# chown -R yarn:yarn /var/log/hadoop-yarn/containers
[root@hc1nn conf]# chown -R yarn:yarn /var/log/hadoop-yarn/apps
[root@hc1nn conf]# chmod 755 /var/log/hadoop-yarn/containers
[root@hc1nn conf]# chmod 755 /var/log/hadoop-yarn/apps
```

You add the following configuration to mapred-site.xml for the Job History server:

```
<property>
  <name>mapreduce.jobhistory.address</name>
  <value>hc1nn:10020</value>
</property>

<property>
  <name>mapreduce.jobhistory.webapp.address</name>
  <value>hc1nn:19888</value>
</property>
```

You also need to set up the staging directory in yarn-site.xml. If you don't specifically create one, YARN will create a directory under /tmp. To create the directory, add the following xml to yarn-site.xml:

```
<property>
    <name>mapreduce.jobhistory.intermediate-done-dir</name>
    <value>/var/lib/hadoop-mapreduce/jobhistory/intermediate/donedir</value>
</property>

<property>
    <name>mapreduce.jobhistory.done-dir</name>
    <value>/var/lib/hadoop-mapreduce/jobhistory/donedir</value>
</property>
```

The directories needed for staging must be created on the file system. You set their ownership and group membership to yarn, then set the permissions:

```
[root@hc1nn conf]# mkdir -p /var/lib/hadoop-mapreduce/jobhistory/intermediate/donedir
[root@hc1nn conf]# mkdir -p /var/lib/hadoop-mapreduce/jobhistory/donedir


[root@hc1nn conf]# chown -R yarn:yarn  /var/lib/hadoop-mapreduce/jobhistory/intermediate/donedir
[root@hc1nn conf]# chown -R yarn:yarn /var/lib/hadoop-mapreduce/jobhistory/donedir


[root@hc1nn conf]# chmod 1777 /var/lib/hadoop-mapreduce/jobhistory/intermediate/donedir
[root@hc1nn conf]# chmod 750 /var/lib/hadoop-mapreduce/jobhistory/donedir
```

Now it's time to start the Hadoop servers. On the Name Node machine (hc1nn), run the following as root to start the HDFS, YARN, and History servers:

```
service hadoop-hdfs-namenode start
service hadoop-mapreduce-historyserver start
service hadoop-yarn-resourcemanager start
service hadoop-yarn-proxyserver start
```

On the Data Node machines (hc1r1m1 to hc1r1m3), run the following as root to start the data node and Node Manager:

```
service hadoop-hdfs-datanode start
service hadoop-yarn-nodemanager start
```

You can make sure that Hadoop is writing to HDFS by executing the Hadoop file system ls command on the /directory:

```
[root@hc1nn conf]# su - hdfs

-bash-4.1$ hadoop fs -ls /
```

Also, ensure that the temporary directory exists on HDFS:

```
-bash-4.1$ hadoop fs -mkdir /tmp
-bash-4.1$
-bash-4.1$ hadoop fs -chmod -R 1777 /tmp
```

The permissions "1777" means that any file system user can write to the directory but cannot remove another user's files.

You need to create the required directories under HDFS; these directories will be owned and used by YARN for logging and history data:

```
[root@hc1nn hadoop-hdfs]# su - hdfs
-bash-4.1$
-bash-4.1$ hadoop fs -mkdir /user/history
-bash-4.1$ hadoop fs -chmod -R 1777 /user/history
-bash-4.1$ hadoop fs -chown yarn /user/history
-bash-4.1$ hadoop fs -mkdir /var/log/hadoop-yarn
-bash-4.1$ hadoop fs -chown yarn:mapred /var/log/hadoop-yarn
```

At this point, check which top-level directories exist on the Hadoop distributed file system (HDFS). Because you have executed a long list (ls -l) and have used a recursive switch (-R), you can also see which subdirectories exist. The following listing shows permissions (drwxrwxrw), plus details like ownership and group membership (hdfs hadoop):

```
-bash-4.1$ hadoop fs -ls -R /
drwxrwxrwt   - hdfs hadoop        0 2014-03-23 14:58 /tmp
drwxr-xr-x   - hdfs hadoop        0 2014-03-23 14:55 /user
drwxrwxrwt   - yarn hadoop        0 2014-03-23 14:55 /user/history
drwxr-xr-x   - hdfs hadoop        0 2014-03-23 14:56 /var
drwxr-xr-x   - hdfs hadoop        0 2014-03-23 14:56 /var/log
drwxr-xr-x   - yarn mapred        0 2014-03-23 14:56 /var/log/hadoop-yarn
```

So, you have the /tmp and /user directories, and the /var/log directory exists with a subdirectory for YARN. Now, you need to create home directories for the Map Reduce users on HDFS. In this case there is only the hadoop account, so you change user (su) to the Linux hdfs account, then use the hadoop file system command mkdir to create the directory and use chown to set its ownership to hadoop:

```
[root@hc1nn sysconfig]# su - hdfs
-bash-4.1$
-bash-4.1$ hadoop fs -mkdir  /user/hadoop
-bash-4.1$ hadoop fs -chown hadoop /user/hadoop
```

Last step is to set up the Map Reduce user environment in the Bash shell by setting some environmental options in the Bash configuration file .bashrc. As in the Hadoop V1 installation, this allows you to set environment variables like JAVA_HOME and HADOOP_MAPRED_HOME in the Bash shell. Each time the Linux account is accessed and a Bash shell is created, these variables will be pre-defined.

```
[hadoop@hc1nn ~]$ tail .bashrc

#######################################################
# Set Hadoop related env variables

# set JAVA_HOME (we will also set a hadoop specific value later)
export JAVA_HOME=/usr/lib/jvm/jre-1.6.0-openjdk

export HADOOP_MAPRED_HOME=/usr/lib/hadoop-mapreduce
```

At this point you have completed the configuration of your installation and you are ready to start the servers. Remember to monitor the logs under /var/log for server errors; when the servers start, they state the location where they are logging to.

You start the HDFS servers and monitor the logs for errors:

```
[root@hc1nn init.d]# cd /etc/init.d
[root@hc1nn init.d]# ls -ld hadoop-hdfs-*
-rwxr-xr-x. 1 root root 4469 Feb 26 23:18 hadoop-hdfs-namenode

[root@hc1nn init.d]# service hadoop-hdfs-namenode start
Starting Hadoop namenode:                               [  OK  ]
starting namenode, logging to /var/log/hadoop-hdfs/hadoop-hdfs-namenode-hc1nn.out
```

If you start the name node on the master and the data nodes on the slaves, check that the necessary ports are open in the firewall. For example, I check that port 8020 is open in the firewall configuration. I also show how the Iptables (Linux kernel firewall) service can be restarted. (I do not provide an in-depth study of the firewall configuration, as that leads us into the realm of systems administration, which is a separate field.)

```
[root@hc1nn sysconfig]# cd /etc/sysconfig
[root@hc1nn sysconfig]# grep 8020 iptables
-A INPUT -m state --state NEW -m tcp -p tcp --dport 8020 -j ACCEPT
```

So, this result tells the firewall to accept tcp-based requests on port 8020.

If the ports are not open, then Iptables needs to be updated with an entry similar to the last line above and the server restarted, as shown next. (If you are unsure about this, consult your systems administrator.)

```
[root@hc1nn sysconfig]# service iptables restart
iptables: Setting chains to policy ACCEPT: filter          [  OK  ]
iptables: Flushing firewall rules:                         [  OK  ]
iptables: Unloading modules:                               [  OK  ]
iptables: Applying firewall rules:                         [  OK  ]
```

You are now ready to start the YARN servers. First, start the Resource Manager on the master, then start the Node managers on the data nodes. Finally, check the logs for errors. The sequence of commands you need for doing this is as follows:

```
[root@hc1nn init.d]# service hadoop-yarn-resourcemanager start
Starting Hadoop resourcemanager:                           [  OK  ]
starting resourcemanager, logging to /var/log/hadoop-yarn/yarn-yarn-resourcemanager-hc1nn.out

[root@hc1r1m1 init.d]# service hadoop-yarn-nodemanager start
Starting Hadoop nodemanager:                               [  OK  ]
starting nodemanager, logging to /var/log/hadoop-yarn/yarn-yarn-nodemanager-hc1r1m1.out
```

Running Another Map Reduce Job Test

You have the Hadoop V2 HDFS and YARN servers running; all servers are up on the data nodes and the Node manager. You have checked the logs and found no errors. So, you are ready to attempt a test of Map Reduce. Try issuing the word-count job on the Poe data, as was done earlier for Hadoop V1 :

1. Switch user to the Linux hadoop account:

    ```
    [root@hc1nn ~]$  su - hadoop
    ```

2. Next, create the working directory for the data on HDFS:

    ```
    [hadoop@hc1nn ~]$ hdfs dfs -mkdir -p /user/hadoop/edgar/edgar
    ```

3. Now, copy the same edgar test data to the working directory on HDFS:

    ```
    [hadoop@hc1nn edgar]$  hdfs dfs -copyFromLocal *.txt /user/hadoop/edgar/edgar
    ```

4. And finally, run the Map Reduce word-count job:

```
[hadoop@hc1nn ~]$ hadoop jar /usr/lib/hadoop-mapreduce/hadoop-mapreduce-examples.jar wordcount /
user/hadoop/edgar/edgar /user/hadoop/edgar-results

14/03/23 16:34:46 INFO jvm.JvmMetrics: Initializing JVM Metrics with processName=JobTracker, sessionId=
14/03/23 16:34:48 INFO util.ProcessTree: setsid exited with exit code 0
14/03/23 16:34:48 INFO mapred.Task: Using ResourceCalculatorPlugin : org.apache.hadoop.util.LinuxR
                                    esourceCalculatorPlugin@18b4587
14/03/23 16:34:48 INFO mapred.MapTask: Processing split: hdfs://hc1nn/user/hadoop/edgar/edgar/10947-8.
                                    txt:0+632294
14/03/23 16:34:48 INFO mapred.MapTask: Map output collector class = org.apache.hadoop.mapred.
                                    MapTask$MapOutputBuffer
14/03/23 16:34:48 INFO mapred.MapTask: io.sort.mb = 100
```

```
14/03/23 16:34:48 INFO mapred.MapTask: data buffer =
........
14/03/23 16:34:56 INFO mapred.JobClient:      Spilled Records=256702
14/03/23 16:34:56 INFO mapred.JobClient:      CPU time spent (ms)=0
14/03/23 16:34:56 INFO mapred.JobClient:      Physical memory (bytes) snapshot=0
14/03/23 16:34:56 INFO mapred.JobClient:      Virtual memory (bytes) snapshot=0
14/03/23 16:34:56 INFO mapred.JobClient:      Total committed heap usage (bytes)=1507446784
```

Notice that the Hadoop jar command is very similar to that used in V1. You have specified an example jar file to use, from which you will execute the word-count function. An input and output data directory on HDFS has also been specified. Also, the run time is almost the same.

Okay, the Map Reduce job has finished, so you take a look at the output. In the edgar-results directory, there is a _SUCCESS file to indicate a positive outcome and a part-r-00000 file that contains the reduced data:

```
[hadoop@hc1nn ~]$ hadoop fs -ls /user/hadoop/edgar-results
Found 2 items
-rw-r--r--   2 hadoop hadoop          0 2014-03-23 16:34 /user/hadoop/edgar-results/_SUCCESS
-rw-r--r--   2 hadoop hadoop     769870 2014-03-23 16:34 /user/hadoop/edgar-results/part-r-00000
```

The job was successful; you have part data. To examine the part file data, you need to extract it from HDFS. The Hadoop file system cat command can be used to dump the contents of the part file. This will then be stored in the Linux file system file /tmp/hadoop/part-r-00000:

```
[hadoop@hc1nn ~]$ mkdir -p /tmp/hadoop/
[hadoop@hc1nn ~]$ hadoop fs -cat /user/hadoop/edgar-results/part-r-00000 > /tmp/hadoop/part-r-00000

[hadoop@hc1nn ~]$ wc -l /tmp/hadoop/part-r-00000
67721 /tmp/hadoop/part-r-00000
```

If you use the Linux command wc -l to show the file lines, you'll see that there are 67,721 lines in the extracted file. This is the same result as you received from the Map Reduce word-count job in the V1 example. To list the actual data, you use:

```
[hadoop@hc1nn ~]$ head -20 /tmp/hadoop/part-r-00000
!            1
"            22
"''T         1
"'--         1
"'A          1
"'After      1
"'Although   1
"'Among      2
"'And        2
"'Another    1
"'As         2
"'At         1
"'Aussi      1
"'Be         2
"'Being      1
"'But        1
"'But,'      1
```

```
"'But--still--monsieur----' 1
"'Catherine, 1
"'Comb       1
```

Again, V2 provides a sorted list of words with their counts. The successful test proves that the installed system, both HDFS and Map Reduce, works. For now, you're finished with the configuration, although in later chapters of this book I'll be introducing more Hadoop components.

Like Hadoop V1, Hadoop V2 offers a web interface for monitoring your cluster's nodes. Direct your web browser to `http://hc1nn:8088/cluster/nodes,` and you can see all of your active data nodes along with information relating to status and storage. The nodes themselves are actually http links, so you can click on then to drill down further.

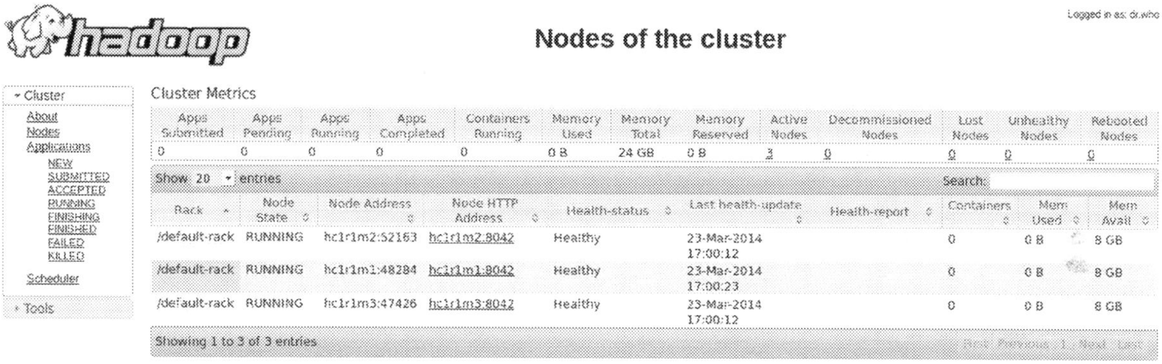

Figure 2-6. *Hadoop V2 UI cluster nodes*

Hadoop Commands

Hadoop offers many additional command-line options. In addition to the shell commands you've already used in this chapter's examples, I'll cover some other essential commands here, but only give a brief introduction to get you going. The following sections will introduce Hadoop shell, user and administration commands. Where possible, I've given a working example for each command. For a complete guide, see the Hadoop site, `https://hadoop.apache.org`.

Hadoop Shell Commands

The Hadoop shell commands are really user commands; specifically, they are a subset related to the file system. Each command is invoked using the hadoop keyword, followed by the fs option, which stands for "file system." Each subcommand is passed as an argument to the fs option. File paths are specified as uniform resource identifiers, or URIs.

A file on the HDFS can be specified as hdfs:///dir1/dir2/file1, whereas the same file on the Linux file system can be specified as file:///dir1/dir2/file1. If you neglect to offer a scheme (hdfs or file), then Hadoop assumes you mean the HDFS.

If you are familiar with Linux or Unix shell commands, then you will find these Hadoop file system commands similar. You can list files on HDFS using the ls command:

```
[hadoop@hc1nn ~]$ hadoop fs -ls  /user/hadoop/edgar/
Found 1 items
drwxr-xr-x   - hadoop hadoop          0 2014-03-23 16:32 /user/hadoop/edgar/edgar
```

To perform a recursive listing, add the –R option; this means that the ls command will list the topmost directory and all subdirectories:

```
[hadoop@hc1nn ~]$ hadoop fs -ls -R /user/hadoop/edgar/
drwxr-xr-x   - hadoop hadoop          0 2014-03-23 16:32 /user/hadoop/edgar/edgar
-rw-r--r--   2 hadoop hadoop     632294 2014-03-23 16:32 /user/hadoop/edgar/edgar/10947-8.txt
-rw-r--r--   2 hadoop hadoop     559342 2014-03-23 16:32 /user/hadoop/edgar/edgar/15143-8.txt
-rw-r--r--   2 hadoop hadoop      66409 2014-03-23 16:32 /user/hadoop/edgar/edgar/17192-8.txt
-rw-r--r--   2 hadoop hadoop     550284 2014-03-23 16:32 /user/hadoop/edgar/edgar/2147-8.txt
-rw-r--r--   2 hadoop hadoop     579834 2014-03-23 16:32 /user/hadoop/edgar/edgar/2148-8.txt
-rw-r--r--   2 hadoop hadoop     596745 2014-03-23 16:32 /user/hadoop/edgar/edgar/2149-8.txt
-rw-r--r--   2 hadoop hadoop     487087 2014-03-23 16:32 /user/hadoop/edgar/edgar/2150-8.txt
-rw-r--r--   2 hadoop hadoop     474746 2014-03-23 16:32 /user/hadoop/edgar/edgar/2151-8.txt
```

You can create directories with mkdir; this example will create a directory on HDFS called "test" under the / root node. Once it has been created, the ls command shows that it exists and is owned by the user hadoop:

```
[hadoop@hc1nn ~]$ hadoop fs -mkdir /test

[hadoop@hc1nn ~]$ hadoop fs -ls /
Found 5 items
drwxr-xr-x   - hadoop hadoop          0 2014-03-24 18:18 /test
```

The chown and chmod commands change ownership and permissions, respectively. If you know Unix commands, then these will be familiar. Their syntax is:

```
[hadoop@hc1nn ~]$ hadoop fs  -chown hdfs:hdfs /test
[hadoop@hc1nn ~]$ hadoop fs  -chmod 700  /test
[hadoop@hc1nn ~]$ hadoop fs  -ls /
Found 5 items
drwx------   - hdfs   hdfs            0 2014-03-24 18:18 /test
```

The chown command has changed the ownership of the HDFS /test directory to user/group hdfs/hdfs. The chmod command has changed the directory permissions to 700 or rwx --- ---. That is read/write/execute for the owner (hdfs), and there's no access for the group or any other user.

You can copy a file to and from the local file system into HDFS by using the copyFromLocal argument:

```
[hadoop@hc1nn ~]$ hadoop fs -copyFromLocal ./test_file.txt /test/test_file.txt
[hadoop@hc1nn ~]$
[hadoop@hc1nn ~]$ hadoop fs -ls /test
Found 1 items
-rw-r--r--   2 hadoop hdfs          504 2014-03-24 18:24 /test/test_file.txt
```

The example above shows that a Linux file system file ./test_file.txt was copied into HDFS to be stored under /test/test_file.txt. The next example shows how copyToLocal can be used to copy a file from HDFS to the Linux file system:

```
[hadoop@hc1nn ~]$ hadoop fs -copyToLocal /test/test_file.txt ./test_file2.txt
[hadoop@hc1nn ~]$ ls -l ./test_file*
-rwxr-xr-x. 1 hadoop hadoop 504 Mar 24 18:25 ./test_file2.txt
-rw-rw-r--. 1 hadoop hadoop 504 Mar 24 18:24 ./test_file.txt
```

In the above example, the HDFS file /test/test_file.txt has been copied to the Linux file system as ./test_file2.txt. To move a file or directory on HDFS, you use the mv command:

```
[hadoop@hc1nn ~]$ hadoop fs -mv /test/test_file.txt /test/test_file3.txt
[hadoop@hc1nn ~]$ hadoop fs -ls /test
Found 1 items
-rw-r--r--   2 hadoop hdfs         504 2014-03-24 18:24 /test/test_file3.txt
```

The HDFS file /test/test_file.txt has been moved to the HDFS file /test/test_file3.txt. You can recursively delete in HDFS by using rm -r:

```
[hadoop@hc1nn ~]$ hadoop fs -rm -r /test

[hadoop@hc1nn ~]$ hadoop fs -ls /
Found 4 items
drwxrwxrwt   - hdfs   hadoop          0 2014-03-23 14:58 /tmp
drwxr-xr-x   - hdfs   hadoop          0 2014-03-23 16:06 /user
drwxr-xr-x   - hdfs   hadoop          0 2014-03-23 14:56 /var
```

The example above has deleted the HDFS directory /test and all of its contents. To determine the space usage in HDFS, you use the du (disk usage) command:

```
[hadoop@hc1nn ~]$ hadoop fs -du -h /
0       /tmp
4.5 M   /user
0       /var
```

The -h option just makes the numbers humanly readable. This last example shows that only the HDFS file system /user directory is using any space.

Hadoop User Commands

This section introduces some Hadoop user commands that you can use to check the health of the HDFS, determine the Hadoop version, and carry out large-scale distributed data copies. Hadoop fsck offers the ability to determine whether the file system is healthy. It displays the total data size plus file and directory volumes. It also offers information like the replication factor and corrupted blocks. Hadoop distcp provides the functionality to move very large volumes of data between clusters.

The following example of the fsck command shows that the file system "/" is healthy. No corrupted or under-replicated blocks are listed. By default, there should be two copies of each block saved (the default replication factor value was 2). If the HDFS had failed in this area, it would be shown in the report as "Under-replicated blocks" with a value greater than zero.

```
[hadoop@hc1nn ~]$ hadoop fsck    /

Connecting to namenode via http://hc1nn:50070
FSCK started by hadoop (auth:SIMPLE) from /192.168.1.107 for path / at Mon Mar 24 18:42:09 NZDT 2014
..........Status: HEALTHY
 Total size: 4716611 B
 Total dirs: 14
 Total files:10
 Total blocks (validated): 9 (avg. block size 524067 B)
 Minimally replicated blocks:    9 (100.0 %)
 Over-replicated blocks:         0 (0.0 %)
 Under-replicated blocks:        0 (0.0 %)
 Mis-replicated blocks:          0 (0.0 %)
 Default replication factor:     2
 Average block replication:      2.0
 Corrupt blocks:                 0
 Missing replicas:               0 (0.0 %)
 Number of data-nodes:           3
 Number of racks:                1
FSCK ended at Mon Mar 24 18:42:09 NZDT 2014 in 9 milliseconds

The filesystem under path '/' is HEALTHY
```

You can use the job command to list jobs, although this example shows no jobs currently running:

```
[hadoop@hc1nn lib]$ hadoop job -list all

14/03/24 18:53:49 INFO service.AbstractService:
Service:org.apache.hadoop.yarn.client.YarnClientImpl is inited.
14/03/24 18:53:50 INFO service.AbstractService:
Service:org.apache.hadoop.yarn.client.YarnClientImpl is started.
Total jobs:0

JobId  State  StartTime  UserName  Queue   Priority  UsedContainers  RsvdContainers  UsedMem
RsvdMem NeededMem    AM info
```

To determine which version of Hadoop you are using, as well as the checksum and dates, you use the version command:

```
[hadoop@hc1nn lib]$ hadoop version
Hadoop 2.0.0-cdh4.6.0
Subversion git://centos32-6-slave.sf.cloudera.com/data/1/jenkins/workspace/generic-package-
centos32-6/topdir/BUILD/hadoop-2.0.0-cdh4.6.0/src/hadoop-common-project/hadoop-common -r
8e266e052e423af592871e2dfe09d54c03f6a0e8
Compiled by jenkins on Wed Feb 26 01:59:02 PST 2014
From source with checksum a9d36604dfb55479c0648f2653c69095
This command was run using /usr/lib/hadoop/hadoop-common-2.0.0-cdh4.6.0.jar
```

The version output above shows that you are using Cloudera CDH 4.6, which is actually Hadoop version 2.0.0. It also shows when the release was built and by whom.

Distcp offers you the ability to copy within a cluster or between clusters. It uses Map Reduce to do this, and it is designed for large-scale copying. I can provide only a theoretical example here because I have only a small four-node cluster for writing this book with a single name node. Check the website hadoop.apache.org for more information. The syntax for distcp is as follows:

```
[hadoop@hc1nn lib]$ hadoop distcp \
                    hdfs://hc1nn:8020/test2 \
                    hdfs://hc2nn:8020/test3
```

This example shows the contents of cluster one HDFS directory /test2 being copied to cluster two directory /test3. The URIs used here (hdfs://hc1nn:8020/test2 and hdfs://hc2nn:8020/test3) use a scheme of hdfs, but they also refer to a hostname and port. These are the hosts and port numbers for the name nodes for the two clusters.

Hadoop Administration Commands

This discussion of Hadoop administration commands will let you sample some of the full set of commands available. For complete details, check hadoop.apache.org. These commands will enable you to format the HDFS, manage upgrades, set logging levels, and save configuration information.

You already used one of the administration commands (-format) when you formatted the file system earlier. Take a second look:

```
hadoop namenode -format
```

The format command starts the name node, executes its command, and then shuts the name node down again. The name node is the centralized place on the HDFS where metadata concerning files in the file system are stored. If the Hadoop file system is running when this command is executed, then HDFS data is lost.

I won't run the upgrade command here, but you can use it after releasing a new version of Hadoop, as follows:

```
hadoop namenode -upgrade
```

This upgrade command will create new working directories on the data nodes for the new Hadoop version. The previous checkpoint version is kept to allow for a rollback to the previous version of software in case the upgrade doesn't work out.

If you need to roll back to the previous version of Hadoop, you can use hadoop namenode -rollback.) This rollback command will cause Hadoop to revert to the previous version of the working directories.

On the other hand, to finalize the upgrade and remove the old version, you use:

```
hadoop namenode -finalize
```

Be sure that the upgrade has worked before you remove the option to roll back.

You can get and set daemon log levels with daemonlog:

```
[hadoop@hc1nn ~]$ hadoop daemonlog -getlevel hc1nn:8088
```

```
org.apache.hadoop.mapred.JobTracker
Connecting to
http://hc1nn:8088/logLevel?log=org.apache.hadoop.mapred.JobTracker
Submitted Log Name: org.apache.hadoop.mapred.JobTracker
Log Class: org.apache.commons.logging.impl.Log4JLogger
Effective level: INFO
```

By changing getLevel to setLevel, you can set the log levels. Some possible values are all, info, debug, and error. The level and volume of the information you receive will vary, with all supplying everything and error just giving an error message.

You can use the dfsadmin report option to get information and administer Hadoop, as follows:

```
[hadoop@hc1nn ~]$ hadoop dfsadmin -report

Configured Capacity: 158534062080 (147.65 GB)
Present Capacity: 141443452928 (131.73 GB)
DFS Remaining: 141433782272 (131.72 GB)
DFS Used: 9670656 (9.22 MB)
DFS Used%: 0.01%
Under replicated blocks: 0
Blocks with corrupt replicas: 0
Missing blocks: 0


-----------------------------------------------
Datanodes available: 3 (3 total, 0 dead)

Live datanodes:
Name: 192.168.1.106:50010 (hc1r1m2)
Hostname: hc1r1m2
Decommission Status : Normal
Configured Capacity: 52844687360 (49.22 GB)
DFS Used: 4157440 (3.96 MB)
Non DFS Used: 6079811584 (5.66 GB)
DFS Remaining: 46760718336 (43.55 GB)
DFS Used%: 0.01%
DFS Remaining%: 88.49%
Last contact: Wed Mar 26 18:26:18 NZDT 2014

Name: 192.168.1.104:50010 (hc1r1m1)
Hostname: hc1r1m1
Decommission Status : Normal
Configured Capacity: 52844687360 (49.22 GB)
DFS Used: 3698688 (3.53 MB)
Non DFS Used: 3301863424 (3.08 GB)
DFS Remaining: 49539125248 (46.14 GB)
DFS Used%: 0.01%
DFS Remaining%: 93.74%
Last contact: Wed Mar 26 18:26:20 NZDT 2014
```

```
Name: 192.168.1.102:50010 (hc1r1m3)
Hostname: hc1r1m3
Decommission Status : Normal
Configured Capacity: 52844687360 (49.22 GB)
DFS Used: 1814528 (1.73 MB)
Non DFS Used: 7708934144 (7.18 GB)
DFS Remaining: 45133938688 (42.03 GB)
DFS Used%: 0.00%
DFS Remaining%: 85.41%
Last contact: Wed Mar 26 18:26:19 NZDT 2014
```

This report shows the HDFS capacity and usage, as well as the same information on each of the data nodes. It also shows information like under-replicated blocks, which gives an indication of potential data loss in the event of server failure.

You can save the name node data structures to file by using the hadoop dfsadmin metasave command:

```
[hadoop@hc1nn ~]$ hadoop dfsadmin -metasave metaFile1

Created metasave file metaFile1 in the log directory of namenode hdfs://hc1nn
```

In this example, you could have specified where this file was saved via the attribute hadoop.log.dir. In any event, having run this command, you can now check the contents of the file via the Linux ls and more commands:

```
[hadoop@hc1nn hadoop-hdfs]$ ls -l /var/log/hadoop-hdfs/metaFile1
-rw-r--r--. 1 hdfs hdfs 658 Mar 26 18:29 /var/log/hadoop-hdfs/metaFile1
```

The ls command shows the location of the metadata file that you have created, metaFile1. It also shows that it is owned by the Linux user hdfs. You can examine the contents of that metadata file by using the Linux more command:

```
[hadoop@hc1nn hadoop-hdfs]$ more /var/log/hadoop-hdfs/metaFile1
25 files and directories, 9 blocks = 34 total
Live Datanodes: 3
Dead Datanodes: 0
Metasave: Blocks waiting for replication: 0
Mis-replicated blocks that have been postponed:
Metasave: Blocks being replicated: 0
Metasave: Blocks 0 waiting deletion from 0 datanodes.
Metasave: Number of datanodes: 3
192.168.1.106:50010 IN 52844687360(49.22 GB) 4157440(3.96 MB) 0.00%
46760718336(
43.55 GB) Wed Mar 26 18:29:18 NZDT 2014
192.168.1.104:50010 IN 52844687360(49.22 GB) 3698688(3.53 MB) 0.00%
49539125248(
46.14 GB) Wed Mar 26 18:29:20 NZDT 2014
192.168.1.102:50010 IN 52844687360(49.22 GB) 1814528(1.73 MB) 0.00%
45133869056(
42.03 GB) Wed Mar 26 18:29:19 NZDT 2014
```

The last three lines of this example metadata file relate to data node capacity and data usage. Instead of refering to the data nodes by their server names, though, their IP addesses have been used. For instance, in my example cluster, the ip address 192.168.1.102 relates to the datanode hc1r1m3. The file also shows that there are three live data nodes and none that are dead.

A full explanation of these administration commands is beyond the scope of this chapter, but by using the `dfsadmin` command you can manage quotas, control the upgrade, refresh the nodes, and enter safe mode. Check the Hadoop site `hadoop.apache.org` for full information.

Summary

In this chapter you have been introduced to both Hadoop V1 and V2 in terms of their installation and use. It is hoped you can see that, by using the CDH stack release, the installation process and use of Hadoop are much simplified.

In the course of this chapter you have installed Hadoop V1 manually via a download package from the Hadoop site. You have then installed V2 and YARN via CDH packages and the `yum` command. Servers for HDFS and YARN are started as Linux services in V2 rather than as scripts, as in V1. Also, in the CDH release logs, binaries and configuration functions were separated into their own, specific directories.

You have been shown the same Map Reduce task as run on both versions of Hadoop. Task run times were comparable between V1 and V2. However, V2 offers the ability to have a larger production cluster than does V1. (In the following chapters you will look at Map Reduce programming in Java and Pig).

You have also configured Hadoop V2 across a mini cluster with name nodes and data nodes on different servers. You have installed and used ZooKeeper, setting up a quorum and using the client. (In the next chapter, HBase—the Hadoop database—will be discussed and that calls upon ZooKeeper).

Lastly, you have looked at the command set for file system and for user and administration commands. True, it was only a brief look, but further information is available at the Hadoop website.

CHAPTER 3

Collecting Data with Nutch and Solr

Many companies collect vast amounts of data from the web by using web crawlers such as Apache Nutch. Available for more than ten years, Nutch is an open-source product provided by Apache and has a large community of committed users. An Apache Lucene open-source search platform, Solr can be used in connection with Nutch to index and search the data that Nutch collects. When you combine this functionality with Hadoop, you can store the resulting large data volume directly in a distributed file system.

In this chapter, you will learn a number of methods to connect various releases of Nutch to Hadoop. I will demonstrate, though architectural examples, what can be accomplished by using the various tools and data. Specifically, the chapter's first architectural example uses Nutch 1.8 configured to implicitly use the local Hadoop installation. If Hadoop is available, Nutch will use it for storage, providing you with the benefits of distributed and resiliant storage. It does not, however, give you much control over the selection of storage. Nutch will use either Hadoop, if it is available, or the file system.

In the second architectural example, employing Nutch 2.x, you will be able to specify the storage used via Gora. By explicitly selecting the storage method in the configuration options, you can gain greater control. This example uses the HBase database, which still employs Hadoop for distributed storage. You then have the option of choosing a different storage mechanism at a later date by altering the configuration.

Remember, although these examples are using small amounts of data, the architectures can scale to a high degree to meet your big data-collection needs.

The Environment

Before we begin, you need to understand a few details about the environment in which we'll be working. This chapter demonstrates the use of Nutch with Hadoop V1.2.1 because I could not get Nutch to build against Hadoop V2 at the time of this writing. (Subsequently, I learned of a version of Nutch developed for YARN, but deadline constraints prevented me from implementing it here.) Although in Chapter 2 you installed Cloudera CDH4 on the CentOS Linux server hc1nn, at this point you'll need to switch back to using Hadoop V1. You'll manage this via a number of steps that are explained in the following sections. A shortage of available machines is the only reason I have installed multiple versions of Hadoop on a single cluster. This kind of multiple Hadoop installation is not appropriate for project purposes.

Stopping the Servers

The Hadoop Cloudera CDH4 cluster servers may still be running, so they need to be stopped on all nodes in the Hadoop cluster. Because these servers are Linux services, you need to stop them as the Linux root user. You carry out the following steps on all servers in the cluster—in this case, hc1nn, hc1r1m1, hc1r1m2, and hc1r1m3.

First, change the user to the root account with the Linux su (switch user) command:

```
[hadoop@hc1nn ~]$ su -
```

Next, issue the command sequence:

```
[root@hc1nn ~]# cd /etc/init.d/

[root@hc1nn init.d]# ls hadoop*mapreduce*
hadoop-0.20-mapreduce-jobtracker  hadoop-mapreduce-historyserver

[root@hc1nn init.d]#  ls hadoop*yarn*
hadoop-yarn-proxyserver  hadoop-yarn-resourcemanager

[root@hc1nn init.d]#  ls hadoop*hdfs*
hadoop-hdfs-namenode
```

The Linux cd (change directory) command moves the current path to /etc/init.d/, and the ls command displays the Map Reduce, Yarn, and HDFS Hadoop services. (The * character is a wildstar value that matches all text to the end of the string.) For each of the services displayed, execute the following command:

```
service <service name> stop
```

For instance, the stop command stops the proxy server, although it was not running:

```
[root@hc1nn init.d]# service hadoop-yarn-proxyserver stop
Stopping Hadoop proxyserver:                         [  OK  ]
no proxyserver to stop
```

Remember to stop the non-HDFS services before the HDFS services and stop Map Reduce and YARN before HDFS. Once you have done this on all the servers in this small cluster, then Hadoop V2 CDH4 will be stopped. You are still logged in as root, however, so use the Linux exit command to return control back to the Linux hadoop account session:

```
[root@hc1nn init.d]# exit
logout
[hadoop@hc1nn ~]$
```

Changing the Environment Scripts

In Chapter 2, you worked with two versions of Hadoop with two different environment configurations. The environment file used to hold these configurations was the Linux hadoop user's $HOME/.bashrc file. During the creation of this book, I needed to switch between Hadoop versions frequently, and so I created two separate versions of the bashrc file on each server in the cluster, as follows:

```
[hadoop@hc1nn ~]$ pwd
/home/hadoop
```

```
[hadoop@hc1nn ~]$ ls -l .bashrc*
lrwxrwxrwx. 1 hadoop hadoop   16 Jun 30 17:59 .bashrc -> .bashrc_hadoopv2
-rw-r--r--. 1 hadoop hadoop 1586 Jun 18 17:08 .bashrc_hadoopv1
-rw-r--r--. 1 hadoop hadoop 1588 Jul 27 11:33 .bashrc_hadoopv2
```

The Linux pwd command shows that the current location is the Linux hadoop user's home directory /home/hadoop/. The Linux ls command produces a long listing that shows a symbolic link called .bashrc, which points to either a Hadoop V1 or a V2 version of the bashrc configuration file. Currently it is pointing to V2, so you need to change it back to V1. (I will not explain the contents of the files, as they are listed in Chapter 2).

Delete the symbolic link named .bashrc by using the Linux rm command, then re-create it to point to the V1 file by using the Linux ln command with a –s (symbolic) switch:

```
[hadoop@hc1nn ~]$ rm .bashrc
[hadoop@hc1nn ~]$ ln -s  .bashrc_hadoopv1 .bashrc
```

```
[hadoop@hc1nn ~]$ ls -l .bashrc*
lrwxrwxrwx  1 hadoop hadoop   16 Nov 12 18:32 .bashrc -> .bashrc_hadoopv1
-rw-r--r--. 1 hadoop hadoop 1586 Jun 18 17:08 .bashrc_hadoopv1
-rw-r--r--. 1 hadoop hadoop 1588 Jul 27 11:33 .bashrc_hadoopv2
```

That creates the correct environment configuration file for the Linux hadopop account, but how does it now take effect? Either log out using the exit command and log back in, or use the following:

```
[hadoop@hc1nn ~]$ . ./.bashrc
```

This means that the .bashrc is executed in the current shell (denoted by the first " . " character). The ./ specifies that the .bashrc file is sourced from the current directory. Now, you are ready to start the Hadoop V1 servers.

Starting the Servers

The Hadoop V1 environment has been configured, and the V2 Hadoop servers have already been stopped. Now, you change to the proper directory and start the servers:

```
[hadoop@hc1nn ~]$ cd $HADOOP_PREFIX/bin
[hadoop@hc1nn hadoop]$ pwd
/usr/local/hadoop/bin/

[hadoop@hc1nn bin]$ ./start-dfs.sh
[hadoop@hc1nn bin]$ ./start-mapred.sh
```

These commands change the directory to the /usr/local/hadoop/bin/ directory using the HADOOP_PREFIX variable. The HDFS servers are started using the start-dfs.sh script, followed by the Map Reduce servers with start-mapred.sh. At this point, you can begin the Nutch work, using Hadoop V1 on this cluster.

Architecture 1: Nutch 1.x

This first example illustrates how Nutch, Solr, and Hadoop work together. You will learn how to download, install, and configure Nutch 1.8 and Solr, as well as how to set up your environment and build Nutch. With the prep work finished, I'll walk you through running a sample Nutch crawl using Solr and then storing the data on the Hadoop file system.

In Nutch 1.8, the crawl class has been replaced by the crawl script. The crawl script, which will be described later, runs the whole Nutch crawl for you, as well as storing data in Hadoop. Then, you will learn how to check that your data has been processed by Solr.

Nutch Installation

For this first example, you will download and install Nutch 1.8 from the Nutch website (nutch.apache.org). From the Downloads page, choose the source version of Nutch 1.8. The -src in the file name means that the source files are included in the software package as well as the binaries. As in the previous chapter, you download a gzipped tar file (.tar.gz), then unpack it using the gunzip command, followed by the tar xvf command. (In the tar command, x stands for "extract," v for "verbose," and the file name to process is specified after the f option.)

```
-rw-rw-r--. 1 hadoop hadoop   2757572 Apr  1 18:12 apache-nutch-1.8-src.tar.gz

[hadoop@hc1nn Downloads]$ gunzip apache-nutch-1.8-src.tar.gz
[hadoop@hc1nn Downloads]$ tar xvf apache-nutch-1.8-src.tar
```

This leaves the raw unpacked Nutch package software extracted under the directory apache-nutch-1.8-src, as shown here:

```
[hadoop@hc1nn Downloads]$ ls -ld apache-nutch-1.8-src
drwxrwxr-x. 7 hadoop hadoop 4096 Apr  1 18:13 apache-nutch-1.8-src
```

Using the mv command, you move this release to a better location and then set the ownership to the Linux hadoop user with the chown command. Note that I used the -R switch, which recursively changes ownership on subdirectories and files under the topmost directory, apache-nutch-1.8-src:

```
[root@hc1nn Downloads]# mv apache-nutch-1.8-src /usr/local
[root@hc1nn Downloads]# cd /usr/local
[root@hc1nn Downloads]# chown -R hadoop:hadoop apache-nutch-1.8-src
[root@hc1nn Downloads]# ln -s apache-nutch-1.8-src    nutch
```

Now the Nutch installation has been moved to /usr/local/ and a symbolic link has been created to point to the installed software, called "nutch." That means that the environment can use this alias to point to the installed software directory. If a new release of Nutch is required in the future, simply change this link to point to it; the environment will not need to be changed.

```
[root@hc1nn local]# ls -ld *nutch*
drwxrwxr-x. 7 hadoop hadoop 4096 Apr  1 18:13 apache-nutch-1.8-src
lrwxrwxrwx. 1 root    root     20 Apr  1 18:16 nutch -> apache-nutch-1.8-src
```

Next, you will set up the configuration files. The first step is to create symbolic links to the Hadoop configuration files in the Nutch build. This avoids the need to copy changes in the Hadoop configuration to the Nutch build each time such a change occurs. Create the links as follows:

```
[root@hc1nn Downloads]# cd /usr/local/nutch/conf
[root@hc1nn Downloads]# ln -s /usr/local/hadoop/conf/core-site.xml    core-site.xml
[root@hc1nn Downloads]# ln -s /usr/local/hadoop/conf/hdfs-site.xml    hdfs-site.xml
[root@hc1nn Downloads]# ln -s /usr/local/hadoop/conf/hadoop-env.sh    hadoop-env.sh
[root@hc1nn Downloads]# ln -s /usr/local/hadoop/conf/mapred-site.xml mapred-site.xml
[root@hc1nn Downloads]# ln -s /usr/local/hadoop/conf/masters          masters
[root@hc1nn Downloads]# ln -s /usr/local/hadoop/conf/slaves           slaves
```

Then you can check that these links exist by creating a long Linux listing using ls -l:

```
[hadoop@hc1nn conf]$ ls -l

lrwxrwxrwx. 1 hadoop hadoop  36 Apr  5 14:15 core-site.xml -> /usr/local/hadoop/conf/core-site.xml
lrwxrwxrwx. 1 hadoop hadoop  36 Apr  5 14:16 hadoop-env.sh -> /usr/local/hadoop/conf/hadoop-env.sh
lrwxrwxrwx. 1 hadoop hadoop  36 Apr  5 14:16 hdfs-site.xml -> /usr/local/hadoop/conf/hdfs-site.xml
lrwxrwxrwx. 1 hadoop hadoop  38 Apr  5 14:16 mapred-site.xml -> /usr/local/hadoop/conf/mapred-site.xml
lrwxrwxrwx. 1 hadoop hadoop  38 Apr  5 14:16 masters -> /usr/local/hadoop/conf/masters
lrwxrwxrwx. 1 hadoop hadoop  38 Apr  5 14:16 slaves -> /usr/local/hadoop/conf/slaves
```

Next, you make some additions to the nutch-site.xml configuration file, as well as to the Hadoop core-site. xml and mapred-site.xml files. When adding the code snippets, place each new property (identified by the opening <property> and closing </property> tags) between the configuration tags in the appropriate file. You can find these files (or links to them) in the Nutch configuration directory /usr/local/nutch/conf.

First, make the nutch-site.xml file changes. These define the name of your Nutch agent and the location of the plug-ins folders, a source of extra modules:

```
<configuration>

  <property>
    <name>http.agent.name</name>
    <value>NutchHadoopCrawler</value>
  </property>

  <property>
    <name>plugin.folders</name>
    <value>/usr/local/nutch/build/plugins</value>
  </property>

</configuration>
```

Next, make those changes that are for the Hadoop core component (core-site.xml) in the Nutch configuration directory /usr/local/nutch/conf/ to enable gzip compression with Hadoop:

```
<property>
  <name>io.compression.codecs</name>
  <value>org.apache.hadoop.io.compress.GzipCodec,org.apache.hadoop.io.compress.DefaultCodec,org.
  apache.hadoop.io.compress.BZip2Codec,org.apache.hadoop.io.compress.SnappyCodec</value>
</property>
```

Place the changes for the Hadoop Map Reduce component in the mapred-site.xml file in the Nutch configuration directory /usr/local/nutch/conf/. These specify the memory limitations and the maximum attempt limits for both Map and Reduce tasks, thereby helping to prevent a runaway task in terms of Map Reduce looping or memory use:

```
<property>
  <name>mapred.child.java.opts</name>
  <value>-Xmx1024m</value>
</property>
```

```
<property>
  <name>mapreduce.map.maxattempts</name>
  <value>4</value>
</property>

<property>
  <name>mapreduce.reduce.maxattempts</name>
  <value>4</value>
</property>

<property>
  <name>mapred.job.map.memory.mb</name>
  <value>4000</value>
</property>

<property>
  <name>mapred.job.reduce.memory.mb</name>
  <value>3000</value>
</property>
```

Each time your Hadoop configuration changes, you need to rebuild your Nutch release with the Apache Ant tool (which I'll describe in a moment). The build is carried out from within the build subdirectory of the Nutch release, and it creates release-specific .jar and .job files, which are used during the Nutch crawl. The .jar file contains the released Nutch Java components built into a single file. The .job file contains all of the classes and plug-ins needed to run a Nutch job.

From the Nutch build directory of the existing Nutch 1.8, you can list the .jar and .job files:

```
[hadoop@hc1nn build]$ pwd
/usr/local/nutch/build

[hadoop@hc1nn build]$ ls -l *.jar *.job
-rw-r--r--. 1 hadoop hadoop   556673 Apr  5 15:35 apache-nutch-1.8.jar
-rw-r--r--. 1 hadoop hadoop 79105966 Apr  5 18:42 apache-nutch-1.8.job
```

The Nutch Ant build re-creates these files and copies them to their runtime directories, as shown here (the NUTCH_HOME path variable is defined below):

```
job file -> $NUTCH_HOME/runtime/deploy

jar file -> $NUTCH_HOME/runtime/local/lib
```

Note that I am using the Linux hadoop account to define the environment for Nutch. This is convenient because it is also the user account that owns the Hadoop installation (see Chapter2). However, this user's shell configuration file ($HOME/.bashrc) needs to be extended to add environmental variables for Nutch. To do so, add the following text at the end of the file, then log out of the Linux hadoop account and log back in to pick up the changes:

```
#######################################################
# Set up Nutch variables

export NUTCH_HOME=/usr/local/nutch
export NUTCH_CONF_DIR=$NUTCH_HOME/conf
```

```
export CLASSPATH=.:$NUTCH_HOME/runtime/local/lib
export CLASSPATH=$CLASSPATH:$NUTCH_HOME/conf

export PATH=$PATH:$NUTCH_HOME/bin
```

You can now build Nutch using Ant. Move to the Nutch installation directory (/usr/local/nutch) using the NUTCH_HOME variable you just set up. Issue the ant command to start the build. The build output will be copied to the session window:

```
[hadoop@hc1nn nutch]# cd $NUTCH_HOME
[hadoop@hc1nn nutch]# ant
Buildfile: build.xml

.....

BUILD SUCCESSFUL
Total time: 11 minutes 15 seconds
```

It can take quite a while to run the build, depending on the volume of changes and the age of the build; for instance, this example took me more than 11 minutes. Checking the job file in the Nutch build directory shows (by date and size) when the Nutch job file has been re-created:

```
[hadoop@hc1nn home]$ cd $NUTCH_HOME/build
[hadoop@hc1nn build]$ ls -l *.job
-rw-r--r--. 1 hadoop hadoop 79104396 Apr  1 18:34 apache-nutch-1.8.job
```

With Nutch installed and ready, you can move on to installing Solr, in preparation for using these tools together.

Solr Installation

To begin your installation, download Solr from the Solr website (https://lucene.apache.org/solr). For example, I selected the zipped file to download and easily unpacked it using the Linux unzip command:

```
[root@hc1nn Downloads]# ls -l solr-4.7.0.zip
-rw-rw-r--. 1 hadoop hadoop 157644303 Mar 29 13:08 solr-4.7.0.zip
```

The release unpacks to a directory called solr-4.7.0 in the same location as the .zip file.

```
[root@hc1nn Downloads]# unzip solr-4.7.0.zip

[root@hc1nn Downloads]# ls -ld solr-4.7.0
drwxr-xr-x. 7 root root 4096 Feb 22 08:39 solr-4.7.0
```

Move the release to /usr/local/ (the same location as your Nutch and Hadoop software) and change its ownership and group membership to hadoop. (Remember to use the recursive -R flag with the Linux chown command.) As with the other software releases, a symbolic link called "solr" is created to point to the installed release.

```
[root@hc1nn Downloads]# mv solr-4.7.0 /usr/local
[root@hc1nn Downloads]# cd /usr/local
[root@hc1nn local]# chown -R hadoop:hadoop solr-4.7.0
[root@hc1nn local]# ln -s solr-4.7.0 solr
```

A quick check of the installation under /usr/local/ shows the solr installation directory as owned by Hadoop and the solr link pointing to it:

```
[root@hc1nn local]# ls -ld *solr*
lrwxrwxrwx. 1 root    root      10 Mar 29 13:11 solr -> solr-4.7.0
drwxr-xr-x. 7 hadoop hadoop 4096 Feb 22 08:39 solr-4.7.0
```

At this point you have Solr installed in the correct location, and you are ready to configure it. You can set up a variable in the hadoop user's Bash shell to point to the Solr installation. Add the following text to the bottom of the Linux hadoop account configuration file $HOME/.bashrc. This will define the Bash shell environment variable SOLR_HOME to be /usr/local/solr.

```
##########################################################
# Set up Solr variables

export SOLR_HOME=/usr/local/solr
```

Next, configure Solr to integrate it with Nutch. Some of the Nutch configuration files need to be copied to the Solr configuration directory; copy the files schema.xml and schema-solr4.xml across:

```
[hadoop@hc1nn ~]$ cd $NUTCH_HOME/conf
[hadoop@hc1nn conf]$ cp schema.xml $SOLR_HOME/example/solr/collection1/conf
[hadoop@hc1nn conf]$ cp schema-solr4.xml $SOLR_HOME/example/solr/collection1/conf
```

These schema files define the field types and fields that the documents being indexed will contain. Solr uses the information in the schema files to help it parse and index the data that it processes.

Next, add a few extra fields at the end of the <fields> section of schema.xml:

```
<!-- fields for Nutch -->
<field name="_version_" type="long" indexed="true" stored="true"/>
<field name="text" type="string" indexed="true" stored="true"/>
```

The filter factory algorithm currently listed in the file is the EnglishPorterFilterFactory, which has been deprecated. To replace it, you need to specify the SnowballPorterFilterFactory; originally devised by by Martin Porter, the algorithm is used as a filter to prepare document tokens before they are processed by Solr. Look for this line:

```
<filter class="solr.EnglishPorterFilterFactory" protected="protwords.txt"/>
```

And replace it with this one:

```
<filter class="solr.SnowballPorterFilterFactory" protected="protwords.txt" language="English"/>
```

Now try starting Solr to test that it will work:

```
[hadoop@hc1nn conf]$ cd $SOLR_HOME/example/
```

```
[hadoop@hc1nn example]$ java -jar start.jar &
```

The & symbol as the end of the line means that the Solr job you are running will run in the background. Look for any errors in the output that are displayed in the session window.

You can also check the log file under $SOLR_HOME/example/logs/. If there are no errors, then try connecting to the Solr admin client at:

```
http://localhost:8983/solr/admin/
```

The Solr admin client (Figure 3-1) contains logging and administrative functions, as well as the core selector that you will use shortly to examine the results of Solr indexing.

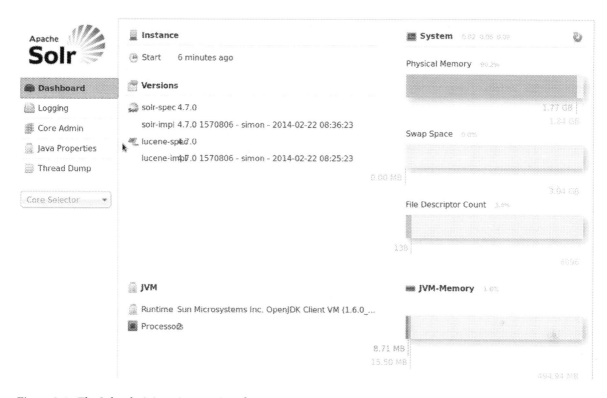

Figure 3-1. *The Solr administration user interface*

Now that you have Solr running without error and you can connect to its admin client, you are ready to run Nutch and do a simple web crawl using Hadoop as the storage mechanism.

Running Nutch with Hadoop 1.8

You are ready to run the first crawl using Nutch and Hadoop. For doing this, I have created a seed file on the Linux file system containing the initial website to crawl. I have also added a single URL to the file, my own website.[1]

```
[hadoop@hc1nn nutch]$ ls -l urls
total 4
-rw-rw-r--. 1 hadoop hadoop 19 Apr  5 13:14 seed.txt
```

---

[1]I own the site and it's contents, so there are no issues with processing the site contents and displaying them here.

This is a small-scale example, so you will start with a single URL to follow. In a real situation, you would populate the crawl database with a large volume of URLs by using a much larger seed file. You can also download seed file databases from the Internet. The file at http://rdf.dmoz.org/rdf/content.rdf.u8.gz has around 3 million URLs.

No matter the size of your seed file, you need to copy it (seed.txt, in this case) from the Linux file system to HDFS (to the nutch/urls directory):

```
[hadoop@hc1nn nutch]$ hadoop dfs -put urls/seed.txt nutch/urls
[hadoop@hc1nn nutch]$ hadoop dfs -ls nutch/urls
Found 1 items
-rw-r--r--   1 hadoop supergroup         19 2014-04-05 13:19
/user/hadoop/nutch/urls/seed.txt
```

Found in $NUTCH_HOME/runtime/deploy/bin, the Nutch crawl command is actually a shell script that automates the sequence of Nutch operations, as follows:

1. **Inject**. Inserts a URL into the Nutch crawl database.

2. **Generate**. Creates a fetch list from the Nutch database for the crawl. This creates a segment directory within the crawl database for fetch processing.

3. **Fetch**. Runs the fetcher against the segment created in step 2.

4. **Parse**. Processes the results of the fetch.

5. **Update db**. Updates the Nutch crawl database with the results of the parse.

6. **Invertlinks**. Creates a link map, listing incoming links for this URL.

7. **Dedup**. Deletes duplicate documents that are in the index.

8. **Index**. Runs the indexer on the database.

9. **Clean**. Cleans up after the crawl cycle.

For the actual crawl itself, the syntax of the crawl command is:

```
crawl <seedDir> <crawlDir> <solrURL> <numberOfRounds>
```

The seed directory, nutch/urls, will be sourced from HDFS, which is why you copied the URL list to HDFS. The Solr URL gives Nutch a link to the Solr instance you started in the last section. The number of rounds is actually the depth that the crawl will process to.

The crawl script runs a couple of steps to decide whether it should use Hadoop for storage. First, it looks for the job file:

```
mode=local
if [ -f ../*nutch-*.job ]; then
    mode=distributed
fi
```

Then it checks to determine whether it can access Hadoop:

```
# check that hadoop can be found on the path
if [ $mode = "distributed" ]; then
  if [ $(which hadoop | wc -l ) -eq 0 ]; then
    echo "Can't find Hadoop executable. Add HADOOP_HOME/bin to the path or run in local mode."
    exit -1;
  fi
fi
```

Given these checks, if Hadoop is available, it will be used for storage; otherwise, the Linux file system will be used. You can now run the crawl as follows:

```
cd $NUTCH_HOME/runtime/deploy/bin
./crawl nutch/urls crawl http://hc1nn:8983/solr/  2
```

This gives you the Nutch crawl output:

```
14/04/06 16:56:22 INFO crawl.Injector: Injector: starting at 2014-04-06 16:56:22
14/04/06 16:56:22 INFO crawl.Injector: Injector: crawlDb: /user/hadoop/crawl/crawldb
14/04/06 16:56:22 INFO crawl.Injector: Injector: urlDir: nutch/urls
14/04/06 16:56:22 INFO crawl.Injector: Injector: Converting injected urls to crawl db entries.
14/04/06 16:56:26 INFO util.NativeCodeLoader: Loaded the native-hadoop library
14/04/06 16:56:26 INFO mapred.FileInputFormat: Total input paths to process : 1
14/04/06 16:56:26 INFO mapred.JobClient: Running job: job_201404061342_0056
14/04/06 16:56:27 INFO mapred.JobClient:  map 0% reduce 0%
14/04/06 16:56:43 INFO mapred.JobClient:  map 50% reduce 0%
14/04/06 16:56:47 INFO mapred.JobClient:  map 100% reduce 0%
14/04/06 16:56:51 INFO mapred.JobClient:  map 100% reduce 33%
14/04/06 16:56:52 INFO mapred.JobClient:  map 100% reduce 100%
14/04/06 16:56:53 INFO mapred.JobClient: Job complete: job_201404061342_0056


..........................

14/04/06 17:05:53 INFO mapred.JobClient: Counters: 30
14/04/06 17:05:53 INFO mapred.JobClient: Job Counters
14/04/06 17:05:53 INFO mapred.JobClient: Launched reduce tasks=1
14/04/06 17:05:53 INFO mapred.JobClient: SLOTS_MILLIS_MAPS=10036
14/04/06 17:05:53 INFO mapred.JobClient: Total time spent by all reduces waiting after reserving
                                         slots (ms)=0
14/04/06 17:05:53 INFO mapred.JobClient: Total time spent by all maps waiting after reserving slots (ms)=0
14/04/06 17:05:53 INFO mapred.JobClient: Rack-local map tasks=1
14/04/06 17:05:53 INFO mapred.JobClient: Launched map tasks=2
14/04/06 17:05:53 INFO mapred.JobClient: Data-local map tasks=1
14/04/06 17:05:53 INFO mapred.JobClient: SLOTS_MILLIS_REDUCES=8334
14/04/06 17:05:53 INFO mapred.JobClient: File Input Format Counters
14/04/06 17:05:53 INFO mapred.JobClient: Bytes Read=3746
14/04/06 17:05:53 INFO mapred.JobClient: File Output Format Counters
14/04/06 17:05:53 INFO mapred.JobClient: Bytes Written=0
14/04/06 17:05:53 INFO mapred.JobClient: FileSystemCounters
14/04/06 17:05:53 INFO mapred.JobClient: FILE_BYTES_READ=6
```

```
14/04/06 17:05:53 INFO mapred.JobClient: HDFS_BYTES_READ=4096
14/04/06 17:05:53 INFO mapred.JobClient: FILE_BYTES_WRITTEN=246622
14/04/06 17:05:53 INFO mapred.JobClient: Map-Reduce Framework
14/04/06 17:05:53 INFO mapred.JobClient: Map output materialized bytes=12
14/04/06 17:05:53 INFO mapred.JobClient: Map input records=43
14/04/06 17:05:53 INFO mapred.JobClient: Reduce shuffle bytes=12
14/04/06 17:05:53 INFO mapred.JobClient: Spilled Records=0
14/04/06 17:05:53 INFO mapred.JobClient: Map output bytes=0
14/04/06 17:05:53 INFO mapred.JobClient: Total committed heap usage (bytes)=360120320
14/04/06 17:05:53 INFO mapred.JobClient: CPU time spent (ms)=3040
14/04/06 17:05:53 INFO mapred.JobClient: Map input bytes=3431
14/04/06 17:05:53 INFO mapred.JobClient: SPLIT_RAW_BYTES=242
14/04/06 17:05:53 INFO mapred.JobClient: Combine input records=0
14/04/06 17:05:53 INFO mapred.JobClient: Reduce input records=0
14/04/06 17:05:53 INFO mapred.JobClient: Reduce input groups=0
14/04/06 17:05:53 INFO mapred.JobClient: Combine output records=0
14/04/06 17:05:53 INFO mapred.JobClient: Physical memory (bytes) snapshot=408395776
14/04/06 17:05:53 INFO mapred.JobClient: Reduce output records=0
14/04/06 17:05:53 INFO mapred.JobClient: Virtual memory (bytes) snapshot=4121174016
14/04/06 17:05:53 INFO mapred.JobClient: Map output records=0
14/04/06 17:05:53 INFO indexer.CleaningJob: CleaningJob: finished at 2014-04-06 17:05:53, elapsed: 00:00:31
```

This output has been clipped because it is too long to include all of it here. As long as you get to the CleaningJob line, you know that the cycle has completed.

Look for any warnings and errors in this output. Common errors relate to undefined or unexpected document tokens being found while crawling. Updating the schema.xml before starting Solr or attempting the crawl will minimize these. Also, check the Hadoop logs and the Nutch log under:

```
$HADOOP_PREFIX/logs/
$NUTCH_HOME/runtime/local/logs/
```

You can now check the Hadoop file system and see the data being stored there:

```
[hadoop@hc1nn nutch]$ hadoop fs -ls /user/hadoop
Found 2 items
drwxr-xr-x   - hadoop supergroup          0 2014-04-06 14:07
/user/hadoop/crawl
drwxr-xr-x   - hadoop supergroup          0 2014-04-06 11:46
/user/hadoop/nutch
```

The Nutch crawl directory stores the current and old data in subdirectories:

```
[hadoop@hc1nn nutch]$ hadoop fs -ls /user/hadoop/crawl/crawldb

Found 2 items
drwxr-xr-x   - hadoop supergroup          0 2014-04-06 14:16
/user/hadoop/crawl/crawldb/current
drwxr-xr-x   - hadoop supergroup          0 2014-04-06 14:07
/user/hadoop/crawl/crawldb/old
```

The segments directory has a list of the segments that have been processed:

```
[hadoop@hc1nn nutch]$ hadoop fs -ls /user/hadoop/crawl/segments

Found 2 items
drwxr-xr-x   - hadoop supergroup          0 2014-04-06 14:08
/user/hadoop/crawl/segments/20140406140827
drwxr-xr-x   - hadoop supergroup          0 2014-04-06 14:17
/user/hadoop/crawl/segments/20140406141732
```

You now have data in Solr. To see it, go to the Solr admin web page at http://localhost:8983/solr/ and select "collection1" in the core selector drop-down menu (halfway down the left side). Figure 3-2 shows that the sample data has loaded into Solr. Specifically, under the Replication heading on the right, you can see that around 10 KB of data loaded from the short crawl of the Semtech Solutions web page. Although this is a comparatively small sum of data for a large-scale distributed system, it serves to prove that the crawl executed and indexed correctly.

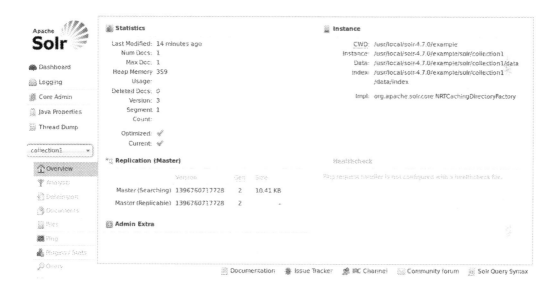

Figure 3-2. *The Solr sample data after processing*

To examine some of the actual data in Solr, you select the Query option (bottom left). An Execute Query option will appear; select it to see the crawl results. Figure 3-3 shows a sample of the data that Solr has indexed from the website at the single URL specified in the seed file.

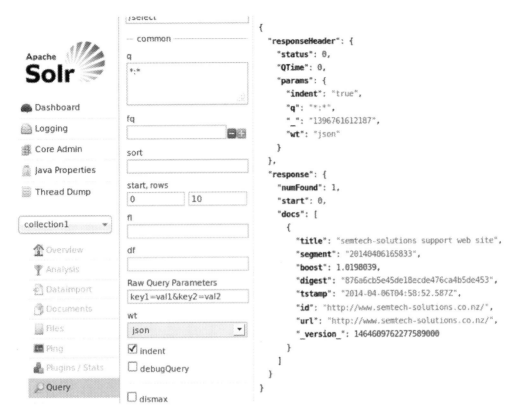

Figure 3-3. Solr with crawl results

So, that's it for architecture example 1. You have connected Hadoop 1.2.1 to Nutch 1.8 and indexed the data using Solr 4.7. Note that during the crawl script, Nutch implicitly checked for Hadoop before using it; otherwise, it would have used the Linux file system for storage. That is an important point to recognize here because, in the second architectural example, using Nutch 2.x, you will explicitly configure Nutch to use HBase, and therefore Hadoop as well.

Architecture 2: Nutch 2.x

In the first architecture example, you used Nutch 1.x. When you executed a crawl, Nutch used Hadoop because it automatically checked whether you were in a distributed environment and it attempted to use Hadoop for storage. The architecture of this next example enables you to specify the storage you will use for your Nutch crawl. Nutch 2.x uses Apache Gora (`gora.apache.org`) to abstract the storage layer. You will also use Apache HBase with Nutch. Using Hadoop and HDFS for storage, Apache HBase (`hbase.apache.org`) offers real-time read/write random access to big data. Should you later need to choose a different storage option, Gora provides the flexibility to do that; you just change the Gora configuration. For instance, you might decide to use the Apache Accumulo database (`accumulo.apache.org`).

Nutch and Solr Configuration

You have already learned how to install Nutch and Solr in the first architecture example, so I will be brief here. To begin, you set up a little configuration in the Linux hadoop account's $HOME/.bashrc for Solr and Nutch.

The following Bash file defines the Hadoop and Java variables; it also creates some useful aliases for Hadoop commands, like fs (saves typing). In addition, it sets up the shell search path (PATH) and defines some variables for Solr and Nutch.

```
[hadoop@hc1r1m2 ~]$ cd $HOME
[hadoop@hc1r1m2 ~]$ ls -l .bashrc*
-rw-r--r--. 1 hadoop hadoop  1142 Apr  5 15:57 .bashrc

[hadoop@hc1r1m2 ~]$ cat .bashrc

####################################################
# Set Hadoop related env variables

export HADOOP_PREFIX=/usr/local/hadoop

# set JAVA_HOME (you will also set a hadoop specific value later)
export JAVA_HOME=/usr/lib/jvm/jre-1.6.0-openjdk

# some handy aliases and functions
unalias fs 2>/dev/null
alias fs="hadoop fs"
unalias hls 2>/dev/null
alias hls="fs -l"
unalias cdh 2>/dev/null
alias cdh="cd $HADOOP_PREFIX"

# add hadoop to the path
export PATH=$HADOOP_PREFIX:$PATH
export PATH=$HADOOP_PREFIX/bin:$PATH
export PATH=$HADOOP_PREFIX/sbin:$PATH

####################################################
# Set up Nutch and Solr variables

export NUTCH_HOME=/usr/local/nutch
export NUTCH_CONF_DIR=$NUTCH_HOME/conf
export SOLR_HOME=/usr/local/solr

export PATH=$PATH:$NUTCH_HOME/bin
```

You now install Nutch 2.x from Apache on the machine hc1r1m2, because that is where ZooKeeper is already installed. For example, I download and unpack the appropriate file from the Nutch website (http://nutch.apache.org/downloads.html):

```
[hadoop@hc1r1m2 Downloads]$ ls -l apache-nutch-2.2.1-src.tar.gz
-rw-rw-r--. 1 hadoop hadoop 3839858 Apr  7 18:32 apache-nutch-2.2.1-src.tar.gz
```

As for example 1, you choose the gzipped tar file; using the gunzip and tar xvf Linux commands, it is easily unpacked. The tar command unpacks the file into the directory named apache-nutch-2.2.1:

```
[hadoop@hc1r1m2 Downloads]$ gunzip apache-nutch-2.2.1-src.tar.gz
[hadoop@hc1r1m2 Downloads]$ tar xvf apache-nutch-2.2.1-src.tar
```

Next, you move it to its home location and set up a symbolic link called "nutch" to simplify the path. Also, you use the Linux chown command with a recursive (-R) switch to change ownership of all files and directories to the Linux hadoop account.

```
[root@hc1r1m2 Downloads]# mv apache-nutch-2.2.1 /usr/local
[root@hc1r1m2 Downloads]# cd /usr/local
[root@hc1r1m2 Downloads]# chown -R hadoop:hadoop apache-nutch-2.2.1
[root@hc1r1m2 local]# ln -s apache-nutch-2.2.1 nutch
[root@hc1r1m2 local]# ls -ld *nutch*

drwxrwxr-x.  7 hadoop hadoop      4096 Apr  7 18:33 apache-nutch-2.2.1
lrwxrwxrwx.  1 root   root          18 Apr  7 18:35 nutch -> apache-nutch-2.2.1
```

Okay, so you have a simple environment set up. Next, you set up the Nutch configuration files:

```
[hadoop@hc1r1m2 ~]$ cd $NUTCH_CONF_DIR ; pwd
/usr/local/nutch/conf
```

Note that the configuration directory for Nutch is under /usr/local/nutch/conf/, as just shown. Extra Nutch configuration properties will now be added to these files for Nutch V2. For example, storage.data.store.class indicates that you are going to use Gora and HBase for storage. (In the Nutch 1.8 architecture, you specified the plug-in folders and agent name options instead.)

```
<property>
  <name>http.agent.name</name>
  <value>NutchHadoopCrawler</value>
</property>

<property>
  <name>storage.data.store.class</name>
  <value>org.apache.gora.hbase.store.HBaseStore</value>
</property>

<property>
  <name>plugin.folders</name>
  <value>/usr/local/nutch/build/plugins</value>
</property>
```

As before, you copy the Nutch configuration files into the Solr configuration directory:

```
[hadoop@hc1r1m2 ~]$ cd $NUTCH_HOME/conf
[hadoop@hc1r1m2 conf]$ cp schema.xml $SOLR_HOME/example/solr/collection1/conf
[hadoop@hc1r1m2 conf]$ cp schema-solr4.xml $SOLR_HOME/example/solr/collection1/conf
```

Now, you need to make some changes to the schema file, schema.xml. As before, find the line that specifies the deprecated EnglishPorterFilterFactory:

```
<filter class="solr.EnglishPorterFilterFactory" protected="protwords.txt"/>
```

Update it to the following to provide filtering functionality for the parsed tokens:

```
<filter class="solr.SnowballPorterFilterFactory" protected="protwords.txt"
    language="English"/>
```

Add the following lines at the end of the <fields> section to define some extra field types in the documents to be parsed. (By making these changes you will avoid Nutch document parsing errors):

```
<!-- fields for Nutch -->
 <field name="_version_" type="long" indexed="true" stored="true"/>
 <field name="text" type="string" indexed="true" stored="true"/>
```

Build Nutch as you did in the last Nutch release:

```
[hadoop@hc1r1m2 nutch]$ pwd
/usr/local/nutch
[hadoop@hc1r1m2 nutch]$ ant

Buildfile: build.xml

....

BUILD SUCCESSFUL
Total time: 1 minute 47 seconds
```

Note: That was a quick build. (As you remember, the last Nutch build took more than 11 minutes). With Nutch built, you are ready to install Apache HBase, the Hadoop-based database, and test it.

HBase Installation

The pieces are moving into place for this second architecture example. Nutch is installed and built, as well as configured to use Gora and HBase. The Gora component was included with the Nutch 2.x release, and Apache ZooKeeper was installed already as part of Chapter2's installation. Now you need to install Apache HBase. To demonstrate its use, I show how to install HBase on a single server.

You can download HBase from the HBase website (hbase.apache.org). After clicking the Downloads option on the left of the page, you may be directed to an alternative mirror site. That's fine—just follow the link. (I downloaded the 0.90.4 release). Again, it is a gzipped tar file that needs to be unpacked.

```
[hadoop@hc1r1m2 Downloads]$ ls -l hbase-0.90.4.tar.gz
-rw-rw-r--. 1 hadoop hadoop 37161251 Apr  8 18:36 hbase-0.90.4.tar.gz

[hadoop@hc1r1m2 Downloads]$ gunzip hbase-0.90.4.tar.gz
[hadoop@hc1r1m2 Downloads]$ tar xvf hbase-0.90.4.tar
```

Move the unpacked release to /usr/local, and change the ownership to the Linux hadoop user recursively with chown -R. Then, create a symbolic link called "hbase" under /usr/local/ to simplify both the path and the environment setup.

```
root@hc1r1m2 Downloads]# mv hbase-0.90.4 /usr/local
[root@hc1r1m2 Downloads]# cd /usr/local
[root@hc1r1m2 local]# chown -R hadoop:hadoop hbase-0.90.4
[root@hc1r1m2 local]# ln -s hbase-0.90.4 hbase
```

```
[root@hc1r1m2 local]# ls -ld *hbase*
lrwxrwxrwx. 1 root    root       12 Apr  8 18:38 hbase -> hbase-0.90.4
drwxrwxr-x. 8 hadoop hadoop 4096 Apr  8 18:36 hbase-0.90.4
```

Next, set up the HBase configuration. You'll find the configuration files under /usr/local/hbase/conf, as shown:

```
[hadoop@hc1r1m2 conf]$ pwd
/usr/local/hbase/conf
[hadoop@hc1r1m2 conf]$ ls
hadoop-metrics.properties  hbase-site.xml    regionservers
hbase-env.sh               log4j.properties
```

You will need to change the contents of the hbase-site.xml file. Specifically, the value of hbase.rootdir needs to point to the name node on the master server. This allows HBase to store data on HDFS. Also, the distributed flag (hbase.cluster.distributed) tells HBase that you are using a cluster.

You also need to specify the HBase master address and port number, as well as the value of the region server port (i/o server) and the fact that it is on a cluster. Other properties to define are the address of the HBase temporary directory, details for HBase access to ZooKeeper, and limits for ZooKeeper operation.

To start, you add the properties that follow to the hbase-site.xml file between the configuration open (<configuration>) and configuration closing (</configuration>) XML tags:

```
<configuration>

  <property>
    <name>hbase.rootdir</name>
    <value>hdfs://hc1nn:54310/hbase</value>
  </property>

  <property>
    <name>hbase.master</name>
    <value>hc1r1m2:60000</value>
  </property>

  <property>
    <name>hbase.master.port</name>
    <value>60000</value>
  </property>

  <property>
    <name>hbase.regionserver.port</name>
    <value>60020</value>
  </property>

  <property>
    <name>hbase.cluster.distributed</name>
    <value>true</value>
  </property>
```

```xml
  <property>
    <name>hbase.tmp.dir</name>
    <value>/var/hbase/</value>
  </property>

  <property>
    <name>hbase.zookeeper.quorum</name>
    <value>hc1r1m1,hc1r1m2,hc1r1m3</value>
  </property>

  <property>
    <name>dfs.replication</name>
    <value>1</value>
  </property>

  <property>
    <name>hbase.zookeeper.property.clientPort</name>
    <value>2181</value>
  </property>

  <property>
    <name>hbase.zookeeper.property.dataDir</name>
    <value>/var/lib/zookeeper</value>
  </property>

  <property>
    <name>zookeeper.session.timeout</name>
    <value>1000000</value>
  </property>

  <property>
    <name>hbase.client.scanner.caching</name>
    <value>6000</value>
  </property>

  <property>
    <name>hbase.regionserver.lease.period</name>
    <value>2500000</value>
  </property>

  <property>
   <name>hbase.zookeeper.property.maxClientCnxns</name>
   <value>0</value>
  </property>

  <property>
    <name>hbase.zookeeper.property.tickTime</name>
    <value>8000</value>
  </property>

</configuration>
```

You also need to tell HBase that it will not be managing the ZooKeeper quorum, because ZooKeeper is already running. To do so, set up the following variable in the Linux hadoop account environment (I have defined this in my environment via an entry in $HOME/.bashrc, as follows):

```
export HBASE_MANAGES_ZK=false
```

You must make sure that Hadoop and HBase are using the same version of Hadoop core libraries, so issue a pair of cd and ls commands to check the versions:

```
[hadoop@hc1r1m2 lib]$ cd /usr/local/hbase/lib
[hadoop@hc1r1m2 lib]$ ls -l hadoop-core-*.jar
-rwxrwxr-x. 1 hadoop hadoop 2707856 Feb 10  2011 hadoop-core-0.20-append-r1056497.jar

[hadoop@hc1r1m2 hadoop]$ cd $HADOOP_PREFIX
[hadoop@hc1r1m2 hadoop]$ ls -l hadoop-core-*.jar
-rw-rw-r--. 1 hadoop hadoop 4203147 Jul 23  2013 hadoop-core-1.2.1.jar
```

As you can see, the versions don't match. The HBase Hadoop core file (hadoop-core-0.20-append-r1056497.jar) is at version 0.20, while the Hadoop jar file (hadoop-core-1.2.1.jar) is at version 1.2.1. Currently, if you tried to run HBase, it would fail with a connect exception.

You already know from the Hadoop installation you carried out in Chapter 2 that the Hadoop installation is version 1.2.1. To work with the Hadoop installation, HBase must use the same version of Hadoop libraries. You copy this library into place, as follows:

```
[hadoop@hc1r1m2 hadoop]$ cp $HADOOP_PREFIX/hadoop-core-*.jar    /usr/local/hbase/lib
[hadoop@hc1r1m2 hadoop]$ cd    /usr/local/hbase/lib
[hadoop@hc1r1m2 hadoop]$ mv hadoop-core-0.20-append-r1056497.jar hadoop-core-0.20-append-r1056497.jar.save
```

Check the version of the Hadoop commons configuration jar file being used by HBase. This library assists with the reading of configuration and preference files. If the version is different, then it can be updated by copying the Hadoop version into place. (This will avoid errors like "NoClassDefFoundError" when you try to run HBase.) Use the command sequence:

```
[root@hc1r1m2 lib]# cd  /usr/local/hadoop/lib/
[root@hc1r1m2 lib]# ls commons-configuration*
 /usr/local/hadoop/lib/commons-configuration-1.6.jar
```

That shows that the Hadoop version of this file is 1.6. You copy it to the HBase library area:

```
[root@hc1r1m2 lib]# cd   /usr/local/hbase/lib
[root@hc1r1m2 lib]# cp /usr/local/hadoop/lib/commons-configuration* .
```

Now, you can try starting HBase by using the start script in the HBase bin directory:

```
[hadoop@hc1r1m2 ~]$ cd /usr/local/hbase
[hadoop@hc1r1m2 hbase]$ ./bin/start-hbase.sh

starting master, logging to
/usr/local/hbase/logs/hbase-hadoop-master-hc1r1m2.out
```

Note that HBase has given the address of its log file, so you can check the logs for errors. The logs files look like this:

```
[hadoop@hc1r1m2 hbase]$ ls -l /usr/local/hbase/logs/
total 28
-rw-rw-r--. 1 hadoop hadoop 11099 Apr  9 18:32 hbase-hadoop-master-hc1r1m2.log
-rw-rw-r--. 1 hadoop hadoop     0 Apr  9 18:32 hbase-hadoop-master-hc1r1m2.out
-rw-rw-r--. 1 hadoop hadoop    78 Apr  9 18:30 hbase-hadoop-regionserver-hc1r1m2.log
-rw-rw-r--. 1 hadoop hadoop   250 Apr  9 18:30 hbase-hadoop-regionserver-hc1r1m2.out
```

A few typical errors reported might be related to ZooKeeper, such as:

```
2014-04-13 14:53:54,827 WARN org.apache.zookeeper.ClientCnxn: Session 0x14558da65420000 for server
null, unexpected error,
closing socket connection and attempting reconnect
java.net.ConnectException: Connection refused
```

This indicates that either Zo1oKeeper is down or there are network issues. Make sure that all of your ZooKeeper servers are up, and check the ZooKeeper logs.

A good way to see that HBase is working is to start a shell and create a table. For example, start the shell as follows:

```
[root@hc1r1m2 bin]# pwd
/usr/local/hbase/bin
[root@hc1r1m2 bin]# ./hbase shell
HBase Shell; enter 'help<RETURN>' for list of supported commands.
Type "exit<RETURN>" to leave the HBase Shell
Version 0.90.4, r1150278, Sun Jul 24 15:53:29 PDT 2011

hbase(main):001:0>
```

Now, create a table named "employer" with a single column named "empname" for the name of the employee and insert some data:

```
hbase(main):001:0> create 'employer', 'empname'
0 row(s) in 0.5650 seconds
```

Insert a single row into the table column created with the employee name "Evans D"; this is row 1:

```
hbase(main):004:0> put 'employer', 'row1', 'empname', 'Evans D'
0 row(s) in 0.0130 seconds
```

Check that this data is accessible from the employee table:

```
hbase(main):005:0> get 'employer', 'row1'
COLUMN                  CELL
 empname:                timestamp=1397028137988, value=Evans D
1 row(s) in 0.0330 seconds

hbase(main):006:0> exit
```

HBase seems to be running okay, but you should check that HBase is storing the data to HDFS. To do so, use the Hadoop file system ls command. HBase should have created a storage directory on HDFS called "/hbase," so you can check that:

```
[hadoop@hc1nn logs]$ hadoop dfs -ls /hbase
Found 5 items
drwxr-xr-x   - hadoop supergroup          0 2014-04-12 19:55 /hbase/-ROOT-
drwxr-xr-x   - hadoop supergroup          0 2014-04-12 19:55 /hbase/.META.
drwxr-xr-x   - hadoop supergroup          0 2014-04-12 19:57 /hbase/.logs
drwxr-xr-x   - hadoop supergroup          0 2014-04-12 19:57 /hbase/.oldlogs
-rw-r--r--   3 hadoop supergroup          3 2014-04-12 19:55 /hbase/hbase.version
```

Gora Configuration

Nutch and Solr are ready, HBase and ZooKeeper are ready, and HBase is storing its data to HDFS. Now it is time to connect Nutch to HBase using the Apache Gora module that was installed with Nutch 2.x. Gora (Gora.apache.org) provides an in-memory data model for big data and data persistence. It allows you to choose where you will store the data that Nutch collects, because it supports a variety of data stores. In this section, you will configure Gora to store Nutch 2.x crawl data to HBase.

You can now set up the Gora connection for Nutch. First, you need to edit the nutch-site.xml file:

```
[hadoop@hc1r1m2 conf]$ pwd
/usr/local/nutch/conf
[hadoop@hc1r1m2 conf]$ vi nutch-site.xml
```

Specifically, you add a property called "storage.data.store.class" to specify that HBase will be the default storage for Nutch Gora. As before, make sure that you add the property to the file so that it sits between the xml open and close configuration tabs:

```
<property>
  <name>storage.data.store.class</name>
  <value>org.apache.gora.hbase.store.HBaseStore</value>
   <description>Default class for storing data</description>
</property>
```

Check the Nutch Ivy configuration. Apache Ivy (http://ant.apache.org/ivy/) is a dependency manager that is integrated with Apache Ant. Intended for Java-based systems, it is mostly used for system build management.

```
[hadoop@hc1r1m2 ivy]$ pwd
/usr/local/nutch/ivy
[hadoop@hc1r1m2 ivy]$ vi ivy.xml
```

Make sure that this line is uncommented so that Ivy is configured to use Gora. This is what the line looks like after the change:

```
<dependency org="org.apache.gora" name="gora-sql" rev="0.3" conf="*->default" />
```

Make sure that the gora.properties file is set up correctly, as shown:

```
[hadoop@hc1r1m2 nutch]$ pwd
/usr/local/nutch
[hadoop@hc1r1m2 nutch]$ vi ./conf/gora.properties
```

Check that Gora is the default data store. The line here should already exist in the file, but it may be commented out. Uncomment or add the line. This will set the default Gora data store to be Apache HBase:

```
gora.datastore.default=org.apache.gora.hbase.store.HBaseStore
```

Remember that each time you change the Nutch configuration, you need to re-compile Nutch. Do so now, so that the Gora changes take effect:

```
[hadoop@hc1r1m2 nutch]$ pwd
/usr/local/nutch
[hadoop@hc1r1m2 nutch]$ ant runtime
Buildfile: build.xml
......
BUILD SUCCESSFUL
Total time: 13 minutes 39 seconds
```

Running the Nutch Crawl

You have managed to start HBase and you know that HBase is storing its data within Hadoop HDFS. You have a ZooKeeper quorum running, and HBase is able to connect to it without error. Solr has been started and is running without error on port 8983. Additionally, Nutch Gora has been configured to use HBase for storage. So now you are ready to run a Nutch crawl, move to the Nutch home directory as shown by the Linux cd command:

```
[hadoop@hc1r1m2 nutch]$ cd $NUTCH_HOME
[hadoop@hc1r1m2 nutch]$ pwd
/usr/local/nutch
```

Now make sure that the seed URL is ready in HDFS. (You know it is ready because you stored it there for the Nutch 1.x crawl.) Checking the contents of the seed file, you can see that it has a single URL line (my website address). You could have put a few million lines into this file for a larger crawl, but you can try that later.

```
[hadoop@hc1r1m2 hadoop]$ hadoop dfs -cat /user/hadoop/nutch/urls/seed.txt
http://www.semtech-solutions.co.nz
```

You can determine the syntax for the crawl by executing the script name without parameters. The error message tells you how it should be run:

```
[hadoop@hc1r1m2 nutch]$ cd runtime/deploy/bin
[hadoop@hc1r1m2 bin]$ ./crawl
Missing seedDir : crawl <seedDir> <crawlID> <solrURL> <numberOfRounds>
```

The crawl is executed in the same format as for Nutch 1.x, and the output is shown as follows:

```
[hadoop@hc1nn bin]$ ./crawl urls crawl1 http://hc1r1m2:8983/solr/   2
```

```
14/04/13 17:28:10 INFO crawl.InjectorJob: InjectorJob: starting at 2014-04-13 17:28:10
14/04/13 17:28:10 INFO crawl.InjectorJob: InjectorJob: Injecting urlDir: nutch/urls
14/04/13 17:28:11 INFO zookeeper.ZooKeeper: Client environment:zookeeper.version=3.3.2-1031432,
                                            built on 11/05/2010 05:32 GMT
14/04/13 17:28:11 INFO zookeeper.ZooKeeper: Client environment:host.name=hc1r1m2
14/04/13 17:28:11 INFO zookeeper.ZooKeeper: Client environment:java.version=1.6.0_30
14/04/13 17:28:11 INFO zookeeper.ZooKeeper: Client environment:java.vendor=Sun Microsystems Inc.
14/04/13 17:28:11 INFO zookeeper.ZooKeeper: Client environment:java.home=/usr/lib/jvm/java-1.6.0-
                                            openjdk-1.6.0.0/jre
14/04/13 17:28:11 INFO zookeeper.ZooKeeper: Client

. . . . . . . . . . . . . . .

14/04/13 17:37:20 INFO mapred.JobClient: Job complete: job_201404131430_0019
14/04/13 17:37:21 INFO mapred.JobClient: Counters: 6
14/04/13 17:37:21 INFO mapred.JobClient: Job Counters
14/04/13 17:37:21 INFO mapred.JobClient: SLOTS_MILLIS_MAPS=54738
14/04/13 17:37:21 INFO mapred.JobClient: Total time spent by all reduces waiting after reserving
                                         slots (ms)=0
14/04/13 17:37:21 INFO mapred.JobClient: Total time spent by all maps waiting after reserving slots (ms)=0
14/04/13 17:37:21 INFO mapred.JobClient: Launched map tasks=8
14/04/13 17:37:21 INFO mapred.JobClient: SLOTS_MILLIS_REDUCES=0
```

You need to monitor all of your logs; that is, you need to monitor the following:

- *ZooKeeper logs*, in this case under /var/log/zookeeper. These allow you to ensure that all servers are up and running as a quorum.

- *Hadoop logs*, in this case under /usr/local/hadoop/logs. Hadoop and MR must be running without error so that HBase can use Hadoop.

- *HBaselogs*, in this case under /usr/local/hbase/logs. You make sure that HBase is running and able to talk to ZooKeeper.

- *Solr output from the Solr session window.* It must be running without error so that it can index the crawl output.

- *Nutch output from the crawl session.* Any errors will appear in the session window.

Each of the components in this architecture must work for the Nutch crawl to work. If you encounter errors, pay particular attention to your configuration. For timeout errors in ZooKeeper, try increasing the tickTime and syncLimit values in your ZooKeeper config files.

Potential Errors

Here are some of the errors that occurred when I tried to use this configuration. They are provided here along with their reasons and solutions. If you encounter them, go back to the step you missed and correct the error.

Consider the first one:

```
2014-04-08 19:05:39,334 ERROR
org.apache.hadoop.hbase.master.HMasterCommandLine: Failed to start master
java.io.IOException: Couldnt start ZK at requested address of 2181, instead
got: 2182. Aborting. Why? Because clients (eg shell) wont be able to find this
```

```
ZK quorum at
org.apache.hadoop.hbase.master.HMasterCommandLine.startMaster(HMasterCommandLine.java:131)
```

This problem was caused by one of the ZooKeeper servers (on hc1r1m1) running under the wrong Linux account. It didn't have file system access. The solution was to shut it down and start as the correct user.

Another error you may encounter is:

```
2014-04-09 18:32:53,741 ERROR
org.apache.hadoop.hbase.master.HMasterCommandLine: Failed to start master
java.io.IOException: Unable to create data directory
/var/lib/zookeeper/zookeeper/version-2
```

Again, this was the same issue as for ZooKeeper in the previous error. The ZooKeeper server (on hc1r1m1) was running under the wrong Linux account. It didn't have file system access. The solution was to shut it down and start as the correct user.

An HBase error that could occur is:

```
org.apache.hadoop.ipc.RemoteException: Server IPC version 7 cannot communicate
with client version 3
```

This was an HBase error. IPC version 4 is for Hadoop 1.0, whereas version 7 is for Hadoop 2.0, so HBase expects Hadoop version 1.x. That is why we are using Hadoop version 1.2.1 with Nutch 2.x.

This error occurred during the Nutch crawl:

```
14/04/12 20:32:30 ERROR crawl.InjectorJob: InjectorJob:
java.lang.ClassNotFoundException: org.apache.gora.hbase.store.HBaseStore
```

The Gora configuration was incorrect. Go back, check the setting, and retry the crawl once you've fixed it.

These errors occurred during a crawl in the HBase logs:

```
2014-04-12 20:52:37,955 INFO org.apache.hadoop.hbase.master.ServerManager:
Waiting on regionserver(s) to checkin

org.apache.hadoop.security.AccessControlException:
org.apache.hadoop.security.AccessControlException: Permission denied:
user=root, access=WRITE, inode=".logs":hadoop:supergroup:rwxr-xr-x
```

They indicate a file system access issue on HDFS for HBase. I had set the HBase directory permissions for the .logs directory incorrectly.

```
[hadoop@hc1nn bin]$ hadoop dfs -ls /hbase/ | grep logs
drwxrwxrwx   - hadoop supergroup          0 2014-04-13 18:00 /hbase/.logs
```

Your permissions for HDFS directories should be fine, but if you encounter a permissions access error, you can use the hadoop dfs -chmod command to set permissions.

This error occurred because I had an error in my /etc/hosts file:

```
14/04/13 12:00:54 INFO mapred.JobClient: Task Id :
attempt_201404131045_0016_m_000000_0, Status : FAILED
java.lang.RuntimeException: java.io.IOException: java.lang.RuntimeException:
org.apache.hadoop.hbase.ZooKeeperConnectionException: HBase is able to connect
```

to ZooKeeper but the connection closes immediately. This could be a sign that the server has too many connections (30 is the default). Consider inspecting your ZK server logs for that error and then make sure you are reusing HBaseConfiguration as often as you can. See HTable's javadoc for more information.

Make sure that your /etc/hosts file entries are defined correctly and that ZooKeeper is working correctly before you move on to start HBase.

As for architecture example 1, you can check the Solr query page for your results. For this example, because the seed URL was from my own site, the query found the apache URLs, as shown in Figure 3-4.

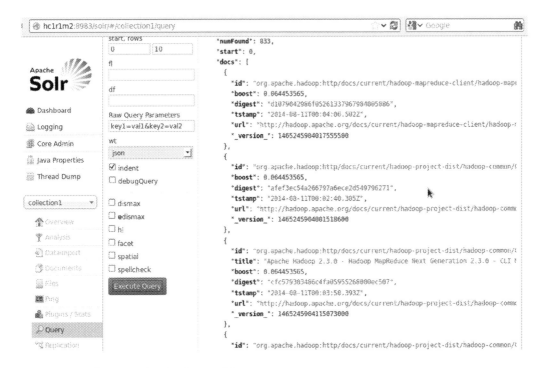

Figure 3-4. *Solr output showing the crawl data*

A Brief Comparison

Of the two architecture examples given in this chapter, the second, using HBase, was the most difficult to use. This may be the future direction that Nutch is taking, but there are a lot more configuration items to take care of. There are more components to worry about and check, plus more potential areas of failure. Having said that, the second architecture example gives you the ability to explicitly choose the storage architecture. If for some reason at a future date you need to use an alternative system to HBase that Gora supports, you will be able to do that.

You have only used Hadoop V1 in both of these examples. If time had allowed, it would have been useful to use Hadoop V2 as well. In that case, you would have needed to rebuild both HBase and Nutch using Hadoop V2 libraries. Nevertheless, it would have been interesting to compare the Nutch processing time using Hadoop V1 and V2.

Summary

In this chapter, you investigated big data collection using Nutch, Solr, Gora, and HBase. You used both Nutch 1.x and 2.x to crawl a seed URL and collect data. Although you crawled only on a small scale, the same process can be used to gather large volumes of data. You used Solr in both cases to index data passed from Nutch. In the second example, you used Apache Gora to determine where Nutch would store its data—in this case, it was HBase. You also looked at two possible approaches for using Nutch and Hadoop. In the first, Nutch implicitly used Hadoop for storage; in the second, Nutch used Apache Gora to explicitly select HBase for storage.

Where do you go from here? The command sequence and examples given in this chapter should enable you to apply these approaches to your own system. Take a logical approach, and make sure that HDFS is working before moving on. Also, make sure that ZooKeeper is working before you attempt HBase. Remember: if you encounter errors, search the web for solutions, because other people may have encountered similar problems. Also, keep trying to think of new ways to approach the problem.

CHAPTER 4

Processing Data with Map Reduce

Hadoop Map Reduce is a system for parallel processing of very large data sets using distributed fault-tolerant storage over very large clusters. The input data set is broken down into pieces, which are the inputs to the Map functions. The Map functions then filter and sort these data chunks (whose size is configurable) on the Hadoop cluster data nodes. The output of the Map processes is delivered to the Reduce processes, which shuffle and summarize the data to produce the resulting output.

This chapter explores Map Reduce programming through multiple implementations of a simple, but flexible word-count algorithm. After first coding the word-count algorithm in raw Java using classes provided by Hadoop libraries, you will then learn to carry out the same word-count function in Pig Latin, Hive, and Perl. Don't worry about these terms yet; you will learn to source, install, and work with this software in the coming sections.

An Overview of the Word-Count Algorithm

The best way to understand the Map Reduce system is with an example. A word-count algorithm is not only the most common and simplest Map Reduce example, but it is also one that contains techniques you can apply to more complex scenarios. To begin, consider Figure 4-1, which breaks the word-count process into steps.

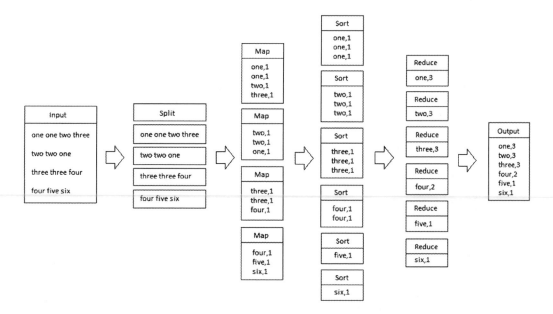

Figure 4-1. *The word-count Map Reduce process*

The input file on the left of Figure 4-1 is split into chunks of data. The size of these splits is controlled by the InputSplit method within the FileInputFormat class of the Map Reduce job. The number of splits is influenced by the HDFS block size, and one mapper job is created for each split data chunk.

Each split data chunk—that is, "one one two three," is sent to a Mapper process on a Data Node server. In the word-count example, the Map process creates a series of key-value pairs where the key is the word—for instance, "one"—and the value is the count of 1. These key-value pairs are then shuffled into lists by key type. The shuffled lists are input to Reduce tasks, which reduce the list volume by summing the values (in this simple example). The Reduce output is then a simple list of summed key-value pairs.

The Map Reduce framework takes care of all other tasks, like scheduling and resources.

Map Reduce Native

Now, it's time to put the word-count algorithm to work, starting with the most basic example: the Map Reduce native version. The term *Map Reduce native* means that the Map Reduce code is written in Java using the functionality provided by the Hadoop core libraries within the Hadoop installation directory. Map Reduce native word-count algorithms are available from the Apache Software Foundation Hadoop 1.2.1 Map Reduce tutorial on the Hadoop website (hadoop.apache.org/docs/r1.2.1/). For instance, I sourced two versions of the Hadoop word-count algorithm and stored them in Java files in a word-count directory, as follows:

```
[hadoop@hc1nn wordcount]$ pwd
/usr/local/hadoop/wordcount
[hadoop@hc1nn wordcount]$ ls *.java
wc-ex1.java  wc-ex2.java
```

The file wc-ex1.java contains the first simple example, while the second Java file contains a second, more complex version.

Java Word-Count Example 1

Consider the Java code for the first word-count example. It follows the basic word-count steps shown in Figure 4-1:

```
01          package org.myorg;
02
03          import java.io.IOException;
04          import java.util.*;
05
06          import org.apache.hadoop.fs.Path;
07          import org.apache.hadoop.conf.*;
08          import org.apache.hadoop.io.*;
09          import org.apache.hadoop.mapred.*;
10          import org.apache.hadoop.util.*;
11
12          public class WordCount
13          {
14
15            public static class Map extends MapReduceBase implements
16            Mapper<LongWritable, Text, Text, IntWritable>
17            {
18              private final static IntWritable one = new IntWritable(1);
19              private Text word = new Text();
20
```

```
21      public void map(LongWritable key, Text value, OutputCollector<Text,
22    IntWritable output, Reporter reporter) throws IOException
23        {
24          String line = value.toString();
25          StringTokenizer tokenizer = new StringTokenizer(line);
26          while (tokenizer.hasMoreTokens())
27          {
28            word.set(tokenizer.nextToken());
29            output.collect(word, one);
30          }
31        }
32      } /* class Map */
33
34      public static class Reduce extends MapReduceBase implements
35          Reducer<Text, IntWritable, Text, IntWritable>
36      {
37        public void reduce(Text key, Iterator<IntWritable> values,
          OutputCollector<Text, IntWritable> output, Reporter reporter)
          throws IOException
39        {
40          int sum = 0;
41          while (values.hasNext())
42          {
43            sum += values.next().get();
44          }
45          output.collect(key, new IntWritable(sum));
46        }
47      } /* class Reduce */
48
49      public static void main(String[] args) throws Exception
50      {
51        JobConf conf = new JobConf(WordCount.class);
52        conf.setJobName("wordcount");
53
54        conf.setOutputKeyClass(Text.class);
55        conf.setOutputValueClass(IntWritable.class);
56
57        conf.setMapperClass(Map.class);
58        conf.setCombinerClass(Reduce.class);
59        conf.setReducerClass(Reduce.class);
60
61        conf.setInputFormat(TextInputFormat.class);
62        conf.setOutputFormat(TextOutputFormat.class);
63
64        FileInputFormat.setInputPaths(conf, new Path(args[0]));
65        FileOutputFormat.setOutputPath(conf, new Path(args[1]));
66
67        JobClient.runJob(conf);
68      }
69
70    } /* class WordCount */
```

Describing the Example 1 Code

In this section, we look at the Java code used in the Java-based simple Map Reduce word-count example. We then proceed to compile it, create a jar file, and run it. This package is called org.myorg, and it is defined at line 1:

```
01       package org.myorg;
```

Lines 6 to 10 import Hadoop functionality for Path, configuration, I/O, Map Reduce, and utilities.

```
06       import org.apache.hadoop.fs.Path;
07       import org.apache.hadoop.conf.*;
08       import org.apache.hadoop.io.*;
09       import org.apache.hadoop.mapred.*;
10       import org.apache.hadoop.util.*;
```

For the details of these APIs, consult https://hadoop.apache.org/docs/stable/api/org/apache/hadoop/. Just select the package name—that is, mapred—and then choose package-summary.html. The fs.Path class provides file system path functionality and the Conf class adds configuration functionality as shown at line 51 in the example. The IO class adds input/output functionality, while the Util class adds utilities like logging and checksums. The Mapred class is the Hadoop V1 Map Reduce class API implimentation. The V1 class names are called *mapred*, while the V2 implementation that you will encounter later in the book uses the term *mapreduce*.

The Map class is defined at line 15:

```
15       public static class Map extends MapReduceBase implements
16         Mapper<LongWritable, Text, Text, IntWritable>
```

Line 25 uses a StringTokenizer to break the input line into words, which are then passed as outputs as key-value pairs—that is, <word,1>.

```
25       StringTokenizer tokenizer = new StringTokenizer(line);
26       while (tokenizer.hasMoreTokens())
27       {
28         word.set(tokenizer.nextToken());
29         output.collect(word, one);
30       }
```

As defined at line 34, the Reduce class accepts shuffled key-value pairs as input:

```
34       public static class Reduce extends MapReduceBase implements
35         Reducer<Text, IntWritable, Text, IntWritable>
```

From line 40 on, the code then totals the values for the key-value pairs with the same key and outputs the totaled key-value pairs; that is, <word,5>:

```
40       int sum = 0;
41       while (values.hasNext())
42       {
43         sum += values.next().get();
44       }
45       output.collect(key, new IntWritable(sum));
```

The main method at line 49 sets up the Map Reduce configuration by defining the type of input. In this case, the input is text.

```
49   public static void main(String[] args) throws Exception
```

The code then defines the Map, Combine, and Reduce classes, as well as specifying the input/output formats:

```
51      JobConf conf = new JobConf(WordCount.class);
52      conf.setJobName("wordcount");
53
54      conf.setOutputKeyClass(Text.class);
55      conf.setOutputValueClass(IntWritable.class);
56
57      conf.setMapperClass(Map.class);
58      conf.setCombinerClass(Reduce.class);
59      conf.setReducerClass(Reduce.class);
60
61      conf.setInputFormat(TextInputFormat.class);
62      conf.setOutputFormat(TextOutputFormat.class);
63
64      FileInputFormat.setInputPaths(conf, new Path(args[0]));
65      FileOutputFormat.setOutputPath(conf, new Path(args[1]));
```

Finally, line 67 runs the job:

```
67      JobClient.runJob(conf);
```

Running the Example 1 Code

To compile this code, I used the Java compiler javac, which was installed with the JDK when Java 1.6 was installed. The compiler expects the Java file name to match the class name, so I renamed the example code as WordCount.Java.

The classes on which this example relies are found in the Hadoop core library that is in the Hadoop release, so I specified that when compiling the code. Also, I placed the compiled output into a subdirectory called wc_classes, which can be used when building an example jar file.

```
[hadoop@hc1nn wordcount]$ cp wc-ex1.java WordCount.java
[hadoop@hc1nn wordcount]$ mkdir wc_classes
[hadoop@hc1nn wordcount]$ javac -classpath $HADOOP_PREFIX/hadoop-core-1.2.1.jar  -d wc_classes
WordCount.java
```

The following recursive listing shows all of the subdirectories and classes from the build of the first example code:

```
[hadoop@hc1nn wordcount]$ ls -R wc_classes
wc_classes:
org

wc_classes/org:
myorg

wc_classes/org/myorg:
WordCount.class  WordCount$Map.class  WordCount$Reduce.class
```

Building the code into a jar library using the `jar` command creates the wordcount1.jar file:

```
[hadoop@hc1nn wordcount]$ jar -cvf ./wordcount1.jar -C wc_classes .
added manifest
adding: org/(in = 0) (out= 0)(stored 0%)
adding: org/myorg/(in = 0) (out= 0)(stored 0%)
adding: org/myorg/WordCount.class(in = 1546) (out= 750)(deflated 51%)
adding: org/myorg/WordCount$Reduce.class(in = 1611) (out= 648)(deflated 59%)
adding: org/myorg/WordCount$Map.class(in = 1938) (out= 798)(deflated 58%)

[hadoop@hc1nn wordcount]$ ls -l *.jar
-rw-rw-r--. 1 hadoop hadoop 3169 Jun 15 15:05 wordcount1.jar
```

This file can now be used to run a word-count task on Hadoop. As in previous Map Reduce runs, the input and output data for the job will be taken from HDFS. To provide the words to count, I copied some data from Edgar Allan Poe books into a directory on HDFS from the Linux file system. The Linux `ls` command shows the text files that will be used:

```
[hadoop@hc1nn wordcount]$ ls $HOME/edgar
10031.txt  15143.txt  17192.txt  2149.txt  932.txt
```

Copying these files to the HDFS directory called /user/hadoop/edgar, using the Hadoop file system `copyFromLocal` command, sets up the data for the word-count job:

```
[hadoop@hc1nn wordcount]$ hadoop dfs -copyFromLocal  $HOME/edgar/* /user/hadoop/edgar

[hadoop@hc1nn wordcount]$ hadoop dfs -ls /user/hadoop/edgar

Found 5 items
-rw-r--r--   1 hadoop supergroup     410012 2014-06-15 15:53 /user/hadoop/edgar/10031.txt
-rw-r--r--   1 hadoop supergroup     559352 2014-06-15 15:53 /user/hadoop/edgar/15143.txt
-rw-r--r--   1 hadoop supergroup      66401 2014-06-15 15:53 /user/hadoop/edgar/17192.txt
-rw-r--r--   1 hadoop supergroup     596736 2014-06-15 15:53 /user/hadoop/edgar/2149.txt
-rw-r--r--   1 hadoop supergroup      63278 2014-06-15 15:53 /user/hadoop/edgar/932.txt
```

By running the word-count example against the data in the input directory (/user/hadoop/edgar), you create the results data in the output directory (/user/hadoop/edgar-results). First, though, make sure the processes are all up before you run the job using jps.

```
[hadoop@hc1nn wordcount]$ jps
1959 SecondaryNameNode
1839 DataNode
4166 TaskTracker
4272 Jps
1720 NameNode
4044 JobTracker
```

This shows that the HDFS processes for the data node and name node are running on hc1nn. Also, the Map Reduce processes for the Task and Job Trackers are running. If you are going to rerun this job, then you will need to delete the HDFS-based results directory by using the Hadoop file system `rmr` command:

```
[hadoop@hc1nn wordcount]$ hadoop dfs -rmr  /user/hadoop/edgar-results
```

You can run the job via the Hadoop jar command. The parameters passed to it are the library file you have just created, the name of the class to run in that library, the input directory on HDFS, and the output directory:

```
[hadoop@hc1nn wordcount]$ hadoop jar ./wordcount1.jar  org.myorg.WordCount  /user/hadoop/edgar  /user/
hadoop/edgar-results


14/06/15 16:04:50 INFO util.NativeCodeLoader: Loaded the native-hadoop library
14/06/15 16:04:50 INFO mapred.FileInputFormat: Total input paths to process : 5
14/06/15 16:04:51 INFO mapred.JobClient: Running job: job_201406151602_0001
14/06/15 16:04:52 INFO mapred.JobClient: map 0% reduce 0%
14/06/15 16:05:02 INFO mapred.JobClient: map 20% reduce 0%
14/06/15 16:05:03 INFO mapred.JobClient: map 40% reduce 0%
14/06/15 16:05:04 INFO mapred.JobClient: map 60% reduce 0%
........................
14/06/15 16:05:19 INFO mapred.JobClient: Combine input records=284829
14/06/15 16:05:19 INFO mapred.JobClient: Reduce input records=55496
14/06/15 16:05:19 INFO mapred.JobClient: Reduce input groups=36348
14/06/15 16:05:19 INFO mapred.JobClient: Combine output records=55496
14/06/15 16:05:19 INFO mapred.JobClient: Physical memory (bytes) snapshot=912035840
14/06/15 16:05:19 INFO mapred.JobClient: Reduce output records=36348
14/06/15 16:05:19 INFO mapred.JobClient: Virtual memory (bytes) snapshot=7949012992
14/06/15 16:05:19 INFO mapred.JobClient: Map output records=284829
```

The job has completed (the output shown above has been trimmed), so you can check the output on HDFS under /user/hadoop/edgar-results/ by using the Hadoop file system ls command:

```
[hadoop@hc1nn wordcount]$ hadoop dfs -ls  /user/hadoop/edgar-results/
Found 3 items
-rw-r--r--   1 hadoop supergroup          0 2014-06-15 16:05 /user/hadoop/edgar-results/_SUCCESS
drwxr-xr-x   - hadoop supergroup          0 2014-06-15 16:04 /user/hadoop/edgar-results/_logs
-rw-r--r--   1 hadoop supergroup     396500 2014-06-15 16:05 /user/hadoop/edgar-results/part-00000
```

These results show a _SUCCESS file, so the job was completed without error. As in previous examples, you use the Hadoop file system cat command to dump the contents of the results file and the Linux head command to limit the job results to the first 10 rows:

```
[hadoop@hc1nn wordcount]$ hadoop dfs -cat /user/hadoop/edgar-results/part-00000 | head -10
!)      1
"''T    1
"'And   1
"'As    1
"'Be    2
"'But--still--monsieur----'     1
"'Catherine,    1
"'Comb  1
"'Come  1
"'Eyes,'        1
```

Well done! You have just compiled and run your own native Map Reduce job from a source file. To create more, you can simply change the algorithm in Java (or write your own) and follow the same process. One change that might be useful is to ignore the white-space and symbol characters when counting the words. The example's output data contains characters like these (" or -). The next example adds these refinements.

Java Word-Count Example 2

Using the same Map and Reduce classes as the first example, you will find this second example adds pattern filtering to the code. You can find the file, wc-ex2.java, in the Apache Software Foundation Hadoop 1.2.1 Map Reduce tutorial at the Hadoop website (hadoop.apache.org/docs/r1.2.1/). Here's the complete listing:

```
01        package org.myorg;
02
03        import java.io.*;
04        import java.util.*;
05
06        import org.apache.hadoop.fs.Path;
07        import org.apache.hadoop.filecache.DistributedCache;
08        import org.apache.hadoop.conf.*;
09        import org.apache.hadoop.io.*;
10        import org.apache.hadoop.mapred.*;
11        import org.apache.hadoop.util.*;
12
13        public class WordCount extends Configured implements Tool
14        {
15
16          /*-------------------------------------------------------------------*/
17          public static class Map extends MapReduceBase
18                implements Mapper < LongWritable, Text, Text, IntWritable >
19          {
20
21            static enum Counters
22            {
23              INPUT_WORDS
24            }
25
26            private final static IntWritable one = new IntWritable(1);
27            private Text word = new Text();
28
29            private boolean caseSensitive = true;
30            private Set < String > patternsToSkip = new HashSet < String > ();
31
32            private long numRecords = 0;
33            private String inputFile;
34
35            /*-------------------------------------------------------------------*/
36            public void configure(JobConf job)
37            {
38              caseSensitive = job.getBoolean("wordcount.case.sensitive", true);
39              inputFile = job.get("map.input.file");
40
41              if (job.getBoolean("wordcount.skip.patterns", false))
42              {
43                Path[] patternsFiles = new Path[0];
44                try
45                {
46                  patternsFiles = DistributedCache.getLocalCacheFiles(job);
47                }
```

```
48        catch (IOException ioe)
49        {
50          System.err.println("Caught exception while getting cached
51            files: " + StringUtils.stringifyException(ioe));
52        }
53        for (Path patternsFile: patternsFiles)
54        {
55          parseSkipFile(patternsFile);
56        }
57      }
58    }
59    /*----------------------------------------------------------------*/
60    private void parseSkipFile(Path patternsFile)
61    {
62      try
63      {
64        BufferedReader fis = new BufferedReader(new
65         FileReader(patternsFile.toString()));
66        String pattern = null;
67        while ((pattern = fis.readLine()) != null)
68        {
69          patternsToSkip.add(pattern);
70        }
71      }
72      catch (IOException ioe)
73      {
74        System.err.println("Caught exception while parsing cached file '"
75          + patternsFile + "' : " + StringUtils.stringifyException(ioe));
76      }
77    }
78    /*----------------------------------------------------------------*/
79    public void map(LongWritable key, Text value, OutputCollector < Text,
80     IntWritable > output, Reporter reporter) throws IOException
81    {
82      String line = (caseSensitive) ? value.toString() :
83       value.toString().toLowerCase();
84
85      for (String pattern: patternsToSkip)
86      {
87        line = line.replaceAll(pattern, "");
88      }
89
90      StringTokenizer tokenizer = new StringTokenizer(line);
91      while (tokenizer.hasMoreTokens())
92      {
93        word.set(tokenizer.nextToken());
94        output.collect(word, one);
95        reporter.incrCounter(Counters.INPUT_WORDS, 1);
96      }
97
```

```
98          if ((++numRecords % 100) == 0)
99          {
100             reporter.setStatus("Finished processing " + numRecords +
101              " records " + "from the input file: " + inputFile);
102          }
103        }
104
105     } /* class Map */
106
107     /*-----------------------------------------------------------------*/
108     public static class Reduce extends MapReduceBase implements Reducer
109     < Text, IntWritable, Text, IntWritable >
110     {
111         public void reduce(Text key, Iterator < IntWritable > values,
112          OutputCollector
113            < Text, IntWritable > output, Reporter reporter) throws
114          IOException
115          {
116            int sum = 0;
117            while (values.hasNext())
118            {
119              sum += values.next().get();
120            }
121            output.collect(key, new IntWritable(sum));
122          }
123      } /* class Reduce */
124      /*-----------------------------------------------------------------*/
125     public int run(String[] args) throws Exception
126     {
127       JobConf conf = new JobConf(getConf(), WordCount.class);
128       conf.setJobName("wordcount");
129
130       conf.setOutputKeyClass(Text.class);
131       conf.setOutputValueClass(IntWritable.class);
132
133       conf.setMapperClass(Map.class);
134       conf.setCombinerClass(Reduce.class);
135       conf.setReducerClass(Reduce.class);
136
137       conf.setInputFormat(TextInputFormat.class);
138       conf.setOutputFormat(TextOutputFormat.class);
139
140       List < String > other_args = new ArrayList < String > ();
141       for (int i = 0; i < args.length; ++i)
142       {
143         if ("-skip".equals(args[i]))
144         {
145           DistributedCache.addCacheFile(new Path(args[++i]).toUri(), conf);
146           conf.setBoolean("wordcount.skip.patterns", true);
147         }
```

```
148              else
149              {
150                 other_args.add(args[i]);
151              }
152          }
153
154          FileInputFormat.setInputPaths(conf, new Path(other_args.get(0)));
155          FileOutputFormat.setOutputPath(conf, new Path(other_args.get(1)));
156
157          JobClient.runJob(conf);
158          return 0;
159      }
160      /*------------------------------------------------------------------*/
161      public static void main(String[] args) throws Exception
162      {
163          int res = ToolRunner.run(new Configuration(), new WordCount(), args);
164          System.exit(res);
165      }
166
167  } /* class word count*/
```

Describing the Example 2 Code

Take a closer look at the code for the simpler example, given earlier. Note that line 1 defines the package name as org. myorg and lines 6 through 11 import the Hadoop functionality for Path, configuration, I/O, Map Reduce, and utilities. New to this second example is the cache definition, which is used to store the configurations pattern file (which will be described later):

```
07        import org.apache.hadoop.filecache.DistributedCache;
```

Line 13 defines the main WordCount class:

```
13        public class WordCount extends Configured implements Tool
```

Meanwhile, the Map class is defined at line 17:

```
17        public static class Map extends MapReduceBase
18                implements Mapper < LongWritable, Text, Text, IntWritable >
```

This class now has a configure method defined at line 36, which offers case-sensitivity and pattern-skipping functionality:

```
36            public void configure(JobConf job)
```

The parseSkipFile method at line 60 parses the pattern file for the pattern-skipping functionality just mentioned. The patternsFile contains a list of patterns that should be removed from the text to be processed when counting words:

```
60            private void parseSkipFile(Path patternsFile)
```

The map method is defined at line 79:

```
79              public void map(LongWritable key, Text value, OutputCollector < Text,
80          IntWritable >  output, Reporter reporter) throws IOException
```

As in the last example, a StringTokenizer (line 90) breaks the line into words, then a while loop outputs the words as key-value pairs where the key is the word and the value is 1:

```
90              StringTokenizer tokenizer = new StringTokenizer(line);
91              while (tokenizer.hasMoreTokens())
92              {
93                word.set(tokenizer.nextToken());
94                output.collect(word, one);
95                reporter.incrCounter(Counters.INPUT_WORDS, 1);
96              }
```

The Reduce class is defined at line 108:

```
108             public static class Reduce extends MapReduceBase implements Reducer
109          < Text, IntWritable, Text, IntWritable >
```

The reduce method totals the values for similar words and outputs the key-value pair beginning at line 117:

```
117                 while (values.hasNext())
118                 {
119                   sum += values.next().get();
120                 }
121                 output.collect(key, new IntWritable(sum));
```

There is now a run method (starting at line 125) that contains the functionality from example 1's main method. It sets the Map Reduce and I/O format classes:

```
133             conf.setMapperClass(Map.class);
134             conf.setCombinerClass(Reduce.class);
135             conf.setReducerClass(Reduce.class);
136
137             conf.setInputFormat(TextInputFormat.class);
138             conf.setOutputFormat(TextOutputFormat.class);
```

The new run method parses the skip command line option, saves the pattern file name, and sets the skip patterns option to True. Processing of the skip file can be seen at line 143 via the –skip command line option:

```
143             if ("-skip".equals(args[i]))
144             {
145               DistributedCache.addCacheFile(new Path(args[++i]).toUri(), conf);
146               conf.setBoolean("wordcount.skip.patterns", true);
147             }
```

The run method also sets the input and output paths for the job:

```
154              FileInputFormat.setInputPaths(conf, new Path(other_args.get(0)));
155              FileOutputFormat.setOutputPath(conf, new Path(other_args.get(1)));
```

Finally, line 157 runs the job:

```
157              JobClient.runJob(conf);
```

Running the Example 2 Code

To run this second Java example, you copy its file to the WordCount.java file so that the file name matches the Java class.

```
[hadoop@hc1nn wordcount]$ cp wc-ex2.java WordCount.java
```

Then, you remove the contents of the wc_classes directory and re-create it to receive the Java build output. Use the Linux rm command for the Remove with r for "recursive"and f for "force switches." Use the Linux mkdir command to re-create the directory:

```
[hadoop@hc1nn wordcount]$ rm -rf wc_classes
[hadoop@hc1nn wordcount]$ mkdir wc_classes
```

You build the WordCount java file by specifying an output directory called wc_classes:

```
[hadoop@hc1nn wordcount]$  javac -classpath $HADOOP_PREFIX/hadoop-core-1.2.1.jar  -d wc_classes
WordCount.java
```

Then, you list the contents of the wc_classes directory recursively to ensure that the org.myorg directory structure exists and contains the newly compiled classes:

```
[hadoop@hc1nn wordcount]$ ls -R wc_classes
wc_classes:
org

wc_classes/org:
myorg

wc_classes/org/myorg:
WordCount.class   WordCount$Map.class   WordCount$Map$Counters.class   WordCount$Reduce.class
```

You build these classes into a jar library called wordcount1.jar, so that the resulting jar file can be used for a Hadoop Map Reduce job run. Use the Linux jar command for this (which operates in a similar manner to tar) by using the options C for "create," v for "verbose," and f to specify the file to create:

```
[hadoop@hc1nn wordcount]$ jar -cvf ./wordcount1.jar -C wc_classes.

added manifest
adding: org/(in = 0) (out= 0)(stored 0%)
adding: org/myorg/(in = 0) (out= 0)(stored 0%)
adding: org/myorg/WordCount.class(in = 2671) (out= 1289)(deflated 51%)
adding: org/myorg/WordCount$Reduce.class(in = 1611) (out= 648)(deflated 59%)
adding: org/myorg/WordCount$Map$Counters.class(in = 983) (out= 504)(deflated 48%)
adding: org/myorg/WordCount$Map.class(in = 4661) (out= 2217)(deflated 52%)
```

```
[hadoop@hc1nn wordcount]$ ls -l *.jar
-rw-rw-r--. 1 hadoop hadoop 5799 Jun 21 17:19 wordcount1.jar
```

The test data from the first example is still available on HDFS under the directory /user/hadoop/edgar; this is shown by using the Hadoop file system ls command:

```
[hadoop@hc1nn wordcount]$ hadoop dfs -ls /user/hadoop/edgar
Found 5 items
-rw-r--r--   1 hadoop supergroup    410012 2014-06-19 11:59 /user/hadoop/edgar/10031.txt
-rw-r--r--   1 hadoop supergroup    559352 2014-06-19 11:59 /user/hadoop/edgar/15143.txt
-rw-r--r--   1 hadoop supergroup     66401 2014-06-19 11:59 /user/hadoop/edgar/17192.txt
-rw-r--r--   1 hadoop supergroup    596736 2014-06-19 11:59 /user/hadoop/edgar/2149.txt
-rw-r--r--   1 hadoop supergroup     63278 2014-06-19 11:59 /user/hadoop/edgar/932.txt
```

To give this first example a thorough test, I also created a patterns file called patterns.txt that contains a series of unwanted characters. I have dumped the contents of the file shown here by using the Linux cat command. Note that some characters have an Escape character (\) at the start of the line to avoid processing errors for characters that Java might consider to have special meaning. By using an Escape character, you will ensure that these patterns are just treated as text:

```
[hadoop@hc1nn wordcount]$ cat patterns.txt
!
"
'

_
;
\(
\)
\#
\$
\&
\.
\,
\*
\-
\/
\{
\}
```

Copy the patterns.txt onto HDFS into the directory /user/hadoop/java by using the Hadoop file system copyFromLocal command. Using the Hadoop file system ls command, list the patterns.txt file that is now on HDFS:

```
[hadoop@hc1nn wordcount]$  hadoop dfs -copyFromLocal ./patterns.txt /user/hadoop/java/patterns.txt

[hadoop@hc1nn wordcount]$ hadoop dfs -ls /user/hadoop/java
Found 1 items
-rw-r--r--   1 hadoop supergroup        46 2014-06-21 17:29 /user/hadoop/java/patterns.txt
```

Now you are ready to run this extended version of the Java Map Reduce task. The library that was just created is specified via the Hadoop jar option. This is followed by the Class name to be called within that library. Next, a flag is set via the -D option to switch the case sensitivity off. After that, the input data file and output directory names on HDFS are listed. Finally, you specify a skip file to remove any unwanted characters in the data processed:

```
[hadoop@hc1nn wordcount]$ hadoop jar ./wordcount1.jar  org.myorg.WordCount
  -Dwordcount.case.sensitive=false  /user/hadoop/edgar/10031.txt
  /user/hadoop/edgar-results  -skip /user/hadoop/java/patterns.txt
```

The command produces the following Map Reduce task output:

```
14/06/21 17:40:06 INFO util.NativeCodeLoader: Loaded the native-hadoop library
14/06/21 17:40:06 INFO mapred.FileInputFormat: Total input paths to process : 1
14/06/21 17:40:07 INFO mapred.JobClient: Running job: job_201406211041_0004
14/06/21 17:40:08 INFO mapred.JobClient: map 0% reduce 0%
14/06/21 17:40:15 INFO mapred.JobClient: map 50% reduce 0%
14/06/21 17:40:23 INFO mapred.JobClient: map 100% reduce 16%
14/06/21 17:40:30 INFO mapred.JobClient: map 100% reduce 100%
14/06/21 17:40:31 INFO mapred.JobClient: Job complete: job_201406211041_0004
14/06/21 17:40:31 INFO mapred.JobClient: Counters: 32
14/06/21 17:40:31 INFO mapred.JobClient: Job Counters
14/06/21 17:40:31 INFO mapred.JobClient: Launched reduce tasks=1
14/06/21 17:40:31 INFO mapred.JobClient: SLOTS_MILLIS_MAPS=17198
14/06/21 17:40:31 INFO mapred.JobClient: Total time spent by all reduces waiting after reserving
slots (ms)=0
.........................
14/06/21 17:40:31 INFO mapred.JobClient: CPU time spent (ms)=5880
14/06/21 17:40:31 INFO mapred.JobClient: Map input bytes=410012
14/06/21 17:40:31 INFO mapred.JobClient: SPLIT_RAW_BYTES=198
14/06/21 17:40:31 INFO mapred.JobClient: Combine input records=63590
14/06/21 17:40:31 INFO mapred.JobClient: Reduce input records=12581
14/06/21 17:40:31 INFO mapred.JobClient: Reduce input groups=9941
14/06/21 17:40:31 INFO mapred.JobClient: Combine output records=12581
14/06/21 17:40:31 INFO mapred.JobClient: Physical memory (bytes) snapshot=404115456
14/06/21 17:40:31 INFO mapred.JobClient: Reduce output records=9941
14/06/21 17:40:31 INFO mapred.JobClient: Virtual memory (bytes) snapshot=4109373440
14/06/21 17:40:31 INFO mapred.JobClient: Map output records=63590
```

Check the results directory on HDFS by using the Hadoop file system ls command. The existence of a _SUCCESS file shows that the job was a success:

```
[hadoop@hc1nn wordcount]$ hadoop dfs -ls /user/hadoop/edgar-results
Found 3 items
-rw-r--r--   1 hadoop supergroup          0 2014-06-21 17:40 /user/hadoop/edgar-results/_SUCCESS
drwxr-xr-x   - hadoop supergroup          0 2014-06-21 17:40 /user/hadoop/edgar-results/_logs
-rw-r--r--   1 hadoop supergroup     103300 2014-06-21 17:40 /user/hadoop/edgar-results/part-00000
```

Checking the last 10 lines of the results part file using the Hadoop file system cat command and the Linux tail command gives a sorted word count with any unwanted characters removed:

```
[hadoop@hc1nn wordcount]$ hadoop dfs -cat /user/hadoop/edgar-results/part-00000 | tail -10
zanthe    1
zeal      2
zeboin    1
zelo      1
zephyr    1
zimmermann        1
zipped    1
zoar      1
zoilus    3
zone      1
```

Comparing the Examples

The two Map Reduce native word-count examples (wc-ex1.java and wc-ex2.java) show that raw Java code for Map Reduce can create complex functionality. They are not, however, the most efficient approaches. Consider the effect achieved when some simple pattern-filtering options were added. The listing grew from 70 lines in example 1 to 167 lines in example 2. As the code volume increases, so do the cost, complexity, and time to implement. Now, imagine the effect on a more complex algorithm; the resulting code could quickly become even more unwieldy.

The good news is that alternatives are available. In the next sections, I will introduce some other Map Reduce coding tools that offer the ability to code these tasks at a higher level and so reduce code volume. Generally, it is more efficient and cheaper to use less code to achieve your objective. You should write your code at a lower level in Java only if higher level systems like Pig native (including UDFs), which will be described in the next section, do not offer the functionality you need.

So, next you will learn to source, install, and use Apache Pig. You will also code the same word-count algorithm in Pig.

Map Reduce with Pig

As it is able to run in interactive or batch mode, Pig is a higher level programming language for processing large data sets. You will be able to see that fewer lines of code are needed to carry out the same word-count example. Apache Pig can be downloaded from pig.apache.org.

As Pig is a higher-level language, you can concentrate more on the logical flow of data processing and less on the lower-level coding to achieve that processing. Also, Pig integrates well with the visual-object-based ETL and reporting tools for big data that are introduced in Chapters 10 and 11 of this guide. This means that you have a quicker and easier path into the world of data processing using Map Reduce. Although this will be explained later, tools like Talend even help to abstract Map Reduce with its predefined Pig-based functionality.

Installing Pig

For this book's examples, I chose to download Pig release 0.12.1 from pig.apache.org/releases.html because it is compatible with the version of Hadoop I have been using up to this point (1.x). The download and installation are straightforward. From the download page, you select to download Pig 0.8 and later. The Pig website then suggests a mirror site for from which you can download (in my case, it was www.carfab.com). After clicking that link, you're offered a series of Pig versions. Choose 0.12.1, and download the version of the release file that is tarred and gzipped. For example, I used wget to download the release to the server hc1nn, which was convenient because I could download the package straight to a server directory using the website URL:

```
[hadoop@hc1nn ~]$ wget http://www.carfab.com/apachesoftware/pig/pig-0.12.1/pig-0.12.1.tar.gz
```

Next, you unzip the release using the Linux command gunzip, and unpack the tar file using the Linux tar -xvf command:

```
[hadoop@hc1nn ~]$ ls -l pig-0.12.1.tar.gz
-rw-rw-r--. 1 hadoop hadoop 59445085 Apr  5 21:44 pig-0.12.1.tar.gz
 [hadoop@hc1nn ~]$ gunzip pig-0.12.1.tar.gz
 [hadoop@hc1nn ~]$ tar xvf pig-0.12.1.tar

[root@hc1nn hadoop]# ls -ld pig-0.12.1
drwxr-xr-x. 15 hadoop hadoop 4096 Apr  5 21:44 pig-0.12.1
```

You move the release to a location under /usr/local/ using the Linux mv command, then set up a symbolic link under /usr/local/ called pig by using the Linux ln command with a –s (symbolic) switch to simplify the path to the software and the environment. Finally, you use the Linux ls command to create a long listing that displays the link and the install the Pig directory.

```
[root@hc1nn hadoop]# mv pig-0.12.1 /usr/local
[root@hc1nn hadoop]# cd /usr/local/
[root@hc1nn hadoop]# ln -s pig-0.12.1 pig

[root@hc1nn local]# ls -ld pig*
lrwxrwxrwx. 1 root    root      10 Jun 18 12:02 pig -> pig-0.12.1
drwxr-xr-x. 15 hadoop hadoop 4096 Apr  5 21:44 pig-0.12.1
```

To simplify access to and use of Pig, you can add some Pig-related variables to the Linux hadoop user's Bash shell configuration at the bottom of the file $HOME/.bashrc:

```
#######################################################
# Set up Pig variables

export PIG_HOME=/usr/local/pig

export PATH=$PATH:$PIG_HOME/bin
```

Once the installation is in place and the environment is set up, you can test that the Pig binary is available and will run. For example, you can use the Linux type command to check that the Pig Linux command is picked up from the correct location:

```
[hadoop@hc1nn ~]$ type pig
pig is hashed (/usr/local/pig/bin/pig)
```

This shows that the path is good—it is the one that was just installed. Issuing the Pig help command is another a good way to ensure that commands will run without error:

```
[hadoop@hc1nn ~]$ pig -help

Apache Pig version 0.12.1 (r1585011)
compiled Apr 05 2014, 01:41:34

USAGE: Pig [options] [-] : Run interactively in grunt shell.
       Pig [options] -e[xecute] cmd [cmd ...] : Run cmd(s).
       Pig [options] [-f[ile]] file : Run cmds found in file.
  options include:
..............
   -M, -no_multiquery - Turn multiquery optimization off; default is on
   -P, -propertyFile - Path to property file
   -printCmdDebug - Overrides anything else and prints the actual command used to run Pig, including
                    any environment variables that are set by the pig command.
```

The results are good: the pig command is located in /usr/local/pig/bin, and it runs as the help option shows. It is now time to use it.

Running Pig

Pig lets you choose how you wish to work with it. For example, you can direct where Pig looks for data by specifying the local mode or the Map Reduce mode (the default). Local mode takes all data from the local server and the file system, while Map Reduce mode uses Hadoop. In addition, you can run tasks interactively or in batch mode. When working interactively, you issue Pig commands via the Grunt command prompt. For larger scheduled or background tasks, you can use batch mode. For the word-count demonstration, you will use Pig interactively in Map Reduce mode.

To prepare to use Pig, you first need to create a Pig working directory on HDFS:

```
[hadoop@hc1nn edgar]$ hadoop dfs -mkdir /user/hadoop/pig/
```

Then, you copy a text-based data file of Edgar Allan Poe's work into that HDFS-based directory from the Linux file system by using the Hadoop file system command copyFromLocal:

```
[hadoop@hc1nn edgar]$ cd $HOME/edgar
[hadoop@hc1nn edgar]$ ls
10031.txt  15143.txt  17192.txt  2149.txt  932.txt

[hadoop@hc1nn edgar]$ hadoop dfs -copyFromLocal ./10031.txt /user/hadoop/pig
```

A quick check on HDFS shows that the file 10031.txt containing the text is now sitting on HDFS in the directory /user/hadoop/pig:

```
[hadoop@hc1nn edgar]$ hadoop dfs -ls /user/hadoop/pig
Found 1 items
-rw-r--r--   1 hadoop supergroup     410012 2014-06-18 12:29 /user/hadoop/pig/10031.txt
```

You can now start Pig in interactive Map Reduce mode. Without any options, the pig command will result in the interactive Grunt command line after trying to access Hadoop:

```
[hadoop@hc1nn edgar]$ pig
2014-06-18 12:27:10,055 [main] INFO  org.apache.pig.Main - Apache Pig version 0.12.1 (r1585011)
compiled Apr 05 2014, 01:41:34
2014-06-18 12:27:10,056 [main] INFO  org.apache.pig.Main - Logging error messages to: /home/hadoop/
edgar/pig_1403051230051.log
2014-06-18 12:27:10,095 [main] INFO  org.apache.pig.impl.util.Utils - Default bootup file /home/
hadoop/.pigbootup not found
2014-06-18 12:27:10,386 [main] INFO  org.apache.pig.backend.hadoop.executionengine.HExecutionEngine
- Connecting to hadoop file system at: hdfs://hc1nn:54310
2014-06-18 12:27:10,750 [main] INFO  org.apache.pig.backend.hadoop.executionengine.HExecutionEngine
- Connecting to map-reduce job tracker at: hc1nn:54311

grunt>
```

In Pig, the character "--" denotes a comment, meaning text between the -- and the start of the next line is ignored. The semicolon (;) denotes the end of a Pig native statement. The data file is loaded from HDFS into variable A by using the load option:

```
grunt>  A = load '/user/hadoop/pig/10031.txt';  -- load the text file
```

With a single line, you can process each word in the data in variable A into a list of words, place each word from the list in variable B, then add them to variable C. TOKENIZE splits the data on white-space characters. Here's the command you need:

```
grunt> C = foreach A generate  flatten(TOKENIZE((chararray)$0)) as B ; -- get list of words
```

Next, you can group the identical words into a variable D, and create a list of word counts in variable E by using the count option:

```
grunt> D = group C by B ; -- group words
grunt> E = foreach D generate  COUNT(C), group; -- create word count
```

To view the word count, you use the dump command to display the contents of variable E in the session window; this shows the word-count list. (I've listed the last 10 lines here.) As you can see, it's very basic counting:

```
grunt> dump E; -- dump result to session

.......
(1,http://pglaf.org/fundraising.)
(1,it!--listen--now--listen!--the)
(1,http://www.gutenberg.net/GUTINDEX.ALL)
(1,http://www.gutenberg.net/1/0/2/3/10234)
(1,http://www.gutenberg.net/2/4/6/8/24689)
(1,http://www.gutenberg.net/1/0/0/3/10031/)
(1,http://www.ibiblio.org/gutenberg/etext06)
(0,)
```

Lastly, you can store the contents of the count, currently in variable E, on HDFS in /user/hadoop/pig/wc_result:

```
grunt> store E into '/user/hadoop/pig/wc_result' ; -- store the results
grunt> quit ; -- quit interactive session
```

Having quit the Pig interactive session, you can examine the results of this Pig job on HDFS. The Hadoop file system ls command shows a success file (_SUCCESS), a part file (part-r-00000) containing the word-count data, and a logs directory. (I have listed the part file from the word count using the Hadoop file system command cat.) Then, you can use the Linux tail command to view the last 10 lines of the file. Both options are shown here:

```
[hadoop@hc1nn edgar]$ hadoop dfs -ls /user/hadoop/pig/wc_result
Found 3 items
-rw-r--r--   1 hadoop supergroup          0 2014-06-18 13:08 /user/hadoop/pig/wc_result/_SUCCESS
drwxr-xr-x   - hadoop supergroup          0 2014-06-18 13:08 /user/hadoop/pig/wc_result/_logs
-rw-r--r--   1 hadoop supergroup     137870 2014-06-18 13:08 /user/hadoop/pig/wc_result/part-r-00000

[hadoop@hc1nn edgar]$ hadoop dfs -cat /user/hadoop/pig/wc_result/part-r-00000  | tail -10
1       http://gutenberg.net/license
1       Dream'--Prospero--Oberon--and
1       http://pglaf.org/fundraising.
1       it!--listen--now--listen!--the
1       http://www.gutenberg.net/GUTINDEX.ALL
1       http://www.gutenberg.net/1/0/2/3/10234
1       http://www.gutenberg.net/2/4/6/8/24689
1       http://www.gutenberg.net/1/0/0/3/10031/
1       http://www.ibiblio.org/gutenberg/etext06
```

It is quite impressive that, with five lines of Pig commands (ignoring the dump and quit lines), you can run the same word-count algorithm as took 70 lines of Java code. Less code means lower development costs and, we all hope, fewer code-based errors.

While efficient, the interactive Pig example does have a drawback: The commands must be manually typed each time you want to run a word count. Once you're finished, they're lost. The answer to this problem, of course, is to store the Pig script in a file and run it as a batch Map Reduce job. To demonstrate, I placed the Pig commands from the previous example into the wordcount.pig file:

```
[hadoop@hc1nn pig]$ ls -l
total 4
-rw-rw-r--. 1 hadoop hadoop 313 Jun 18 13:24 wordcount.pig

[hadoop@hc1nn pig]$ cat wordcount.pig

01          -- get raw line data from file
02
03          rlines = load '/user/hadoop/pig/10031.txt';
04
05          -- get list of words
06
07          words = foreach rlines generate flatten(TOKENIZE((chararray)$0)) as word;
08
09          -- group the words by word value
10
```

```
11        gwords = group words by word ;
12
13        -- create a word count
14
15        wcount = foreach gwords generate  group, COUNT(words) ;
16
17        -- store the word count
18
19        store wcount into '/user/hadoop/pig/wc_result1' ;
```

I also added comments, line numbers, and meaningful names for the variables. These modifications can help when you're trying to determine what a script is doing. They also help to tie this example to the work in the next section, on Pig user-defined functions.

Instead of invoking the interactive Grunt command line, you invoke Pig with the name of the file containing the Pig script. Pig will use Map Reduce mode by default and so access HDFS. The output will be stored in the HDFS directory /user/hadoop/pig/wc_result1/. So, when the task starts, a Map Reduce job is initiated.

```
[hadoop@hc1nn pig]$ pig wordcount.pig


..................
Counters:
Total records written : 13219
Total bytes written : 137870
Spillable Memory Manager spill count : 0
Total bags proactively spilled: 0
Total records proactively spilled: 0

Job DAG:
job_201406181226_0003

2014-06-18 13:27:49,446 [main] INFO  org.apache.pig.backend.hadoop.executionengine.mapreducelayer.
mapreducelauncher - Success!
```

As mentioned previously, you can use Hadoop and Linux commands to output the word count:

```
[hadoop@hc1nn pig]$ hadoop dfs -cat /user/hadoop/pig/wc_result1/part-r-00000  | tail -10

1       http://gutenberg.net/license
1       Dream'--Prospero--Oberon--and
1       http://pglaf.org/fundraising.
1       it!--listen--now--listen!--the
1       http://www.gutenberg.net/GUTINDEX.ALL
1       http://www.gutenberg.net/1/0/2/3/10234
1       http://www.gutenberg.net/2/4/6/8/24689
1       http://www.gutenberg.net/1/0/0/3/10031/
1       http://www.ibiblio.org/gutenberg/etext06
```

Notice that in both the interactive and batch script versions, the count includes non-alpha-numeric chararacters like ":"(colon), and that the case of the words has not been standardized. For instance, the word *Dream* (with a capital D) is part of the count. In the next section, you will learn how to add greater selectivity to Pig by creating user-defined functions.

Pig User-Defined Functions

Coded in Java, user-defined functions (UDFs) provide custom functionality that you can invoke from a Pig script. For instance, you might create a UDF if you found that you needed to carry out an operation that the standard Pig Latin language did not include. This section will provide an example of just such a function. You will examine the UDF Java code, as well as the method by which it is built into a jar library. You will then use an extended version of the Pig script from the last section that incorporates this UDF. You will learn how to incorporate both the jar file and its classes into a Pig script.

As greater functionality was obtained for earlier Map Reduce jobs, such as removing unwanted characters from the word-count process, the same will be done here. Using Java, you will create a UDF to remove unwanted characters, so that the final word count is more precise. For instance, I have created a UDF build directory on the Linux file system under /home/hadoop/pig/wcudfs that contains a number of files:

```
[hadoop@hc1nn wcudfs]$ pwd
/home/hadoop/pig/wcudfs/

[hadoop@hc1nn wcudfs]$ ls
build_clean_ws.sh  build_lower.sh  CleanWS.java
```

The Java files contain the code for UDFs while the shell scripts (*.sh) are used to build them. The CleanWS.java file contains the following code:

```
01        package wcudfs;
02
03        import java.io.*;
04
05        import org.apache.pig.EvalFunc;
06        import org.apache.pig.data.Tuple;
07        import org.apache.hadoop.util.*;
08
09        public class CleanWS extends EvalFunc<String>
10        {
11          /*-------------------------------------------------------*/
12          @Override
13          public String exec(Tuple input) throws IOException
14          {
15            if (input == null || input.size() == 0)
16              return null;
17            try
18            {
19              String str = (String)input.get(0);
20
21              return str.replaceAll("[^A-Za-z0-9]"," ");
22            }
23            catch(IOException ioe)
24            {
25                System.err.println("Caught exception processing input row : "
26                            + StringUtils.stringifyException(ioe) );
27            }
28
```

```
29            return null;
30          }
31          /*-------------------------------------------------------*/
32
33        } /* class CleanWS */
```

Line 1 defines the package name to be wcudfs

```
01          package wcudfs;
```

Pig and Hadoop functionality for tuples and utilities is imported into the UDF between lines 5 and 7. Line 5 invokes the EvalFunc class and identifies this as an eval type of UDF function:

```
05          import org.apache.pig.EvalFunc;
06          import org.apache.pig.data.Tuple;
07          import org.apache.hadoop.util.*;
```

Line 9 specifies the class name as CleanWS, which extends the EvalFunc class and has a String return type.

```
09          public class CleanWS extends EvalFunc<String>
```

Line 13 onward defines the exec method that will be called to process every tuple in the data:

```
13            public String exec(Tuple input) throws IOException
```

Line 21 changes the return string, removing all characters that are not in the character sets A–Z, a–z, or 0–9 and replacing them with a space character.

```
21              return str.replaceAll("[^A-Za-z0-9]"," ");
```

For example, I built CleanWS as follows:

```
[hadoop@hc1nn wcudfs]$ cat build_clean_ws.sh

javac -classpath $PIG_HOME/pig-0.12.1.jar   -Xlint:deprecation CleanWS.java
```

The Java compiler is called javac; an option is added via the classpath to include the Pig library in the build. The lint:deprecation option uses lint to check the code for deprecated API calls. These scripts build the class files:

```
[hadoop@hc1nn wcudfs]$ ./build_clean_ws.sh
[hadoop@hc1nn wcudfs]$ ls
build_clean_ws.sh  CleanWS.class  CleanWS.java
```

The class file is created as part of the build for the Java file. The class file for the UDF is built into a library that can be used within a Pig script. The library is built using the jar command with the options c (create), v (verbose), and f (file). The next parameter to be created is the library name, followed by the list of classes to be placed in the library:

```
[hadoop@hc1nn wcudfs]$ cd ..
[hadoop@hc1nn pig]$ jar cvf wcudfs.jar wcudfs/*.class
added manifest
adding: wcudfs/CleanWS.class(in = 1318) (out= 727)(deflated 44%)

[hadoop@hc1nn pig]$ ls -l wcudfs.jar
-rw-rw-r--. 1 hadoop hadoop 2018 Jun 24 18:57 wcudfs.jar
```

This is the library containing the UDF classes that will be called in the Pig script register line.

Next, take a look at the updated version of the Pig script wordcount2.pig by using the Linux cat command, which employs the newly created UDF function:

```
[hadoop@hc1nn pig]$ cat wordcount2.pig

01        REGISTER /home/hadoop/pig/wcudfs.jar ;
02
03        DEFINE CleanWS wcudfs.CleanWS() ;
04
05        -- get raw line data from file
06
07        rlines = load '/user/hadoop/pig/10031.txt' AS (rline:chararray);
08
09        -- filter for empty lines
10
11        clines = FILTER rlines BY SIZE(rline) > 0 ;
12
13        -- get list of words
14
15        words = foreach clines generate
16        flatten(TOKENIZE(CleanWS( (chararray) $0  ))) as word ;
17
18        -- group the words by word value
19
20        gword = group words by word ;
21
22        -- create a word count
23
24        wcount = foreach gword generate  group, COUNT(words) ;
25
26        -- store the word count
27
28        store wcount into '/user/hadoop/pig/wc_result1' ;
```

There are some new terms in this script. At line 1, the REGISTER keyword is used to register the word-count UDF library wcudfs.jar for use with this Pig script.

```
01        REGISTER /home/hadoop/pig/wcudfs.jar ;
```

Line 3 uses the DEFINE keyword to refer to the classes of the package within this library that use a single term. For instance, the class CleanWS in the package wcudfs, in the library wcudfs.jar, can now be called as just CleanWS in the code.

```
03        DEFINE CleanWS wcudfs.CleanWS() ;
```

Line 11 introduces the FILTER keyword. Using this filter removes any lines that are empty from the data set. The variable clines is used to contain lines that have more than zero characters. This is accomplished by using a check on the size of the line (rline) and ensuring that the size is greater than zero.

```
11        clines = FILTER rlines BY SIZE(rline) > 0 ;
```

Line 16 calls the user-defined function named CleanWS, which removes unwanted characters from the input text.

```
15          words = foreach clines generate
16              flatten(TOKENIZE(CleanWS( (chararray) $0  ))) as word ;
```

I have created some Bash shell scripts to assist in running this second Pig job. This is just to provide an example of how to speed up a manual job. For instance, instead of having to type the Pig job execution command each time, I can just execute a simple script. Instead of having to manually delete the job results directory for a job rerun, I can run a clean script. Here is the clean_wc.sh script that was used to delete the job results HDFS directory, employing the Linux cat command:

```
[hadoop@hc1nn pig]$ cat clean_wc.sh

01  #!/bin/bash
02
03  # remove the pig script results directory
04
05  hadoop dfs -rmr /user/hadoop/pig/wc_result1
```

The script does this by calling the Hadoop file system rmr command to remove the directory and its contents.

The next script that is run_wc2.sh, which is used to run the job, calls the clean script (at line 5) each time it is run. This single script cleans the results directory on HDFS and runs the wordcount2.pig job:

```
[hadoop@hc1nn pig]$ cat run_wc2.sh

01  #!/bin/bash
02
03  # run the pig wc 2 job
04
05  ./clean_wc.sh
06
07  pig  -stop_on_failure  wordcount2.pig
```

This shell script calls the clean_wc.sh script and then invokes the Pig wordcount2.pig script. The pig command on line 7 is called with a flag (-stop_on_failure), telling it to stop as soon as it encounters an error. The results are listed via the result_wc.sh script:

```
[hadoop@hc1nn pig]$ cat result_wc.sh

01  #!/bin/bash
02
03  # remove the pig script results directory
04
05  hadoop dfs -ls /user/hadoop/pig/wc_result1
07
08  hadoop dfs -cat /user/hadoop/pig/wc_result1/part-r-00000 | tail -10
```

This script lists the contents of the Pig job results directory and then dumps the last 10 lines of the part file within that directory that contains the word-count job data. It does this by using the Hadoop file system cat command and the Linux tail command. So, to run the job, you just execute the run_wc2.sh Bash script:

```
[hadoop@hc1nn pig]$ ./run_wc2.sh

Deleted hdfs://hc1nn:54310/user/hadoop/pig/wc_result1
2014-06-24 19:06:44,651 [main] INFO  org.apache.pig.Main - Apache Pig version 0.12.1 (r1585011)
compiled Apr 05 2014, 01:41:34
2014-06-24 19:06:44,652 [main] INFO  org.apache.pig.Main - Logging error messages to: /home/hadoop/
pig/pig_1403593604648.log
.............................

Input(s):
Successfully read 10377 records (410375 bytes) from: "/user/hadoop/pig/10031.txt"

Output(s):
Successfully stored 9641 records (95799 bytes) in: "/user/hadoop/pig/wc_result1"

Counters:
Total records written : 9641
Total bytes written : 95799
Spillable Memory Manager spill count : 0
Total bags proactively spilled: 0
Total records proactively spilled: 0

Job DAG:
job_201406241807_0002

2014-06-24 19:07:23,252 [main] INFO  org.apache.pig.backend.hadoop.executionengine.mapReduceLayer.
MapReduceLauncher - Success!
```

You then list the results of the job that will output the Pig job results directory and the last 10 lines of the Pig job data file, as explained in the description of the result_wc_.sh script described above:

```
[hadoop@hc1nn pig]$ ./result_wc.sh
Found 3 items
-rw-r--r--   1 hadoop supergroup          0 2014-06-24 19:07 /user/hadoop/pig/wc_result1/_SUCCESS
drwxr-xr-x   - hadoop supergroup          0 2014-06-24 19:06 /user/hadoop/pig/wc_result1/_logs
-rw-r--r--   1 hadoop supergroup      95799 2014-06-24 19:07 /user/hadoop/pig/wc_result1/part-r-00000

unexceptionable        1
constitutionally       1
misunderstanding       1
tintinnabulation       1
unenforceability       1
Anthropomorphites      1
contradistinction      1
preconsiderations      1
undistinguishable      1
transcendentalists     1
```

Notice that all of the unwanted characters have now been removed from the output. Given this simple example, you will be able to build your own UDF extentions to Pig. You are now able to use Apache Pig to code your Map Reduce jobs and expand its functionality via UDFs. You can also gain the greater flexibility of reduced code volume by using your own UDF libraries.

In the next section, you will tackle the same Map Reduce job using Apache Hive, the big-data data warehouse. A similar word-count algorithm will be presented using Hive QL, Hive's SQL-like query language. All of these methods and scripts for creating Map Reduce jobs are presented to give you a sample of each approach. The data that you want to use and the data architecture that you choose will govern which route you take to create your jobs. The ETL tools that you choose will also affect your approach—for instance, Talend and Pentaho, to be discussed in Chapter 10, can integrate well with Pig functionality.

▓ **Note**   For more information on Apache Pig and Pig Latin, see the Apache Software Foundation guide at
`http://pig.apache.org/docs/r0.12.1/start.html`.

Map Reduce with Hive

This next example involves installing Apache Hive from `hive.apache.org`. Hive is a data warehouse system that uses Hadoop for storage. It is possible to interrogate data on HDFS by using an SQL-like language called HiveQL. Hive can represent HDFS-based data via the use of external tables (described in later chapters) or relational data where there are relationships between data in different Hive tables.

In this section, I will explain how to source and install Hive, followed by a simple word-count job on the same data as used previously.

Installing Hive

When downloading and installing Hive, be sure to choose the version compatible with the version of Hadoop you are using in conjunction with this book. For this section's examples, I chose version 0.13.1, which is compatible with Hadoop version 1.2.1 used earlier. As before, I used wget from the Linux command line to download the tarred and gzipped release from a suggested mirror site:

```
[hadoop@hc1nn Downloads]$ wget http://apache.mirror.quintex.com/hive/hive-0.13.1/apache-hive-0.13.1-
bin.tar.gz

[hadoop@hc1nn Downloads]$ ls -l apache-hive-0.13.1-bin.tar.gz
-rw-rw-r--. 1 hadoop hadoop 54246778 Jun  3 07:31 apache-hive-0.13.1-bin.tar.gz
```

As before, you unpack the software using the Linux commands gunzip and tar:

```
[hadoop@hc1nn Downloads]$ gunzip apache-hive-0.13.1-bin.tar.gz
[hadoop@hc1nn Downloads]$ tar xvf apache-hive-0.13.1-bin.tar

[hadoop@hc1nn Downloads]$ ls -ld apache-hive-0.13.1-bin
drwxrwxr-x. 8 hadoop hadoop 4096 Jun 18 17:03 apache-hive-0.13.1-bin
```

Next, you move the release to /usr/local/ and create a symbolic link to the release to simplify the path and environment:

```
[root@hc1nn Downloads]# mv apache-hive-0.13.1-bin /usr/local
[root@hc1nn Downloads]# cd /usr/local
[root@hc1nn local]# ln -s apache-hive-0.13.1-bin hive

[root@hc1nn local]# ls -ld *hive*
drwxrwxr-x. 8 hadoop hadoop 4096 Jun 18 17:03 apache-hive-0.13.1-bin
lrwxrwxrwx. 1 root   root     22 Jun 18 17:05 hive -> apache-hive-0.13.1-bin
```

You update the user environment for the Linux account hadoop via its $HOME/.bashrc. For instance, I added the following to the end of the file:

```
#######################################################
# Set up Hive variables

export HIVE_HOME=/usr/local/hive

export PATH=$PATH:$HIVE_HOME/bin
```

Once installed, Hive needs to use several HDFS-based directories, including a temporary directory and a warehouse directory. You first check that these exist and are group writeable:

```
[hadoop@hc1nn bin]$ hadoop fs -mkdir       /tmp
[hadoop@hc1nn bin]$ hadoop fs -chmod g+w   /tmp
[hadoop@hc1nn bin]$ hadoop fs -mkdir       /user/hive/warehouse
[hadoop@hc1nn bin]$ hadoop fs -chmod g+w   /user/hive/warehouse
```

After making one last check for the proper release via the Linux type command, which shows that the hive command is accessible, you can start Hive by using the hive command, which results in the Hive> prompt:

```
[hadoop@hc1nn bin]$ type hive
hive is hashed (/usr/local/hive/bin/hive)

[hadoop@hc1nn bin]$ hive

Logging initialized using configuration in jar:file:/usr/local/apache-hive-0.13.1-bin/lib/hive-
common-0.13.1.jar!/hive-log4j.properties

hive>
```

The hive command starts the Hive command line interface (CLI), readying Hive for some Hive QL.

Hive Word-Count Example

With the Hive CLI running, you're ready to do a word count on HDFS data using HiveQL. Notice that you must terminate each command with a semicolon. To create a table to hold the data file lines, use the CREATE TABLE command. The first parameter is the table's name—in this case, "rawdata." Next, you pass the column names and their type. Here, you specify a single column named "line," of type STRING:

```
hive> CREATE TABLE rawdata (line STRING);
```

You load the text files under /user/hadoop/edgar/ on HDFS into the Hive table rawdata by using the LOAD DATA statement.

```
hive> LOAD DATA INPATH '/user/hadoop/edgar/' INTO TABLE rawdata ;
```

The data in the rawdata table is converted to the word-count table via a CREATE TABLE with a sub SELECT. (I have taken the liberty of adding line numbers to the script that follows to better explain this.) You have created the rawdata table and populated it with data via a LOAD DATA command. The Hive QL script that will carry out the word count now is as follows:

```
hive>
```

```
01  > CREATE TABLE wordcount AS
02  >   SELECT
03  >     word,
04  >     count(1) AS count
05  >   FROM
06  >     (SELECT
07  >       EXPLODE(SPLIT(line,' ')) AS word
08  >     FROM
09  >       rawdata
10  >     ) words
11  >   GROUP BY word
12  >   ORDER BY word ;
```

Lines 6 to 10 create a derived table called words that takes data from the rawdata table. It does this by splitting the rawdata.line column into a column called word in the derived table at line 7. The rawdata.line free text is split by space characters so that the derived table column words.word contains a list of words.

The rest of the Hive QL then groups these words together (line 11), counts the instances of each (line 4), and orders the list (line 12) that is output.

When you check the word-count table with a SELECT COUNT(*) command, you find there are over 36 thousand rows:

```
hive> SELECT COUNT(*) FROM wordcount;
```

```
OK
36511
```

You can narrow your results by using SELECT, as well. Selecting the data from the word count where the count is greater than 1,500 instances gives a short list of the most frequently occurring words in the data.

```
hive> SELECT
    >     word,
    >     count
    > FROM
    >     wordcount
    > WHERE
    >     count > 1500
    > ORDER BY
    >     count ;
```

```
he        1585
for       1614
at        1621
his       1795
had       1839
it        1918
my        1921
as        1950
with      2563
that      2726
was       3119
I         4532
in        5149
a         5649
to        6230
and       7826
of        10538
the       18128
```

This is an SQL-like example that uses HiveQL to run a word count. It is easy to install and use, and Hive provides a powerful HiveQL interface to the data. Not including the COUNT(*) line, the word-count job took just three statements. Each of the statements issued to the Hive CLI was passed on to Hadoop as a Map Reduce task.

Whether you employ Hive for Map Reduce will depend on the data you are using, its type, and the relationships between it and other data streams you might wish to incorporate. You have to view your use of Hive QL in terms of your ETL chains—that is, the sequence of steps that will transform your data. You might find that Hive QL doesn't offer the functionality to process your data; in that case, you would choose either Pig Latin or Java.

Map Reduce with Perl

Additionally, you can use the library-based data streaming functionality provided with Hadoop. The important point to note is that this approach allows you to process streams of data using Hadoop libraries. With the Hadoop streaming functionality, you can create Map Reduce jobs from many executable scripts, including Perl, Python, and Bash. It is best used for textual data, as it allows data streaming between Hadoop and external systems.

In this example, you will run a Perl-based word-count task. (I present this example in Perl simply because I am familiar with that language.) The streaming library can be found within the Hadoop release as a jar file. This is the library within Hadoop that provides the functionality for users to write their own scripts and have Hadoop use them for Map Reduce:

```
[hadoop@hc1nn hadoop]$ pwd
/usr/local/hadoop
[hadoop@hc1nn hadoop]$ ls -l contrib/streaming/hadoop-*streaming*.jar
-rw-rw-r--. 1 hadoop hadoop 107399 Jul 23  2013 contrib/streaming/hadoop-streaming-1.2.1.jar
```

First, we need a Perl working directory on HDFS called /user/hadoop/perl which will be used for result data for the Map Reduce run:

```
[hadoop@hc1nn python]$ hadoop dfs -mkdir /user/hadoop/perl
```

I have already created a number of scripts in the Linux file system directory /home/hadoop/perl:

```
[hadoop@hc1nn perl]$ ls
mapper.pl    test1.sh  wc_clean.sh    wordcount.sh
reducer.pl   test2.sh  wc_output.sh
```

The file names ending in .pl are Perl scripts, while those ending in .sh are shell scripts used either to test the perl scripts or to run them. The Map function is in the file mapper.pl and looks like this:

```
[hadoop@hc1nn perl]$ cat mapper.pl
```

```
01  #!/usr/bin/perl
02
03  my $line;
04  my @words = ();
05  my $word;
05
06  # process input line by line
07
08  foreach $line ( <STDIN> )
09  {
10    # strip new line from string
11
12    chomp( $line );
13
14    # strip line into words using space
15
16    @words = split( ' ', $line );
17
18    # now print the name value pairs
19
20    foreach  $word (@words)
21    {
22      # convert word to lower case
23
24      $word = lc( $word ) ;
25
26      # remove unwanted characters from string
27
28      $word =~ s/!//g ;   # remove ! character from word
29      $word =~ s/"//g ;   # remove " character from word
30      $word =~ s/'//g ;   # remove ' character from word
31      $word =~ s/_//g ;   # remove _ character from word
32      $word =~ s/;//g ;   # remove ; character from word
33      $word =~ s/\(//g ; # remove ( character from word
34      $word =~ s/\)//g ; # remove ) character from word
35      $word =~ s/\#//g ; # remove # character from word
36      $word =~ s/\$//g ; # remove $ character from word
37      $word =~ s/\&//g ; # remove & character from word
38      $word =~ s/\.//g ; # remove . character from word
39      $word =~ s/\,//g ; # remove , character from word
40      $word =~ s/\*//g ; # remove * character from word
```

115

```
41      $word =~ s/\-//g ; # remove - character from word
42      $word =~ s/\///g ; # remove / character from word
43      $word =~ s/\{//g ; # remove { character from word
44      $word =~ s/\}//g ; # remove } character from word
45      $word =~ s/\}//g ; # remove } character from word
46
47      # only print the key,value pair if the key is not
48      # empty
49
50      if ( $word ne "" )
51      {
52        print "$word,1\n" ;
53      }
54
55    }
56
57  }
```

This script takes text file lines from STDIN at line 8, the Linux standard input stream; breaks the input down into lines, then into words at line 16; and strips the words of unwanted characters between lines 28 and 45. It then prints a series of key-value pairs as word,1 at line 52. Look at the Reduce script in the Perl file reducer.pl:

[hadoop@hc1nn perl]$ cat reducer.pl

```
01  #!/usr/bin/perl
02
03  my $line;
04  my @lineparams = ();
05  my $oldword,$word,$value,$sumval;
06
07  # the reducer is going to receive a key,value pair from stdin and it
08  # will need to sum up the values. It will need to split the name and
09  # value out of the comma separated string.
10
11  $oldword = "" ;
12
13  foreach $line ( <STDIN> )
14  {
15    # strip new line from string
16
17    chomp( $line );
18
19    # split the line into the word and value
20
21    @lineparams = split( '\,', $line );
22
23    $word  = $lineparams[0];
24    $value = $lineparams[1];
25
26    # Hadoop sorts the data by value so just sum similar word values
27
```

```
28    if ( $word eq $oldword )
29    {
30      $sumval += $value ;
31    }
32    else
33    {
34      if ( $oldword ne "" )
35      {
36        print "$oldword,$sumval\n" ;
37      }
38      $sumval = 1 ;
39    }
40
41    # now print the name value pairs
42
43    $oldword = $word ;
44 }
45
46 # remember to print last word
47
48 print "$oldword,$sumval\n" ;
```

The reducer.pl Perl script that receives data from the mapper.pl script splits its STDIN (standard input) line into the key-value pair of word,1 (at line 21). It then groups similar words and increments their count between lines 28 and 39. Lastly, it outputs key-value pairs as word,count at lines 36 and 48.

You already have some basic text files on HDFS under the directory /user/hadoop/edgar on which you can run the Perl word-count example. Check the data using the Hadoop file system ls command to be sure that it is ready to use:

```
[hadoop@hc1nn python]$ hadoop dfs -ls /user/hadoop/edgar
Found 5 items
-rw-r--r--   1 hadoop supergroup     410012 2014-06-15 15:53 /user/hadoop/edgar/10031.txt
-rw-r--r--   1 hadoop supergroup     559352 2014-06-15 15:53 /user/hadoop/edgar/15143.txt
-rw-r--r--   1 hadoop supergroup      66401 2014-06-15 15:53 /user/hadoop/edgar/17192.txt
-rw-r--r--   1 hadoop supergroup     596736 2014-06-15 15:53 /user/hadoop/edgar/2149.txt
-rw-r--r--   1 hadoop supergroup      63278 2014-06-15 15:53 /user/hadoop/edgar/932.txt
```

The test1.sh shell script tests the Map function on the Linux command line to ensure that it works, giving a single word count—that is, a count of 1 for each word in the string:

```
[hadoop@hc1nn perl]$ cat test1.sh

01   #!/bin/bash
02
03   # test the mapper
04
05   echo "one one one two three" | ./mapper.pl

[hadoop@hc1nn perl]$ ./test1.sh
one,1
one,1
one,1
two,1
three,1
```

Okay, that works. The input of five words separated by spaces is outputted as five key-value pairs of the words with a value of 1. Now, you test the Reduce function with test2.sh:

```
[hadoop@hc1nn perl]$ cat test2.sh

01  #!/bin/bash
02
03  # test the mapper
04
05  echo "one one one two three" | ./mapper.pl | ./reducer.pl
```

This script pipes the output from the Map function shown above into the Reduce function:

```
[hadoop@hc1nn perl]$ ./test2.sh
one,3
two,1
three,1
```

The Reduce function sums the values of the similar words correctly: three instances of the word *one* followed by one each of *two* and *three*. Now, it is time to run the Hadoop streaming Map Reduce job by using these Perl scripts. You create three scripts to help with this:

```
[hadoop@hc1nn perl]$ ls w*
wc_clean.sh  wc_output.sh  wordcount.sh
```

The script wc_clean.sh is used to delete the contents of the results directory on HDFS so that the Map Reduce job can be rerun:

```
[hadoop@hc1nn perl]$ cat wc_clean.sh

01  #!/bin/bash
02
03  # Clean the hadoop perl run data directory
04
05  hadoop dfs -rmr /user/hadoop/perl/results_wc
```

This uses the Hadoop file system rmr command to delete the directory and its contents.

The script wc_output.sh is used to display the results of the job:

```
[hadoop@hc1nn perl]$ cat wc_output.sh

01  #!/bin/bash
02
03  # List the results directory
04
05  hadoop dfs -ls /user/hadoop/perl/results_wc
06
07  # Cat the last ten lines of the part file
08
09  hadoop dfs -cat /user/hadoop/perl/results_wc/part-00000 | tail -10
```

It lists the files in the results directory on HDFS and dumps the last 10 lines of the results part file using the Hadoop file system cat command and the Lunix tail command.

The script wordcount.sh runs the Map Reduce task by using the Map and Reduce Perl scripts:

```
[hadoop@hc1nn perl]$ cat wordcount.sh

01  #!/bin/bash
02
03  # Now run the Perl based word count
04
05  cd $HADOOP_PREFIX
06
07  hadoop jar contrib/streaming/hadoop-*streaming*.jar \
08      -file    /home/hadoop/perl/mapper.pl    \
09      -mapper  /home/hadoop/perl/mapper.pl \
10      -file    /home/hadoop/perl/reducer.pl   \
11      -reducer /home/hadoop/perl/reducer.pl \
12      -input   /user/hadoop/edgar/* \
13      -output  /user/hadoop/perl/results_wc
```

The \ characters allow you to make your Hadoop command line more readable by breaking a single command line over multiple lines. The -file options make a file executable within Hadoop. The -mapper and -reducer options identify the Map and Reduce functions for the job. The -input option gives the path on HDFS to the input text data. The -output option specifies where the job output will be placed on HDFS.

The Hadoop jar parameter allows the command line to specify which library file to use—in this case, the streaming library. Using the last three scripts for cleaning, running, and outputting the results makes the Map Reduce task quickly repeatable; you do not need to retype the commands! The output is a Map Reduce job, as shown below:

```
[hadoop@hc1nn perl]$ ./wordcount.sh

packageJobJar: [/home/hadoop/perl/mapper.pl, /home/hadoop/perl/reducer.pl, /app/hadoop/tmp/hadoop-
unjar5199336797215175827/] [] /tmp/streamjob5502063820605104626.jar tmpDir=null
14/06/20 13:35:56 INFO util.NativeCodeLoader: Loaded the native-hadoop library
14/06/20 13:35:56 INFO mapred.FileInputFormat: Total input paths to process : 5
14/06/20 13:35:57 INFO streaming.StreamJob: getLocalDirs(): [/app/hadoop/tmp/mapred/local]
14/06/20 13:35:57 INFO streaming.StreamJob: Running job: job_201406201237_0010
14/06/20 13:35:57 INFO streaming.StreamJob: To kill this job, run:
14/06/20 13:35:57 INFO streaming.StreamJob: /usr/local/hadoop-1.2.1/libexec/../bin/hadoop job
                                            -Dmapred.job.tracker=hc1nn:54311 -kill job_201406201237_0010
14/06/20 13:35:57 INFO streaming.StreamJob: Tracking URL: http://hc1nn:50030/jobdetails.
                                            jsp?jobid=job_201406201237_0010
14/06/20 13:35:58 INFO streaming.StreamJob: map 0%    reduce 0%
14/06/20 13:36:06 INFO streaming.StreamJob: map 20%   reduce 0%
14/06/20 13:36:08 INFO streaming.StreamJob: map 60%   reduce 0%
14/06/20 13:36:13 INFO streaming.StreamJob: map 100% reduce 0%
14/06/20 13:36:15 INFO streaming.StreamJob: map 100% reduce 33%
14/06/20 13:36:19 INFO streaming.StreamJob: map 100% reduce 100%
14/06/20 13:36:22 INFO streaming.StreamJob: Job complete: job_201406201237_0010
14/06/20 13:36:22 INFO streaming.StreamJob: Output: /user/hadoop/perl/results_wc
```

Looking on HDFS, you will find a results_wc directory under /user/hadoop/perl that contains the output of the word-count task. As in previous examples, it is the part file that contains the result. When you dump the part file to the session by using the Hadoop file system cat command and the Linux tail command, you limit the results to the last 10 lines, with the following resulting data:

```
[hadoop@hc1nn perl]$ ./wc_output.sh

Found 3 items
-rw-r--r--   1 hadoop supergroup        0 2014-06-20 13:36 /user/hadoop/perl/results_wc/_SUCCESS
drwxr-xr-x   - hadoop supergroup        0 2014-06-20 13:35 /user/hadoop/perl/results_wc/_logs
-rw-r--r--   1 hadoop supergroup   249441 2014-06-20 13:36 /user/hadoop/perl/results_wc/part-00000

zephyr,1
zero,1
zigzag,2
zimmermann,5
zipped,1
zoar,1
zoilus,3
zone,1
zones,1
zoophytes,1
```

The words have been sorted and their values totaled, and many of the unwanted characters have been removed from the words. This last example shows the wide-ranging possibility of using scripts for Map Reduce jobs with Hadoop streaming. No Java code was needed and no code was compiled.

Summary

In this chapter you have investigated Map Reduce programming by using one example implemented in several ways. That is, by using a single algorithm, you can better compare the different approaches.

A Java-based approach, for example, gives a low-level means to Map Reduce development. The downside is that code volumes are large, and so costs and potential error volume increase. On a positive note, using low-level Hadoop-based APIs gives a wide range of functionality for your data processing.

In contrast, the Apache Pig examples involved a high-level Pig native code API. This resulted in a lower code volume and therefore lower costs and quicker times. You can also extend the functionality of Pig by writing user-defined functions (UDFs) in Java. Pig can be a vehicle for processing HDFS-based data, and although there was no time to cover it here, it can also load data to Hive by using a product called HCatalog.

A word-count example was then attempted using Hive, the Hadoop data warehouse. A file was imported into a table and a count of words was created in Hive QL, an SQL-like language. While this is a functional language and quite easy to use, it may not offer the full range of functions that are available when using Pig and UDFs. Although it was quick to implement and needed very little code, choosing this technique depends on the complexity of your task.

Lastly, word count was coded in Perl and called via the Hadoop streaming library. This showed that a third-party language like Python or Perl can be used to create Map Reduce jobs. In this example, unstructured text was employed for the streaming job, making it possible to create user-defined input and output formats. See the Hadoop streaming guide at http://hadoop.apache.org/docs/r1.2.1/streaming.html#Hadoop+Streaming.

These different Map Reduce methods offer the ability to create simple ETL building blocks that can be used to build complex ETL chains. Later chapters will discuss this concept in relation to products like Oozie, Talend, and Pentaho. Therefore, the reader should consider this chapter in conjunction with Chapter 10, which will present big-data visual ETL tools such as Talend and Pentaho; these offer a highly functional approach to ETL job creation using object drag and drop.

CHAPTER 5

Scheduling and Workflow

When you're working with big data in a distributed, parallel processing environment like Hadoop, job scheduling and workflow management are vital for efficient operation. *Schedulers* enable you to share resources at a job level within Hadoop; in the first half of this chapter, I use practical examples to guide you in installing, configuring, and using the Fair and Capacity schedulers for Hadoop V1 and V2. Additionally, at a higher level, *workflow* tools enable you to manage the relationships between jobs. For instance, a workflow might include jobs that source, clean, process, and output a data source. Each job runs in sequence, with the output from one forming the input for the next. So, in the second half of this chapter, I demonstrate how workflow tools like Oozie offer the ability to manage these relationships.

An Overview of Scheduling

The default Hadoop scheduler was FIFO—first in, first out. It did not support task preemption, which is the ability to temporarily halt a task and allow another task to access those resources. Apache, though, offers two extra schedulers for Hadoop—Capacity and Fair—that you can use in place of the default.

Although both schedulers are available in Hadoop V1 and V2, the functions they offer depend on the Hadoop version you're using. To decide which scheduler is right for your applications, take a look at the prime features of each, detailed in the following sections, or consult the Apache Software Foundation website at hadoop.apache.org/docs for in-depth information.

The Capacity Scheduler

Capacity handles large clusters that are shared among multiple organizations or groups. In this multi-tenancy environment, a cluster can have many job types and multiple job priorities. Some key features of Capacity are as follows:

Organization: As it is designed for situations in which clusters need to support multi-tenancy, its resource sharing is more stringent so as to meet capacity, security, and resource guarantees.

Capacity: Resources are allocated to queues and are shared among the jobs on that queue. It is possible to set soft and hard limits on queue-based resources.

Security: In a multi-tenancy cluster, security is a major concern. Capacity uses access control lists (ACLs) to manage queue-based job access. It also permits per-queue administration, so that you can have different settings on the queues.

Elasticity: Free resources from under-utilized queues can be assigned to queues that have reached their capacities. When needed elsewhere, these resources can then be reassigned, thereby maximizing utilization.

Multi-tenancy: In a multi-tenancy environment, a single user's rogue job could possibly soak up multiple tenants' resources, which would have a serious impact on job-based service-level agreements (SLAs). The Capacity scheduler provides a range of limits for these multiple jobs, users, and queues so as to avoid this problem.

Resource-based Scheduling: Capacity uses an algorithm that supports memory-based resource scheduling for jobs that are resource intensive.

Hierarchical Queues: When used with Hadoop V2, Capacity supports a hierarchy of queues, so that under-utilized resources are first shared among subqueues before they are then allocated to other cluster tenant queues.

Job Priorities: In Hadoop V1, the scheduler supports scheduling by job priority.

Operability: Capacity enables you to change the configuration of a queue at runtime via a console that permits viewing of the queues. In Hadoop V2, you can also stop a queue to let it drain.

The Fair Scheduler

Fair aims to do what its name implies: share resources fairly among all jobs within a cluster that is owned and used by a single organization. Over time, it aims to share resources evenly to job pools. Some key aspects of Fair are:

Organization: This scheduler organizes jobs into pools, with resources shared among the pools. Attributes, like priorities, act as weights when the resources are shared.

Resource Sharing: You can specify a minimum level of resources to a pool. If a pool is empty, then Fair shares the resources of other pools.

Resource Limits: With Fair, you can specify concurrent job limits by user and pool so as to limit the load on the cluster.

Scheduling in Hadoop V1

Now that you have a sense of each scheduler's strengths, you're ready to see them put to work. This section demonstrates job scheduling in a Hadoop V1 environment. You'll learn how to configure the Capacity and Fair schedulers, and you'll see that the libraries necessary to use them are already supplied with Hadoop V1.2.1, just waiting for you to plug them in.

V1 Capacity Scheduler

As mentioned, the library used by the Hadoop Capacity scheduler is included in the V1.2.1 release within the lib directory of the installation, as you can see:

```
[hadoop@hc1nn lib]$ pwd
/usr/local/hadoop/lib
[hadoop@hc1nn lib]$ ls -l hadoop-capacity-scheduler*
-rw-rw-r--. 1 hadoop hadoop 58461 Jul 23  2013 hadoop-capacity-scheduler-1.2.1.jar
```

To use the library, you plug it into the configuration by adding the following property to the mapred-site.xml file in the conf directory of the installation:

```
<property>
  <name>mapred.jobtracker.taskScheduler</name>
  <value>org.apache.hadoop.mapred.CapacityTaskScheduler</value>
  <description>Plugin the Capcity scheduler</description>
</property>
```

As described earlier, Capacity is designed for multiple tenancy and is queue-based, with queues shared among cluster tenants. A configuration file for the queue configuration, called capacity-scheduler.xml, is supplied in the conf directory as well. The file contains the configuration for the default queue:

```
[hadoop@hc1nn conf]$ pwd
/usr/local/hadoop/conf
[hadoop@hc1nn conf]$ ls -l capacity-scheduler.xml
-rw-rw-r--. 1 hadoop hadoop 7457 Jul 23  2013 capacity-scheduler.xml
```

To demonstrate an example of a Hadoop V1 Capacity scheduling queue, I have set up a new queue called "tqueue" in this file, with the following configuration. The configuration file, which you can copy, shows the attributes that can be set for the queue. The default queue already exists; the new tqueue queue is added:

```
<!-- Set up test queue -->

<property>
  <name>mapred.capacity-scheduler.queue.tqueue.capacity</name>
  <value>50</value>
</property>

<property>
  <name>mapred.capacity-scheduler.queue.tqueue.maximum-capacity</name>
  <value>100</value>
</property>

<property>
  <name>mapred.capacity-scheduler.queue.tqueue.supports-priority</name>
  <value>true</value>
</property>

<property>
  <name>mapred.capacity-scheduler.queue.tqueue.minimum-user-limit-percent</name>
  <value>20</value>
</property>

<property>
  <name>mapred.capacity-scheduler.queue.tqueue.user-limit-factor</name>
  <value>1</value>
</property>

<property>
  <name>mapred.capacity-scheduler.queue.tqueue.maximum-initialized-active-tasks</name>
  <value>200000</value>
</property>

<property>
  <name>mapred.capacity-scheduler.queue.tqueue.maximum-initialized-active-tasks-per-user</name>
  <value>100000</value>
</property>
```

```
<property>
  <name>mapred.capacity-scheduler.queue.tqueue.init-accept-jobs-factor</name>
  <value>10</value>
</property>
```

You can specify access controls, such as which users can submit jobs and which can administer queues, in the conf directory file mapred-queue-acls.xml. For complete details, see the configuration guide supplied by the Apache Software Foundation, available at hadoop.apache.org/docs/r1.2.1/; click Map Reduce (on the left), and then select Capacity Scheduler.

Next, in mapred-queue-acls.xml, you need to specify the list of queues via the mapred.queue.names property:

```
<property>
  <name>mapred.queue.names</name>
  <value>default,tqueue</value>
</property>
```

Make sure that the sum of the queue capacity values in the configuration file for all the queues is 100. For instance, both the default and tqueue queues in my configuration file have a capacity value of 50 (see below).

```
<property>
  <name>mapred.capacity-scheduler.queue.default.capacity</name>
  <value>50</value>
</property>
```

```
<property>
  <name>mapred.capacity-scheduler.queue.tqueue.capacity</name>
  <value>50</value>
</property>
```

After you add the configuration for the Capacity scheduler, it will become visible within the system when the Map Reduce servers are restarted. They can be checked via the Job Tracker user interface. For example, to check the jobs on the name node hc1nn server, use the URL http://hc1nn:50030/jobtracker.jsp.

Figure 5-1 shows a list of queues, along with their attributes, for the Capacity scheduler on the name node hc1nn server. Notice the scheduling configuration of the two queues—default and tqueue.

Scheduling Information

Queue Name	State	Scheduling Information
default	running	Queue configuration Capacity Percentage: 50.0% User Limit: 100% Priority Supported: NO ──────── Map tasks Capacity: 4 slots Used capacity: 0 (0.0% of Capacity) Running tasks: 0 ──────── Reduce tasks Capacity: 4 slots Used capacity: 0 (0.0% of Capacity) Running tasks: 0 ──────── Job info Number of Waiting Jobs: 0 Number of Initializing Jobs: 0 Number of users who have submitted jobs: 0
tqueue	running	Queue configuration Capacity Percentage: 50.0% User Limit: 20% Priority Supported: YES ──────── Map tasks Capacity: 4 slots Maximum capacity: 8 slots Used capacity: 0 (0.0% of Capacity) Running tasks: 0 ──────── Reduce tasks Capacity: 4 slots Maximum capacity: 8 slots Used capacity: 0 (0.0% of Capacity) Running tasks: 0 ──────── Job info Number of Waiting Jobs: 0 Number of Initializing Jobs: 0 Number of users who have submitted jobs: 0

Figure 5-1. *Capacity scheduler queue list*

V1 Fair Scheduler

As was the case for the Capacity scheduler, the library you need to use the Fair scheduler is included with the Hadoop V1.2.1 installation; it is within the lib directory:

```
[hadoop@hc1nn lib]$ pwd
/usr/local/hadoop/lib
[hadoop@hc1nn lib]$ ls -l  hadoop-fairscheduler*
-rw-rw-r--. 1 hadoop hadoop 70409 Jul 23  2013 hadoop-fairscheduler-1.2.1.jar
```

However, you need to modify the property mapred.jobtracker.taskScheduler in the file mapred-site.xml within the conf directory, as follows:

```
<property>
  <name>mapred.jobtracker.taskScheduler</name>
  <value>org.apache.hadoop.mapred.FairScheduler</value>
  <description>Plugin the Fair scheduler</description>
</property>
```

To give greater control, I add some properties to the mapred-site.xml file to switch off pre-emption by setting mapred.fairscheduler.preemption to False and I disallow unspecified pool names by setting mapred.fairscheduler.allow.undeclared.pools to False. Also, I assign pool property names to queue names by using mapred.fairscheduler.poolnameproperty. Finally, I use the mapred.queue.names property to define a list of allowed queue names that could be used in the configuration file, all as follows:

```
<property>
  <name>mapred.fairscheduler.preemption</name>
  <value>false</value>
</property>

<property>
  <name>mapred.fairscheduler.allow.undeclared.pools</name>
  <value>false</value>
</property>

<property>
  <name>mapred.fairscheduler.poolnameproperty</name>
  <final>true</final>
  <value>mapred.job.queue.name</value>
</property>

<property>
  <name>mapred.queue.names</name>
  <final>true</final>
  <value>high_pool,low_pool,default</value>
</property>
```

To see the full configuration guide, go to the Apache Software Foundation website (hadoop.apache.org/docs/r1.2.1/), click Map Reduce, and then select Fair Scheduler.

Like the configuration for the Capacity scheduler, you can add access control in the mapred-queue-acls.xml file for the Fair scheduler to specify user and administration access to each queue. For example, here I grant the hadoop user access to the high_pool and administration access to that queue, as follows:

```
<property>
  <name>mapred.queue.high_pool.acl-submit-job</name>
  <value>hadoop</value>
</property>

<property>
  <name>mapred.queue.low_pool.acl-submit-job</name>
  <value>smitha</value>
</property>
```

```
<property>
  <name>mapred.queue.default.acl-submit-job</name>
  <value>jonesb</value>
</property>

<property>
  <name>mapred.queue.high_pool.acl-administer-jobs</name>
  <value>hadoop</value>
</property>

<property>
  <name>mapred.queue.low_pool.acl-administer-jobs</name>
  <value>smitha</value>
</property>

<property>
  <name>mapred.queue.default.acl-administer-jobs</name>
  <value>jonesb</value>
</property>
```

There is already a configuration file provided for this scheduler in the installation configuration directory:

```
[hadoop@hc1nn conf]$ pwd
/usr/local/hadoop/conf
[hadoop@hc1nn conf]$ ls -l fair-scheduler.xml
-rw-rw-r--. 1 hadoop hadoop 327 Jul 23  2013 fair-scheduler.xml
```

I add the following configuration to this file to specify the fair scheduler pools high_pool, low_pool, and default queues, as well as their attributes, such as the minimum and maximum number of Map and Reduce function instances, as follows:

```
<pool name="high_pool">
  <minMaps>10</minMaps>
  <minReduces>10</minReduces>
  <maxMaps>50</maxMaps>
  <maxReduces>50</maxReduces>
  <maxRunningJobs>1000</maxRunningJobs>
  <weight>3</weight>
</pool>

<pool name="low_pool">
  <minMaps>10</minMaps>
  <minReduces>10</minReduces>
  <maxMaps>50</maxMaps>
  <maxReduces>50</maxReduces>
  <maxRunningJobs>1000</maxRunningJobs>
  <weight>1</weight>
</pool>
```

```
<pool name="default">
  <minMaps>10</minMaps>
  <minReduces>10</minReduces>
  <maxMaps>50</maxMaps>
  <maxReduces>50</maxReduces>
  <maxRunningJobs>1000</maxRunningJobs>
  <weight>1</weight>
</pool>
```

Here, I define three pools (high_pool, low_pool, and default), each with the same configuration for the Map Reduce minimum and maximum limits. They all have a maximum of 1,000 running jobs, but the high pool has three times the weighting, so the high_pool will get three times the share of the cluster than the other pools will get.

To show how the Fair scheduler works, I will run a Pig-based job as an example. But before running the job example, I restart the Map Reduce servers as I did for the Capacity scheduler earlier, so as to pick up the changes to the configuration. Using the mapred.fairscheduler.pool property, I specify the name of the queue that my Pig Latin job will be placed on when I issue the command line, using the -D switch, as follows.

```
[hadoop@hc1nn pig]$ pig -Dmapred.fairscheduler.pool=high_pool wordcount2.pig
```

Note: If I did not specify the pool to be used by using the –D option, I would encounter an error because, by default, Fair assumes that the queue name matches the Linux account name. Given that I am running the job using the Linux hadoop account, Fair would have looked for a queue named "hadoop," which does not exist. The error message I would receive is an example of an UndeclaredPoolException error:

```
Failed Jobs:
JobId   Alias     Feature Message Outputs
N/A     clines,gword,rlines,wcount,words         GROUP_BY,COMBINER
Message: org.apache.hadoop.ipc.RemoteException: org.apache.hadoop.mapred.UndeclaredPoolException:
Pool name: 'hadoop' is invalid. Add pool name to the fair scheduler allocation file. Valid pools
are: high_pool, low_pool
```

However, proceeding with the Fair scheduler example, I can see that the scheduler has started without error from the Job Server's log in the install logs directory. I look for the following line:

```
2014-06-29 13:41:27,882 INFO org.apache.hadoop.mapred.FairScheduler: Successfully configured
FairScheduler
```

By checking the Job Tracker user interface, I can learn more about the job details. Figure 5-2 shows a compound image of the Pig Latin job that I submitted to the high_pool pool. For instance, the top table is taken from the list of running jobs from the Job Tracker user interface. It shows that the hadoop user has submitted the Pig Latin job wordcount2.pig, which is running.

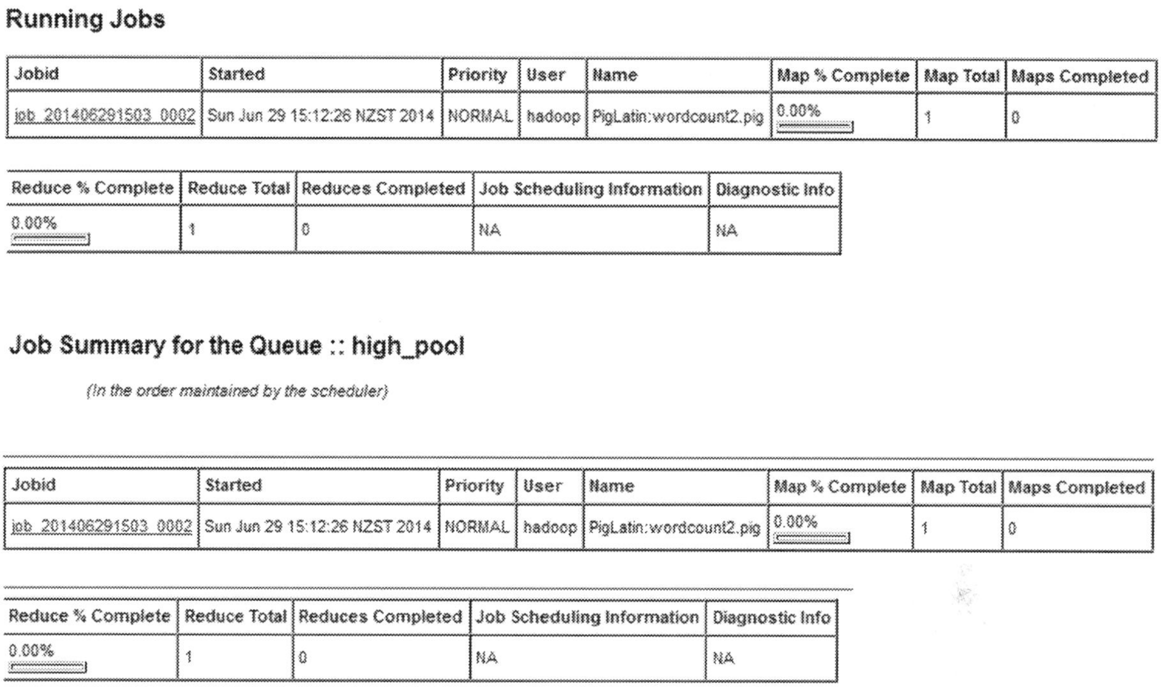

Running Jobs

Jobid	Started	Priority	User	Name	Map % Complete	Map Total	Maps Completed
job_201406291503_0002	Sun Jun 29 15:12:26 NZST 2014	NORMAL	hadoop	PigLatin:wordcount2.pig	0.00%	1	0

Reduce % Complete	Reduce Total	Reduces Completed	Job Scheduling Information	Diagnostic Info
0.00%	1	0	NA	NA

Job Summary for the Queue :: high_pool

(In the order maintained by the scheduler)

Jobid	Started	Priority	User	Name	Map % Complete	Map Total	Maps Completed
job_201406291503_0002	Sun Jun 29 15:12:26 NZST 2014	NORMAL	hadoop	PigLatin:wordcount2.pig	0.00%	1	0

Reduce % Complete	Reduce Total	Reduces Completed	Job Scheduling Information	Diagnostic Info
0.00%	1	0	NA	NA

Figure 5-2. *Fair scheduler job queue*

The second table is taken from the high_pool queue, and it shows that the job was submitted to this queue with Normal priority, using the command line switch -Dmapred.fairscheduler.pool=high_pool.

Now that you've examined the schedulers in Hadoop V1, it's time to take a look at those same schedulers in Hadoop V2. You can use the environment section in Chapter 4 to find out how to switch between versions on a single cluster of servers. You will then see the similarities of the V1 and V2 schedulers, but also note the more advanced interfaces that V2 offers.

Scheduling in Hadoop V2

In this section, I show how to configure the Capacity and Fair plug-in schedulers for Hadoop V2. You can find full details at the Apache Software Foundation website (hadoop.apache.org/docs) by selecting your Hadoop version, then the desired scheduling options from the menu on the left. In comparison to V1, the V2 user interface offers better functionality and generally looks much more presentable. The configuration is similar, however, and the files that need to be set up are the same.

V2 Capacity Scheduler

You need to set up the Capacity scheduler to work with YARN. You begin by configuring YARN so that its scheduler class is defined as Capacity, rather than as the default FIFO scheduler. At the end of the yarn-site.xml file, which is located under /etc/hadoop/conf, you add the following property to define Capacity:

```
<!-- add scheduler configuration -->

<property>
  <name>yarn.resourcemanager.scheduler.class</name>
  <value>org.apache.hadoop.yarn.server.resourcemanager.scheduler.capacity.CapacityScheduler</value>
</property>
```

As you remember, Hadoop V2 can support hierarchical queues for each cluster client. This way, free resources can be shared across a client's queues before being offered to other queues. To demonstrate how this works, I define some queues in the Capacity scheduler's configuration file (capacity-scheduler.xml, found in /etc/hadoop/conf). Specifically, I define and create three new queues (client1, client2, and client3) and their subqueues, as follows:

```
<property>
  <name>yarn.scheduler.capacity.root.queues</name>
  <value>client1,client2,client3</value>
</property>

<property>
  <name>yarn.scheduler.capacity.root.client1.queues</name>
  <value>client1a,client1b</value>
</property>

<property>
  <name>yarn.scheduler.capacity.root.client2.queues</name>
  <value>client2a,client2b,client2c</value>
</property>

<property>
  <name>yarn.scheduler.capacity.root.client3.queues</name>
  <value>client3a,client3b</value>
</property>
```

In addition, for each queue I create, I define its properties in the capacity-scheduler.xml file. The XML that follows is an example of that needed for one of the queues. Note that this must be repeated for each parent and child queue; therefore, I must remember to change the name of each attribute to match the queue name—that is, yarn.scheduler.capacity.root.**client1**.capacity.

As in the V1 example, the sum of the capacity values for queues in the configuration file must be 100. Also, the sum of the capacity values for the child queues within a parent queue must be 100. For instance, the children of client1 are named client1.client1a. This XML sets up the attributes for queue client1 in terms of capacity, state, and access. In this case, the * values mean "anyone," so access has been left open:

```
<property>
  <name>yarn.scheduler.capacity.root.client1.capacity</name>  <value>100</value>
</property>
```

```
<property>
  <name>yarn.scheduler.capacity.root.client1.user-limit-factor</name> <value>1</value>
</property>
<property>
  <name>yarn.scheduler.capacity.root.client1.maximum-capacity</name> <value>100</value>
</property>
<property>
  <name>yarn.scheduler.capacity.root.client1.state</name> <value>RUNNING</value>
</property>
<property>
  <name>yarn.scheduler.capacity.root.client1.acl_submit_applications</name> <value>*</value>
</property>
<property>
  <name>yarn.scheduler.capacity.root.client1.acl_administer_queue</name> <value>*</value>
</property>
```

Now, I get YARN to refresh its queue configuration by using the yarn rmadmin command with a -refreshQueues option. This causes YARN to reread its configuration files and so pick up the changes that have been made:

```
[hadoop@hc1nn conf]$ yarn rmadmin –refreshQueues
```

My reconfigured scheduler is ready, and I now submit a word-count job to show the queues in use. To display the queue's functionality, however, I need some test data; therefore, I have created the job's input data in the HDFS directory /usr/hadoop/edgar, as the HDFS file system command shows (and I populate it with data):

```
[hadoop@hc1nn edgar]$ hdfs dfs -ls /usr/hadoop/edgar
Found 5 items
-rw-r--r--   2 hadoop hadoop     410012 2014-07-01 18:14 /usr/hadoop/edgar/10031.txt
-rw-r--r--   2 hadoop hadoop     559352 2014-07-01 18:14 /usr/hadoop/edgar/15143.txt
-rw-r--r--   2 hadoop hadoop      66401 2014-07-01 18:14 /usr/hadoop/edgar/17192.txt
-rw-r--r--   2 hadoop hadoop     596736 2014-07-01 18:14 /usr/hadoop/edgar/2149.txt
-rw-r--r--   2 hadoop hadoop      63278 2014-07-01 18:14 /usr/hadoop/edgar/932.txt
```

The word-count job will read this data, run a word count, and place the output results in the HDFS directory /usr/hadoop/edgar-results1. The word-count job results look like this:

```
hdfs \
    jar /usr/lib/hadoop-mapreduce/hadoop-mapreduce-examples.jar \
    wordcount \
    -Dmapred.job.queue.name=client1a \
    /usr/hadoop/edgar \
    /usr/hadoop/edgar-results1
```

The backslash characters (\) allow me to spread the command over multiple lines to make it more readable. I've used a -D option to specify the queue in which to place this job (client1a).

Now I examine the scheduler configuration using the Name Node server name hc1nn, with a port value of 8088, taken from the property yarn.resourcemanager.webapp.address in the configuration file yarn-site.xml: http://hc1nn:8088/cluster/scheduler. Figure 5-3 illustrates the resulting hierarchy of job queues, with the currently running word-count Map Reduce job placed in the client1a. child queue.

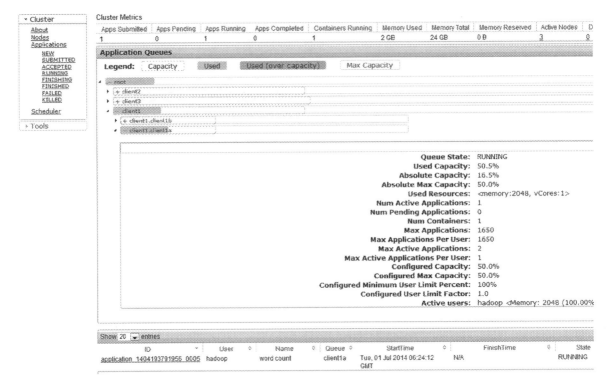

Figure 5-3. *Hadoop V2 Capacity scheduler*

Note that I have cropped Figure 5-3 so that the used queues are visible. You can see the Map Reduce word-count job running on the client1.client1a queue. The Capacity scheduler, therefore, is working on Hadoop V2 because it can accept Hadoop jobs into its queues.

V2 Fair Scheduler

The configuration for the Hadoop V2 Fair scheduler is quite simple and follows the same method as for the V1 configuration. In the yarn-site.xml file within the configuration directory (/etc/hadoop/conf), you define the `yarn.resourcemanager.scheduler.class` property.

```
<property>
  <name>yarn.resourcemanager.scheduler.class</name>
  <value>org.apache.hadoop.yarn.server.resourcemanager.scheduler.fair.FairScheduler</value>
</property>
```

Then, you set up the default Fair scheduler configuration file fair-scheduler.xml in the same directory.

To demonstrate, I have borrowed this file from the V1 configuration:

```
<allocations>

  <pool name="high_pool">
    <minMaps>10</minMaps>
    <minReduces>10</minReduces>
    <maxMaps>50</maxMaps>
    <maxReduces>50</maxReduces>
    <maxRunningJobs>1000</maxRunningJobs>
    <weight>3</weight>
  </pool>

  <pool name="low_pool">
    <minMaps>10</minMaps>
    <minReduces>10</minReduces>
    <maxMaps>50</maxMaps>
    <maxReduces>50</maxReduces>
    <maxRunningJobs>1000</maxRunningJobs>
    <weight>1</weight>
  </pool>

  <pool name="default">
    <minMaps>10</minMaps>
    <minReduces>10</minReduces>
    <maxMaps>50</maxMaps>
    <maxReduces>50</maxReduces>
    <maxRunningJobs>1000</maxRunningJobs>
    <weight>1</weight>
  </pool>

</allocations>
```

You now use the yarn rmadmin command to refresh the YARN scheduler queue configuration. This will cause YARN (as for V2 Capacity) to reread its configuration files and so pick up the changes that you have made:

```
[hadoop@hc1nn conf]$ yarn rmadmin -refreshQueues
```

To demonstrate its use, I create a Map Reduce job and specify a queue value of high_queue so that I can ensure that the submitted job will be processed by the YARN Fair scheduler and will be placed in the right queue. The job I use specifies the queue name using a -D option:

```
hadoop \
    jar /usr/lib/hadoop-mapreduce/hadoop-mapreduce-examples.jar \
    wordcount \
    -Dmapred.job.queue.name=high_pool \
    /usr/hadoop/edgar \
    /usr/hadoop/edgar-results1
```

Figure 5-4 shows the job status in the scheduler user interface. The queue legend now displays Fair Share, and the job details appear in the high_pool queue, which is marked in green to indicate it's being used.

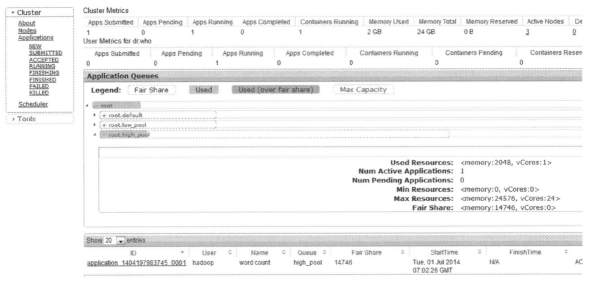

Figure 5-4. *Hadoop V2 Fair scheduler*

As shown in Figure 5-4, the V2 fair scheduler is configured and working correctly.

Using Oozie for Workflow

Capacity, Fair, and similar plug-in schedulers deal with resources allocated to individual jobs over a period of time. However, what about the relationships between jobs and the dependencies between them? That's where *workflow* managers, like Apache's Oozie, come in. This section examines Hadoop job-based workflows and scheduling, and demonstrates how tools like Oozie enable you to manage related jobs as workflows.

A workflow scheduler for Hadoop, Oozie is integrated into many of the Hadoop tools, such as Pig, Hive, Map Reduce, and Streaming. Oozie workflows are defined as *directed acyclical graphs* (DAGs) and are stored as XML.

In this section, I will demonstrate how to install the Oozie component that is part of the Cloudera installation that was used in Chapter 2. I show how to check that it is working, and how to create an example workflow. For further details, check the Apache Software Foundation website at oozie.apache.org.

Installing Oozie

You start by installing the Oozie client and server as root, using Yum, the Linux-based package manager. You install Oozie on the single CentOS 6 Linux server hc1nn, as follows:

```
[root@hc1nn conf]# yum install oozie
```

```
[root@hc1nn conf]# yum install oozie-client
```

Next, you configure Oozie to use YARN by editing the oozie-env.sh file under the directory /etc/oozie/conf:

```
[root@hc1nn conf]# cd /etc/oozie/conf

[root@hc1nn conf]# ls -l oozie-env.sh
-rw-r--r--. 1 root root 1360 May 29 07:20 oozie-env.sh
```

You can configure Oozie to use either YARN or Map Reduce V1, but not both at the same time. The oozie-server-0.20 option is for Map Reduce V1, while setting CATALINA_BASE as oozie-server configures it to use YARN. (In this demonstration, I have chosen to use YARN because this is Hadoop V2, and it is the default).

Because I'm using YARN, I edit the file as follows:

```
#export CATALINA_BASE=/usr/lib/oozie/oozie-server-0.20 ## use MRv1
export CATALINA_BASE=/usr/lib/oozie/oozie-server           ## use Yarn
```

Note that I left the MRv1 line in the file but commented it out so that I know how to configure Oozie later if I want to switch back to use Map Reduce version 1. I have also elected to use the default Derby database that comes with the Oozie installation. If you need to use an alternative database, such as Oracle or MySql, you can. Check the Oozie website at oozie.apache.org for details.

To proceed, you set up Oozie and run it as the Linux hadoop user. This is the account where you previously stored your scripts and set up your Hadoop-based configuration. So, you run the ooziedb.sh script with parameters create and -run to set up the oozie database:

```
[hadoop@hc1nn conf]# /usr/lib/oozie/bin/ooziedb.sh create -run


Validate DB Connection
DONE
Check DB schema does not exist
DONE
Check OOZIE_SYS table does not exist
DONE
Create SQL schema
DONE
Create OOZIE_SYS table
DONE

Oozie DB has been created for Oozie version '3.3.2-cdh4.7.0'

The SQL commands have been written to: /tmp/ooziedb-3196886912358277569.sql
```

This sets up the Oozie Derby database under the directory /var/lib/oozie. In this example, the directory is owned by the Oozie Linux user. If you plan to use a different account to run Oozie, make sure it has access to this directory and its contents. (I used the su [switch user] command to switch the user to the Linux root account. I then changed the directory to /var/lib/oozie and used the Linux chmod command to recursively set permissions to 777. This is because the Linux hadoop account needs to run Oozie and so access this Derby database instance).

```
[hadoophc1nn ~]$ su -
[root@hc1nn oozie]$ cd /var/lib/oozie
[root@hc1nn oozie]$ ls -l
drwxr-xr-x. 2 oozie  oozie  4096 Jul  9 17:44 oozie-db
[root@hc1nn oozie]$ chmod -R 777  *
[root@hc1nn oozie]$ exit
```

To use the Oozie web-based console, it has to be configured. You download and install the file ext-2.2.zip onto Hadoop. You use wget to download the file:

```
[hadoop@hc1nn Downloads]$ wget  http://extjs.com/deploy/ext-2.2.zip

[hadoop@hc1nn Downloads]$ ls -l ext-2.2.zip
-rw-rw-r--. 1 oozie oozie 6800612 Oct 24  2008 ext-2.2.zip
```

You unpack the file with the Linux unzip command, and then use the Linux ls command to check the unpacked directory ext-2.2:

```
[hadoop@hc1nn Downloads]$ unzip ext-2.2.zip

[hadoop@hc1nn Downloads]$ ls -ld ext-2.2
drwxr-xr-x. 9 oozie oozie 4096 Aug  4  2008 ext-2.2
```

This unzipped package needs to be moved to the /var/lib/oozie directory, which is the location where Oozie will look for it when it runs. In this example, I move the installation directory using the root account. The commands that follow show that the Linux su (switch user) command switches to the root account. The Linux cd (change directory) command moves it to the Downloads directory. The Linux mv (move) command moves the Oozie web console package to the /var/lib/oozie/ directory:

```
[hadoophc1nn ~]$ su -
[root@hc1nn ~]$ cd /home/Hadoop/Downloads
[root@hc1nn Downloads]$ mv  ext-2.2  /var/lib/oozie/
[root@hc1nn Downloads]$ cd  /var/lib/oozie/

[root@hc1nn Downloads]# ls -l /var/lib/oozie
total 8
drwxrwxr-x. 9 oozie  oozie  4096 Aug  4  2008 ext-2.2
drwxr-xr-x. 2  oozie  oozie  4096 Jul  9 17:44 oozie-db

[root@hc1nn Downloads]# exit
```

You unpack the Oozie shared library in a temporary directory under /tmp so that you can install it onto HDFS. This library provides the functionality for tools like Pig, Hive, and Sqoop when used in workflows. There is a version for YARN and one for Map Reduce V1. For this example, I use the YARN version because that is what my Hadoop V2 CDH4 cluster is using.

```
 [hadoop@hc1nn ~]$ ls -l /usr/lib/oozie/oozie-sharelib*.gz

-rwxrwxrwx. 1 root root 84338242 May 29 07:02 /usr/lib/oozie/oozie-sharelib-mr1.tar.gz
-rwxrwxrwx. 1 root root 84254668 May 29 07:02 /usr/lib/oozie/oozie-sharelib-yarn.tar.gz
```

You choose the file that you wish to use; my example needs the YARN version. The Linux mkdir command creates a directory under /tmp, called "ooziesharelib." The Linux cd command then moves it to that directory. The tar command extracts the Oozie tarred and gzipped file oozie-sharelib-yarn.tar.gz; the x option means "extract" while the z option unzips the tar file:

```
[hadoop@hc1nn ~]$ mkdir /tmp/ooziesharelib
[hadoop@hc1nn ~]$ cd /tmp/ooziesharelib
[hadoop@hc1nn ooziesharelib]$ tar xzf /usr/lib/oozie/oozie-sharelib-yarn.tar.gz

[hadoop@hc1nn ooziesharelib]$ ls -ld *
drwxr-xr-x. 3 oozie oozie 4096 May 29 06:59 share
```

Now, you use the Hadoop file system command put to copy the share directory onto HDFS under /user/Oozie workflow:

```
[oozie@hc1nn ooziesharelib]$ hdfs dfs -put share /user/oozie/share
```

It is quite simple to start the Oozie server by using the Linux service command as the root user. You use the Linux su command to switch the user to root, then start the Oozie service:

```
[hadoop@hc1nn ooziesharelib]$ su -
[root@hc1nn ~]$ service oozie start
[root@hc1nn ~]$ exit
```

Finally, you can use the Oozie client as the Linux hadoop user to access Oozie and check the server's status:

```
[hadoop@hc1nn ~]$ oozie admin -oozie http://localhost:11000/oozie -status
System mode: NORMAL
```

```
[hadoop@hc1nn ~]$ oozie admin -oozie http://localhost:11000/oozie -version
Oozie server build version: 3.3.2-cdh4.7.0
```

By setting the OOZIE_URL variable, you can simplify the Oozie client commands. The URL tells the Oozie client the location in terms of the host name and port of the Oozie server, as follows:

```
[hadoop@hc1nn ~]$ export OOZIE_URL=http://localhost:11000/oozie
[hadoop@hc1nn ~]$ oozie admin -version
Oozie server build version: 3.3.2-cdh4.7.0
```

At this point, you can access the Oozie web console via the URL http://localhost:11000/oozie. (I discuss this in more detail following the discussion of workflows in Oozie).

The Mechanics of the Oozie Workflow

In general, the *workflow* is a set of chained actions that call HDFS-based scripts like Pig and Hive. All input comes from HDFS, not from the Linux file system, because Oozie cannot guarantee which cluster nodes will be used to process the workflow. Created as an XML document, an Oozie workflow script contains a series of linked *actions* controlled via pass/fail *control nodes* that determine where the control flow moves next. The fork option, for example, allows actions to be run in parallel. You can configure the script to send notifications of the workflow outcome via email or output message, as well as set action parameters and add tool-specific actions like Pig, Hive, and Java to the workflow.

Oozie Workflow Control Nodes

The workflow control nodes are like traffic cops in a script, directing the flow of work. The start control node defines the starting point for the workflow. Each workflow script can have only one start node, and it must define an existing action.

```
<start to="pig-fork"/>
```

The end control node is also mandatory and indicates the end of the workflow. If the control flow reaches the end control node, it has finished sucessfully.

```
<end name="end"/>
```

Offering the ability to split the flow of work into a series of parallel streams, the fork and join control nodes are used as a pair. For instance, you can use these to run several Pig data-processing scripts in parallel. The join control node will not complete until all of the fork actions have completed.

```
<fork name="pig-fork">
  <path start="pig-manufacturer"/>
  <path start="pig-model"/>
</fork>
```

Optional in the workflow, the kill control node stops the workflow. It is useful for error conditions; if any actions are still running when an error occurs, they will be ended.

```
<kill name="fail">
  <message>Workflow died, error message[${wf:errorMessage(wf:lastErrorNode())}]</message>
</kill>
```

The decision control node uses a switch statement with a series of cases and a default option to decide which control flow to use. The first case to be True is used; otherwise, the default case is used. In the pig-decision switch statement the workflow uses a file system test of file size and will pass control to the end control node if the data required by the workflow is not greater than 1 GB.

```
<decision name="pig-decision">
  <switch>
    <case to="pig-fork">
      ${fs:fileSize("${hdfsRawData}) gt 1 * GB}
    </case>
    <default to="end"/>
  </switch>
</decision>
```

Oozie Workflow Actions

The best way to understand an Oozie workflow action is to examine an example. I present the following Pig Latin–based action to define the values of the JobTracker and name node. The prepare section deletes data for the action. The configuration section then defines the Pig action queue name, and it is followed by a script section that specifies the script to call. Finally, the OK and error options define which nodes to move to, depending on the outcome of the Pig script.

```
<action name="pig-manufacturer">
  <pig>
    <job-tracker>${jobTracker}</job-tracker>
    <name-node>${nameNode}</name-node>
    <prepare>
      <delete path="${hdfsEntityData}/manufacturer"/>
    </prepare>
    <configuration>
      <property>
        <name>mapred.job.queue.name</name>
        <value>${queueName}</value>
      </property>
    </configuration>
```

```
        <script>manufacturer.pig</script>
    </pig>
    <ok to="pig-join"/>
    <error to="fail"/>
</action>
```

Creating an Oozie Workflow

In this example, I examine and run a Pig- and Hive-based Oozie workflow against Oozie. The example uses a Canadian vehicle fuel-consumption data set that is provided at the website data.gc.ca. You can either search for "Fuel Consumption Ratings" to find the data set or use the link http://open.canada.ca/data/en/dataset/98f1a129-f628-4ce4-b24d-6f16bf24dd64.

To begin, I download the English version of each CSV file. For instance, I have downloaded these files using the Linux hadoop account, downloading them to that account's Downloads directory, as the Linux ls command shows:

```
[hadoop@hc1nn Downloads]$ ls

MY1995-1999 Fuel Consumption Ratings.csv  MY2007 Fuel Consumption Ratings.csv
MY2000 Fuel Consumption Ratings.csv       MY2008 Fuel Consumption Ratings.csv
MY2001 Fuel Consumption Ratings.csv       MY2009 Fuel Consumption Ratings.csv
MY2002 Fuel Consumption Ratings.csv       MY2010 Fuel Consumption Ratings.csv
MY2003 Fuel Consumption Ratings.csv       MY2011 Fuel Consumption Ratings.csv
MY2004 Fuel Consumption Ratings.csv       MY2012 Fuel Consumption Ratings.csv
MY2005 Fuel Consumption Ratings.csv       MY2013 Fuel Consumption Ratings.csv
MY2006 Fuel Consumption Ratings.csv       MY2014 Fuel Consumption Ratings.csv
```

I then need to copy these files to an HDFS directory so that they can be used by an Oozie workflow job. To do this, I create some HDFS directories, as follows:

```
[hadoop@hc1nn Downloads]$ hdfs dfs -mkdir /user/hadoop/oozie_wf
[hadoop@hc1nn Downloads]$ hdfs dfs -mkdir /user/hadoop/oozie_wf/fuel
[hadoop@hc1nn Downloads]$ hdfs dfs -mkdir /user/hadoop/oozie_wf/fuel/rawdata
[hadoop@hc1nn Downloads]$ hdfs dfs -mkdir /user/hadoop/oozie_wf/fuel/pigwf
[hadoop@hc1nn Downloads]$ hdfs dfs -mkdir /user/hadoop/oozie_wf/fuel/entity
[hadoop@hc1nn Downloads]$ hdfs dfs -mkdir /user/hadoop/oozie_wf/fuel/entity/manufacturer
[hadoop@hc1nn Downloads]$ hdfs dfs -mkdir /user/hadoop/oozie_wf/fuel/entity/model
```

The Hadoop file system ls command produces a long list that shows the three HDFS subdirectories I've just created and that will be used in this example.

```
[hadoop@hc1nn Downloads]$ hdfs dfs -ls /user/hadoop/oozie_wf/fuel/
Found 3 items
drwxr-xr-x   - hadoop hadoop          0 2014-07-12 18:16 /user/hadoop/oozie_wf/fuel/entity
drwxr-xr-x   - hadoop hadoop          0 2014-07-12 18:15 /user/hadoop/oozie_wf/fuel/pigwf
drwxr-xr-x   - hadoop hadoop          0 2014-07-08 18:16 /user/hadoop/oozie_wf/fuel/rawdata
```

I employ the rawdata directory under /user/hadoop/oozie_wf/fuel/ on HDFS to contain the CSV data that I will use. I use the pigwf directory to contain the scripts for the task. I use the entity directory and its subdirectories to contain the data used by this task.

So, my next step is to upload the CSV files from the Linux file system Downloads directory to the HDFS directory rawdata:

```
[hadoop@hc1nn Downloads]$ hdfs dfs –copyFromLocal  *.csv  /user/hadoop/oozie_wf/fuel/rawdata
```

Now, the workflow data are ready, and the scripts and configuration files that the workflow will use need to be copied into place. For this example, I have created all the necessary files. To begin, I load them to the HDFS pigwf directory using the Hadoop file system copyFromLocal command:

```
[hadoop@hc1nn Downloads]$ cd /home/hadoop/oozie/pig/fuel
[hadoop@hc1nn fuel]$ ls

load.job.properties  model.pig       manufacturer.pig    model.sql       workflow.xml
manufacturer.sql

[hadoop@hc1nn fuel]$ hdfs dfs –copyFromLocal   * /user/hadoop/oozie_wf/fuel/pigwf
```

Next, using the Hadoop file system ls command, I check the contents of the pigwf directory. The listing shows the sizes of the files that were just uploaded:

```
[oozie@hc1nn fuel]$ hdfs dfs -ls /user/oozie/oozie_wf/fuel/pigwf/
Found 6 items
-rw-r--r--   2 oozie oozie      542 2014-07-06 15:48 /user/oozie/oozie_wf/fuel/pigwf/load.job.properties
-rw-r--r--   2 oozie oozie      567 2014-07-08 19:13 /user/oozie/oozie_wf/fuel/pigwf/manufacturer.pig
-rw-r--r--   2 oozie oozie      306 2014-07-12 18:06 /user/oozie/oozie_wf/fuel/pigwf/manufacturer.sql
-rw-r--r--   2 oozie oozie      546 2014-07-08 19:13 /user/oozie/oozie_wf/fuel/pigwf/model.pig
-rw-r--r--   2 oozie oozie      283 2014-07-12 18:06 /user/oozie/oozie_wf/fuel/pigwf/model.sql
-rw-r--r--   2 oozie oozie     2400 2014-07-12 18:15 /user/oozie/oozie_wf/fuel/pigwf/workflow.xml
```

Note that I actually don't need to copy the load.job.properties file to HDFS, as it will be located from the local Linux file system. Having uploaded the files, it is time to explain their contents.

The Workflow Configuration File

The first file is the workflow configuration file, called load.job.properties; this specifies parameters for the workflow. I have listed its contents using the Hadoop file system cat command and have taken the liberty of adding line numbers here and elsewhere to use in explaining the steps:

```
[hadoop@hc1nn fuel]$ hdfs dfs -cat /user/hadoop/oozie_wf/fuel/pigwf/load.job.properties

01   # ---------------------------------------
02   # Workflow job properties
03   # ---------------------------------------
04
05   nameNode=hdfs://hc1nn:8020
06
07   # Yarn resource manager host and port
08   jobTracker=hc1nn:8032
09   queueName=high_pool
10
```

```
11   oozie.libpath=${nameNode}/user/hadoop/share/lib
12   oozie.use.system.libpath=true
13   oozie.wf.rerun.failnodes=true
14
15   hdfsUser=hadoop
16   wfProject=fuel
17   hdfsWfHome=${nameNode}/user/${hdfsUser}/oozie_wf/${wfProject}
18   hdfsRawData=${hdfsWfHome}/rawdata
19   hdfsEntityData=${hdfsWfHome}/entity
20
21   oozie.wf.application.path=${hdfsWfHome}/pigwf
22   oozieWfPath=${hdfsWfHome}/pigwf/
```

The parameters in this file specify the Hadoop name node by server and port. Because YARN is being employed, the Resource Manager is defined via its host and port by using the JobTracker variable. Job Tracker is obviously a Hadoop V1 component name, but this works for YARN. The queue name to be used for this workflow, high_pool, is also specified.

The library path of the Oozie shared library is defined by oozie.libpath, along with the parameter oozie.use.system.libpath. The HDFS user for the job is specified, as is a project name. Finally, the paths are defined for the workflow scripts and entity data that will be produced. The special variable oozie.wf.application.path is used to define the location of the workflow job file.

The workflow.txt file is the main control file for the workflow job. It controls the flow of actions, via Oozie, and manages the subtasks. This workflow file runs two parallel streams of processing to process the data in the HDFS rawdata directory.

The manufacturer.pig script is called to strip manufacturer-based data from the HDFS-based rawdata files. This data is placed in the HDFS-based entity/manufacturer directory. Then the script manufacturer.sql is called to process this data to the Hive data warehouse.

In parallel to this (via a fork option in the xml), the model.pig script is called to strip the vehicle model–based data from the HDFS rawdata files. This data is placed in the HDFS entity/model directory. Then the script model.sql is called to process this data to the Hive data warehouse.

The workflow.xml workflow file has been built using a combination of the workflow elements described earlier (see "The Mechanics of the Oozie Workflow"). I have used the Hadoop file system cat command to display its contents:

```
[hadoop@hc1nn fuel]$ hdfs dfs -cat /user/hadoop/oozie_wf/fuel/pigwf/workflow.xml
```

```
01      <workflow-app name="FuelWorkFlow" xmlns="uri:Oozie workflow:workflow:0.1">
02
03          <start to="pig-fork"/>
04
05          <fork name="pig-fork">
06            <path start="pig-manufacturer"/>
07            <path start="pig-model"/>
08          </fork>
09
10          <action name="pig-manufacturer">
11            <pig>
12              <job-tracker>${jobTracker}</job-tracker>
13              <name-node>${nameNode}</name-node>
14              <prepare>
15                <delete path="${hdfsEntityData}/manufacturer"/>
16              </prepare>
```

```
17          <configuration>
18            <property>
19              <name>mapred.job.queue.name</name>
20              <value>${queueName}</value>
21            </property>
22          </configuration>
23          <script>manufacturer.pig</script>
24        </pig>
25        <ok to="pig-join"/>
26        <error to="fail"/>
27      </action>
28
29      <action name="pig-model">
30        <pig>
31          <job-tracker>${jobTracker}</job-tracker>
32          <name-node>${nameNode}</name-node>
33          <prepare>
34            <delete path="${hdfsEntityData}/model"/>
35          </prepare>
36          <configuration>
37            <property>
38              <name>mapred.job.queue.name</name>
39              <value>${queueName}</value>
40            </property>
41          </configuration>
42          <script>model.pig</script>
43        </pig>
44        <ok to="pig-join"/>
45        <error to="fail"/>
46      </action>
47
48      <join name="pig-join" to="hive-fork"/>
49
50      <fork name="hive-fork">
51        <path start="hive-manufacturer"/>
52        <path start="hive-model"/>
53      </fork>
54
55      <action name="hive-manufacturer">
56        <hive xmlns="uri:Oozie workflow:hive-action:0.2">
57          <job-tracker>${jobTracker}</job-tracker>
58          <name-node>${nameNode}</name-node>
59          <configuration>
60            <property>
61              <name>mapred.job.queue.name</name>
62              <value>${queueName}</value>
63            </property>
64          </configuration>
65          <script>model.sql</script>
66        </hive>
67        <ok to="hive-join"/>
```

```
68            <error to="fail"/>
69        </action>
70
71        <action name="hive-model">
72          <hive xmlns="uri:Oozie workflow:hive-action:0.2">
73            <job-tracker>${jobTracker}</job-tracker>
74            <name-node>${nameNode}</name-node>
75            <configuration>
76              <property>
77                <name>mapred.job.queue.name</name>
78                <value>${queueName}</value>
79              </property>
80            </configuration>
81            <script>model.sql</script>
82          </hive>
83          <ok to="hive-join"/>
84          <error to="fail"/>
85        </action>
86
87        <join name="hive-join" to="end"/>
88
89        <kill name="fail">
90          <message>Workflow died, error message[${wf:errorMessage(wf:lastErrorNode())}]</message>
91        </kill>
92
93        <end name="end"/>
94
95    </workflow-app>
```

The workflow uses a fork control node at line 05 to run the Pig manufacturer and model jobs in parallel. The join control is issued at line 48 when both Pig jobs have finished. The Pig actions at lines 10 and 29 are exactly the same as the example earlier (see "Oozie Workflow Actions"). They set up the actions by defining the name node and the Job Tracker, and then they prepare the job by deleting any previous data. They define the Hadoop-based queue that the job should be sent to, and finally they define the script to be called.

The Hive jobs use the same fork and join structure as the Pig jobs at lines 50 and 87. The Pig join control node at line 48 has a to element that passes control to the Hive fork control node at line 50.

```
48        <join name="pig-join" to="hive-fork"/>
49
50        <fork name="hive-fork">
51          <path start="hive-manufacturer"/>
52          <path start="hive-model"/>
53        </fork>
```

The workflow definition of the Hive actions are similar to the Pig actions. The only major difference is that the <Hive> label is used and that there is no prepare section to delete data.

Each of the four actions has a set of error conditions that determine where control will be passed to:

```
83          <ok to="hive-join"/>
84          <error to="fail"/>
85        </action>
86
```

```
87      <join name="hive-join" to="end"/>
88
89      <kill name="fail">
90        <message>Workflow died, error message[${wf:errorMessage(wf:lastErrorNode())}]</message>
91      </kill>
92
93      <end name="end"/>
```

An error condition in any of the four main actions passes control to the kill control node, called fail. The OK condition just passes the control flow to the next success node in the workflow.

Now I will briefly explain the manufacturer and model pig and sql script contents. I have used the Hadoop file system cat command to display the contents of the manufacturuer.pig file.

```
[hadoop@hc1nn fuel]$ hdfs dfs -cat /user/hadoop/oozie_wf/fuel/pigwf/manufacturer.pig
```

```
01  -- get the raw data from the files from the csv files
02
03  rlines = LOAD '/user/hadoop/oozie_wf/fuel/rawdata/*.csv' USING PigStorage(',') AS
04    ( year:int, manufacturer:chararray, model:chararray, class:chararray, size:float,
      cylinders:int,
05        transmission:chararray, fuel:chararray, cons_cityl100:float, cond_hwyl100:float, cons_
          citympgs:int,
06        cond_hwympgs:int, lyears:int, co2s:int
07    );
08
09  mlist = FOREACH rlines GENERATE manufacturer;
10
11  dlist = DISTINCT mlist ;
12
13  -- save to a new file
14
15  STORE dlist INTO '/user/hadoop/oozie_wf/fuel/entity/manufacturer/' ;
```

The pig script is just stripping the manufacturer information from the rawdata CSV files and storing that data in the HDFS directory under entity/manufacturer. The sql script called manufacturuer.sql then processes that information and stores it in Hive.

```
[hadoop@hc1nn fuel]$ hdfs dfs -cat /user/hadoop/oozie_wf/fuel/pigwf/manufacturer.sql
```

```
01  drop table if exists rawdata2 ;
02
03  create external table rawdata2 (
04    line string
05  )
06  location '/user/hadoop/oozie_wf/fuel/entity/manufacturer/' ;
07
08  drop table if exists manufacturer ;
09
10  create table manufacturer as
11    select distinct line from rawdata2 where line not like '%=%'
12    and line not like '% % %' ;
```

The sql just creates an external Hive table called rawdata2 from the manufacturer HDFS-based files. It then creates a second table in Hive called "manufacturer" by selecting the contents of the rawdata2 table.

The model.pig and sql files are very similar, pulling vehicle model data from the HDFS-based rawdata files and moving it to HDFS. I use the Hadoop file system cat command to display the model.pig file:

```
[hadoop@hc1nn fuel]$ hdfs dfs -cat /user/hadoop/oozie_wf/fuel/pigwf/model.pig

01  -- get the raw data from the files from the csv files
02
03  rlines = LOAD '/user/hadoop/oozie_wf/fuel/rawdata/*.csv' USING PigStorage(',') AS
04    ( year:int, manufacturer:chararray, model:chararray, class:chararray, size:float,
      cylinders:int,
05      transmission:chararray, fuel:chararray, cons_cityl100:float, cond_hwyl100:float, cons_
        citympgs:int,
06      cond_hwympgs:int, lyears:int, co2s:int
07    );
08
09  mlist = FOREACH rlines GENERATE manufacturer,year,model ;
10
11  dlist = DISTINCT mlist ;
12
13  STORE dlist INTO '/user/hadoop/oozie_wf/fuel/entity/model/' using PigStorage(',');
```

Again, it strips vehicle model information from the HDFS-based CSV files in the rawdata directory. It then stores that information in the entity/model HDFS directory. The model.sql script then processes that information to a Hive table:

```
[hadoop@hc1nn fuel]$ hdfs dfs -cat /user/hadoop/oozie_wf/fuel/pigwf/model.sql

01  drop table if exists rawdata2 ;
02
03  create external table rawdata2 (
04    line string
05  )
06  location '/user/hadoop/oozie_wf/fuel/entity/model/' ;
07
08  drop table if exists model ;
09
10  create table model as
11    select
12      distinct split(line,',')
13    from rawdata2
14    where
15      line not like '%=%' ;
```

The Hive QL script creates an external table over the HDFS-based entity/model data, called rawdata2; it then selects that data into a Hive-based table called "model."

The intention of this workflow example is to show that complex ETL ("extract, transform, load") chains of subtasks can be built using Oozie. The tasks can be run in parallel and control can be added to the workflow to set up the jobs and define the end conditions. Having described the workflow, it is now time for me to run the job; the next section explains how the workflow can be run and monitored with the Oozie web-based user interface.

Running an Oozie Workflow

To make the Oozie task invocation simpler, I define the OOZIE_URL variable. This means that when I want to invoke an Oozie job, I have simply invoke the oozie command; I do not have to specify the whole Oozie URL. For the Oozie example, this is as follows:

```
[hadoop@hc1nn fuel]$ export OOZIE_URL=http://localhost:11000/oozie
```

The oozie command is then used with a job parameter and the load.job.properties file. The –submit parameter is used to submit the file, and this returns the Oozie job number 0000000-140706152445409-oozie-oozi-W:

```
oozie job  -config ./load.job.properties  -submit
job: 0000000-140706152445409-oozie-oozi-W
```

The workflow can now be started using this Oozie job number. I issue the oozie command with the parameters job and –start, followed by the job number:

```
oozie job  -start  0000000-140706152445409-oozie-oozi-W
```

I can pause and restart the job with the -suspend and -resume oozie command options. The help option passed to the oozie command gives me a full list of the possible options.

I now access the Oozie console (shown in Figure 5-5) by using the OOZIE_URL value I specified previously.

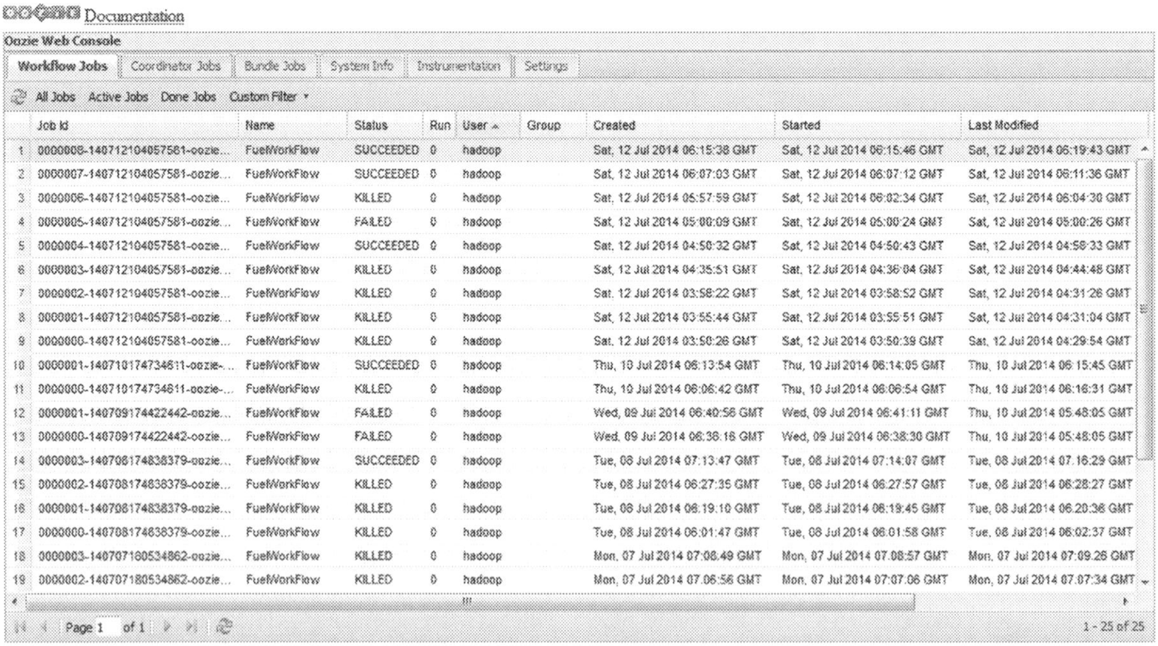

Figure 5-5. *Oozie console job list*

The left column of Figure 5-5 gives a list of job IDs that are issued when each Oozie job is submitted. The next column to the right shows the workflow name—in this case, FuelWorkFlow. The Status column shows the job's last status, which could be Prep, Running, Suspended, Succeeded, Failed, or Killed. Table 5-1 lists the meanings of these statuses.

Table 5-1. *Job Statuses and Meanings*

Status	Meaning
PREP	Job just created; it can be moved to Running or Killed.
RUNNING	Job is being executed; its state can change to Suspended, Succeeded, Killed, or Failed.
SUSPENDED	A running job can be suspended; its state can change to Running or Killed.
SUCCEEDED	The job sucessfully reached its end state.
FAILED	The job encountered an unexpected error.
KILLED	A job action failed or the job was killed by an administrator.

The User and date columns show which Linux user account ran the job and when it was created, started, and modified. The tabs at the top of the display are for workflows, coordinator jobs, and bundling. Coordinator jobs are for scheduling, either via time or event, while bundling allows the grouping of coordinator jobs. The other tabs are for configuration information and metrics.

By selecting an individual workflow job, you can bring up a job-related Oozie window. In Figure 5-6, I have selected the topmost successful job on the list. As you can see, the Job Info tab shows the job's attributes and its actions and status details. The Job Definition tab contains the xml contents of the workflow, while the Job Configuration tab shows configured attributes for the job. The Job Log tab shows the logged output for this job, as Figure 5-7 shows.

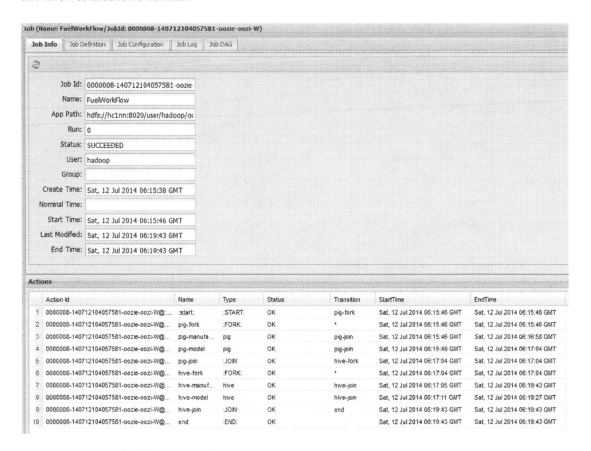

Figure 5-6. *Oozie job information window*

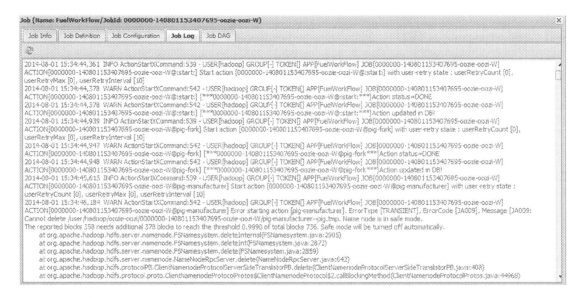

Figure 5-7. *Oozie job log information window*

The Job DAG (directed acyclical graph) tab is interesting because it shows both the structure and the state of the job's actions. Figures 5-8 and 5-9 provide examples of Oozie DAGs. Figure 5-8 has no state or colors shown on it; it is a job that has been submitted but not run. However, the DAG contains enough detail so that you can read the labels. It shows the pig-manufacturer and pig-model tasks running in parallel. The triangular pig-fork and pig-join nodes split the streams of processing. The Hive-based functionality is organized in the same manner, with the manufacturer and model Hive tasks running in parallel. The workflow begins with a start node and terminates at an end node.

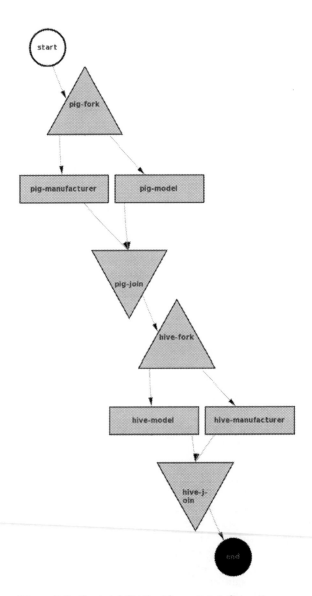

Figure 5-8. *Oozie job DAG with no state information*

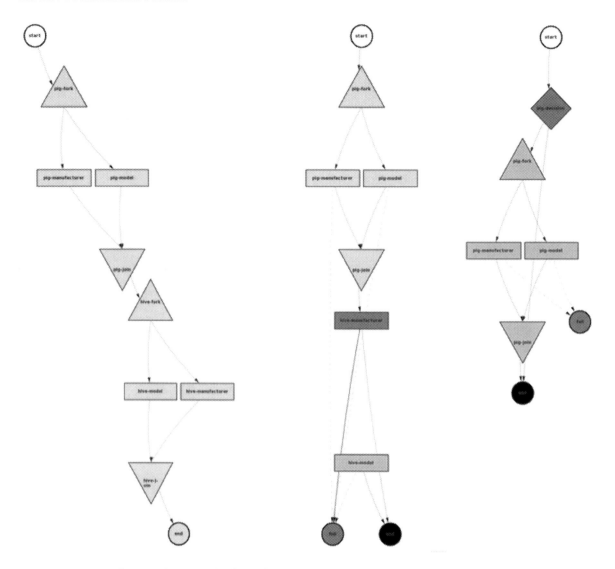

Figure 5-9. *Oozie job DAG information for three jobs*

Figure 5-9 shows three different DAGs. The text is illegible because the figures have been minimized to fit together on a single page, but you can see the different states that nodes in a DAG can achieve.

For instance, in Figure 5-9 the circles show the start and endpoints, while the triangles show the forks and joins. The boxes show the actions to be executed, while the actions between a fork and join show that they will be executed in parallel. The color codes indicate the execution states: gray means that a node has not been executed, green means that that node has executed sucessfully, red means that an error has occurred, and black is the unreached endpoint. That is, the DAG on the left completed successfully and the two on the right failed.

Scheduling an Oozie Workflow

How do you schedule a workflow to run at a specific time or run after a given event? For instance, you might want the workflow to run at 01:00 each Tuesday morning or each time data arrives. Oozie coordinator jobs exist for this purpose. To continue my example, I updated the workflow properties file to use a coordinator job:

```
oozieWfPath=${hdfsWfHome}/pigwf

# Job Coordination properties

jobStart=2014-07-10T12:00Z
jobEnd=2014-09-10T12:00Z

# Frequency in minutes

JobFreq=10080
jobNZTimeZone=GMT+1200

oozie.coord.application.path=${hdfsWfHome}/pigwf
```

The path to the workflow script is now called oozieWfPath, and the path to the coordinator script is called oozie.coord.application.path, the latter which is the reserved pathname that Oozie expects will be used to identify a cordinator job. I also specify some time-based parameters to the cordinator job, a start time, an end time, and a job frequency in minutes. Lastly, I set the time zone for New Zealand.

I create an XML-based coordinator job file called coordinator.xml, which I copy to the workflow directory in HDFS. The file looks like this:

```
1    <coordinator-app
2
3      name="FuelWorkFlowCoord"
4      frequency="${JobFreq}"
5      start="${jobStart}"
6      end="${jobEnd}"
7      timezone="${jobNZTimeZone}"
8      xmlns="uri:Oozie workflow:coordinator:0.4">
9
10     <action>
11       <workflow>
12         <app-path>${oozieWfPath}/workflow.xml</app-path>
13       </workflow>
14     </action>
15
16   </coordinator-app>
```

This is a time-based coordinator job that will run between the start and end dates for a given frequency using New Zealand time.

I send the coordinator job to Oozie as follows:

```
oozie job  -config ./load.job.properties  -submit
job: 0000000-140713100519754-oozie-oozi-C
```

A coordinator job appears on the Oozie console coordinator tab, as shown in Figure 5-10, indicating that the coordinator job has started and its frequency is 10,080 minutes (24 hours times 60 minutes times 7 days).

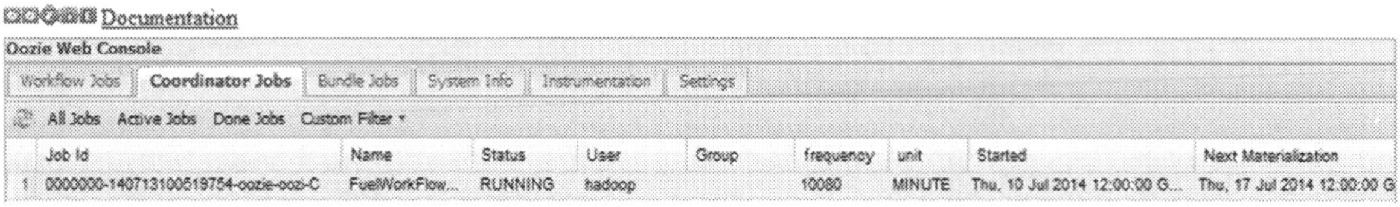

Figure 5-10. *Ozzie coordinator job status*

So, as set, this job will run weekly at this time until its end date. The last column in Figure 5-10 also shows the next time that this job will run, which is a week from the start date. Clicking on the job ID provides a window with its details, as shown in Figure 5-11.

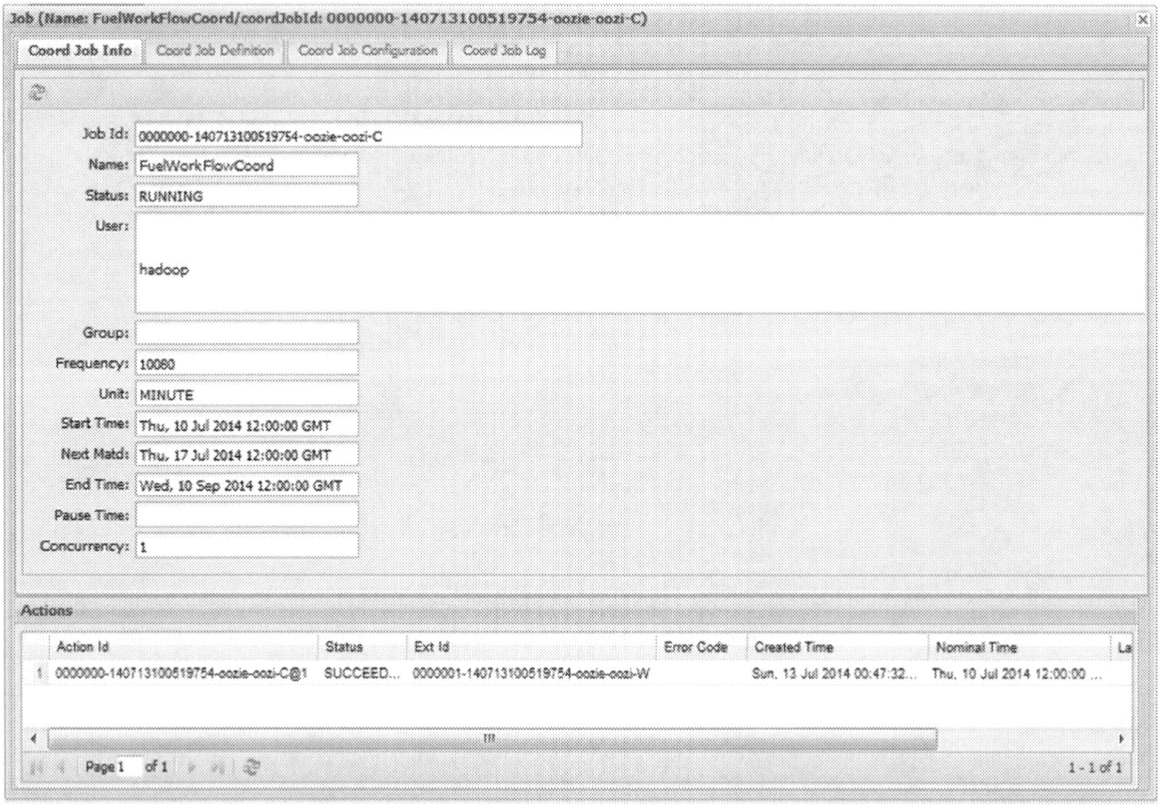

Figure 5-11. *Ozzie coordinator job details window*

The Definition tab in the Figure 5-11 window contains the contents of the job's coordinator.xml file. The Configuration tab contains all of the job's parameters in XML from the configuration file. And the Log tab contains the job's log entries.

As an extension to this coordinator job, I add a dataset requirement and create a new data set variable, as follows:

```
DataJobFreq=1440
```

And the XML now looks like this:

```
 1     <coordinator-app
 2
 3       name="FuelWorkFlowCoord"
 4       frequency="${JobFreq}"
 5       start="${jobStart}"
 6       end="${jobEnd}"
 7       timezone="${jobNZTimeZone}"
 8       xmlns="uri:Oozie workflow:coordinator:0.4">
 9
10       <datasets>
11         <dataset
12
13           name="vehicle"
14           frequency="${DataJobFreq}"
15           initial-instance="${jobStart}"
16           timezone="${jobNZTimeZone}">
17
18           <uri-template>${hdfsRawData}/${YEAR}_${MONTH}_${DAY}_Fuel_Consumption</uri-template>
19         </dataset>
20       </datasets>
21
22       <action>
23         <workflow>
24           <app-path>${oozieWfPath}/workflow.xml</app-path>
25         </workflow>
26       </action>
27
28     </coordinator-app>
```

The data set requirement is added between lines 10 and 20, using the hdfsPathRawData variable from the configuration file and the predefined YEAR, MONTH, and DAY variables. So, if the file for July 13, 2014 (${hdfsRawData}/2014_07_13_Fuel_Consumption) is not available in the rawdata directory, then the job will not run.

This section has provided a brief introduction, via examples, for you to sample Oozie. For a full definition of workflow specification, you should check the Oozie website at oozie.apache.org. Choose the documentation level that matches your Oozie installation; there is a detailed specification there for workflow and coordinator jobs. You could also investigate bundler jobs, which allow you to group coordinator jobs.

Summary

This chapter presented Hadoop-based schedulers and discussed their use for Hadoop V1 and V2. Remember that each scheduler type is meant for a different scenario. The Capacity scheduler enables multiple tenants to share a cluster of resources, while the Fair scheduler enables multiple projects for a single tenant to share a cluster. The aim overall is to share cluster resources appropriately. Keep checking the Hadoop website (`hadoop.apache.org`) for version updates applicable to the scheduling function.

While these schedulers allow the sharing of resources, tools like Oozie offer the ability to schedule jobs that are organized into workflows by time and event. Using an example, this chapter has shown how to create a workflow and how to schedule it. Additionally, the Oozie console was used to examine the job output and status.

As a final suggestion, you might consider investigating workflow schedulers like Azkaban and Luigi as well to give you some idea of comparable functionality. Azkaban uses DAGs like Oozie, and it integrates with Hadoop components like Pig and Hive. Luigi is a simple workflow engine written in Python; at the time of this writing, it integrates with Hive but not with Pig.

## CHAPTER 6

Moving Data

The tools and methods you use to move big data within the Hadoop sphere depend on the type of data to be processed. This is a large category with many data sources, such as relational databases, log data, binary data, and realtime data, among others. This chapter focuses on a few common data types and discusses some of the tools you can use to process them. For instance, in this chapter you will learn to use Sqoop to process relational database data, Flume to process log data, and Storm to process stream data.

You will also learn how this software can be sourced, installed, and used. Finally, I will show how a sample data source can be processed and how all of these tools connect to Hadoop. But I begin with an explanation of the Hadoop file system commands.

Moving File System Data

You can use Hadoop file system commands to move file-based data into and out of HDFS. In all the examples in this book that employ Hadoop file system commands, I have used either a simple file name (myfile.txt) or a file name with a path (/tmp/myfile.txt). However, a file may also be defined as a Uniform Resource Identifier (URI). The URI contains the file's path, name, server, and a source identifier. For instance, for the two URIs that follow, the first file is on HDFS while the second is on the file system. They also show that the files in question are located on the server hc1nn:

```
hdfs://hc1nn/user/hadoop/oozie_wf/fuel/pigwf/manufacturer.pig
file://hc1nn/tmp/manufacturer.pig
```

To indicate the data's source and destination for the move, each command accepts one or more URIs. The Hadoop file system cat command (below) dumps the contents of the HDFS-based file manufacturer.pig to STDOUT (the standard out stream on Linux). The URI is the text in the line that starts at the string "hdfs" and ends with the file type (.pig):

```
hdfs dfs -cat hdfs://hc1nn/user/hadoop/oozie_wf/fuel/pigwf/manufacturer.pig
```

On the other hand, the cat command below dumps the Linux file system file flume_exec.sh to STDOUT (the standard out stream):

```
hdfs dfs -cat file:///home/hadoop/flume/flume_exec.sh
```

Although file or hdfs and the server name can be specified in the URI, they are optional. In this chapter I use only file names and paths.

Now, let's take a closer look at some of the most useful system commands.

The Cat Command

The Hadoop file system cat command copies the contents of the URIs presented to it to STDOUT.

```
hdfs dfs -cat hdfs://hc1nn/user/hadoop/oozie_wf/fuel/pigwf/manufacturer.pig
```

The cat command is useful if you want to run adhoc Linux-based commands against Hadoop-based data. For instance, the following:

```
hdfs dfs -cat hdfs://hc1nn/user/hadoop/oozie_wf/fuel/pigwf/manufacturer.pig | wc -l
```

would give you a line count of this file using the Linux command wc (word count), with a -l switch for the number of lines.

The CopyFromLocal Command

The source for the CopyFromLocal command is the local Linux file system; this command copies files from the local Linux file system to Hadoop.

```
[hadoop@hc1nn flume]$ hdfs dfs -copyFromLocal /home/hadoop/flume /tmp/flume
[hadoop@hc1nn flume]$ hdfs dfs -ls /tmp/flume
Found 6 items
-rw-r--r--   2 hadoop hadoop       1343 2014-07-26 20:09 /tmp/flume/agent1.cfg
-rw-r--r--   2 hadoop hadoop       1483 2014-07-26 20:09 /tmp/flume/agent1.cfg.nl
-rw-r--r--   2 hadoop hadoop         45 2014-07-26 20:09 /tmp/flume/flume_clean_hdfs.sh
-rw-r--r--   2 hadoop hadoop        197 2014-07-26 20:09 /tmp/flume/flume_exec.sh
-rw-r--r--   2 hadoop hadoop        233 2014-07-26 20:09 /tmp/flume/flume_exec.sh.nl
-rw-r--r--   2 hadoop hadoop         42 2014-07-26 20:09 /tmp/flume/flume_show_hdfs.sh
```

In this example, the contents of the Linux file system directory /home/hadoop/flume have been copied to HDFS under /tmp/flume.

The CopyToLocal Command

For the CopyToLocal command, the destination directory must be the local Linux file system. This command copies files from HDFS to the Linux file system:

```
[hadoop@hc1nn flume]$  hdfs dfs -copyToLocal /tmp/flume /tmp/hdfscopy
[hadoop@hc1nn flume]$ ls -l /tmp/hdfscopy
total 24
-rwxr-xr-x. 1 hadoop hadoop 1343 Jul 26 20:13 agent1.cfg
-rwxr-xr-x. 1 hadoop hadoop 1483 Jul 26 20:13 agent1.cfg.nl
-rwxr-xr-x. 1 hadoop hadoop   45 Jul 26 20:13 flume_clean_hdfs.sh
-rwxr-xr-x. 1 hadoop hadoop  197 Jul 26 20:13 flume_exec.sh
-rwxr-xr-x. 1 hadoop hadoop  233 Jul 26 20:13 flume_exec.sh.nl
-rwxr-xr-x. 1 hadoop hadoop   42 Jul 26 20:13 flume_show_hdfs.sh
```

The contents of the /tmp/hdfscopy directory on HDFS have been copied to the Linux file system directory /tmp/hdfscopy.

The Cp Command

The cp command can copy multiple sources to a destination. The sources and destination might be on HDFS or on the local file system. If multiple sources are specified, the destination must be a directory:

```
[hadoop@hc1nn flume]$  hdfs dfs -ls /tmp/flume/agent*
Found 1 items
-rw-r--r--   2 hadoop hadoop        1343 2014-07-26 20:09 /tmp/flume/agent1.cfg
Found 1 items
-rw-r--r--   2 hadoop hadoop        1483 2014-07-26 20:09 /tmp/flume/agent1.cfg.nl

[hadoop@hc1nn flume]$ hdfs dfs -cp /tmp/flume/agent1.cfg /tmp/flume/agent2.cfg

[hadoop@hc1nn flume]$  hdfs dfs -ls /tmp/flume/agent*
Found 1 items
-rw-r--r--   2 hadoop hadoop        1343 2014-07-26 20:09 /tmp/flume/agent1.cfg
Found 1 items
-rw-r--r--   2 hadoop hadoop        1483 2014-07-26 20:09 /tmp/flume/agent1.cfg.nl
Found 1 items
-rw-r--r--   2 hadoop hadoop        1343 2014-07-26 20:19 /tmp/flume/agent2.cfg
```

This example shows that an HDFS-based file agent1.cfg is being copied to another HDFS-based file agent2.cfg.

The Get Command

The get command copies HDFS-based files to the local Linux file system. The get command is similar to copyToLocal, except that copyToLocal must copy to a local Linux file system based file.

```
[hadoop@hc1nn tmp]$ hdfs dfs -get /tmp/flume/agent2.cfg
[hadoop@hc1nn tmp]$ ls -l ./agent2.cfg
-rwxr-xr-x. 1 hadoop hadoop 1343 Jul 26 20:23 ./agent2.cfg
```

This example copies the HDFS-based file agent2.cfg to the local Linux directory (".").

The Put Command

The put command copies a single file or multiple source files and writes them to a destination. This command can also read from STDIN (the input file stream). The put command is similar to copyFromLocal, except that copyFromLocal must copy from a local Linux file system based file.

```
[hadoop@hc1nn tmp]$ ps -ef | hdfs dfs -put - /tmp/ps/list.txt

[hadoop@hc1nn tmp]$  hdfs dfs -cat /tmp/ps/list.txt | head -10

UID        PID  PPID  C STIME TTY          TIME CMD
root         1     0  0 13:48 ?        00:00:01 /sbin/init
root         2     0  0 13:48 ?        00:00:00 [kthreadd]
root         3     2  0 13:48 ?        00:00:00 [migration/0]
root         4     2  0 13:48 ?        00:00:03 [ksoftirqd/0]
```

```
root          5     2  0 13:48 ?        00:00:00 [migration/0]
root          6     2  0 13:48 ?        00:00:00 [watchdog/0]
root          7     2  0 13:48 ?        00:00:00 [migration/1]
root          8     2  0 13:48 ?        00:00:00 [migration/1]
root          9     2  0 13:48 ?        00:00:02 [ksoftirqd/1]
```

This example takes input from STDIN, which is sourced from a full-process listing. The Hadoop file system put command places the contents in list.txt on HDFS. The Hadoop file system cat command then dumps the contents of the list.txt file and the Linux head command is used to limit the output to the first 10 lines.

The Mv Command

The mv command allows files to be moved from a source to a destination, but not across a file system:

```
[hadoop@hc1nn tmp]$ hdfs dfs -ls /tmp/flume/agent*
Found 1 items
-rw-r--r--   2 hadoop hadoop        1343 2014-07-26 20:09 /tmp/flume/agent1.cfg
Found 1 items
-rw-r--r--   2 hadoop hadoop        1483 2014-07-26 20:09 /tmp/flume/agent1.cfg.nl
Found 1 items
-rw-r--r--   2 hadoop hadoop        1343 2014-07-26 20:19 /tmp/flume/agent2.cfg


[hadoop@hc1nn tmp]$  hdfs dfs -mv /tmp/flume/agent2.cfg /tmp/flume/agent3.cfg


[hadoop@hc1nn tmp]$ hdfs dfs -ls /tmp/flume/agent*
Found 1 items
-rw-r--r--   2 hadoop hadoop        1343 2014-07-26 20:09 /tmp/flume/agent1.cfg
Found 1 items
-rw-r--r--   2 hadoop hadoop        1483 2014-07-26 20:09 /tmp/flume/agent1.cfg.nl
Found 1 items
-rw-r--r--   2 hadoop hadoop        1343 2014-07-26 20:19 /tmp/flume/agent3.cfg
```

This example shows that the HDFS file agent2.cfg has been moved to the file agent3.cfg.

The Tail Command

You can use the Hadoop file system tail command to dump the end of the file to STDOUT. Adding the -f switch (tail -f) enables you to continuously dump the contents of a file as it changes. The example that follows dumps the end of the HDFS-based file list.txt.

```
[hadoop@hc1nn tmp]$ hdfs dfs -tail /tmp/ps/list.txt

0 ?          00:00:00 pam: gdm-password
root       4494  1348  0 14:00 ?        00:00:00 sshd: hadoop [priv]
hadoop     4514  4494  0 14:01 ?        00:00:01 sshd: hadoop@pts/0
hadoop     4515  4514  0 14:01 pts/0    00:00:00 -bash
```

```
root      4539  1348  0 14:01 ?          00:00:00 sshd: hadoop [priv]
hadoop    4543  4539  0 14:01 ?          00:00:01 sshd: hadoop@pts/1
hadoop    4544  4543  0 14:01 pts/1      00:00:00 -bash
postfix  16083  1459  0 20:29 ?          00:00:00 pickup -l -t fifo -u
hadoop   16101  4544  0 20:31 pts/1      00:00:00 ps -ef
hadoop   16102  4544  0 20:31 pts/1      00:00:00 /usr/lib/jvm/java-1.6
```

As you can see, the system commands enable you to move data to and from Hadoop from the Linux file system, but what happens if data is in another source such as a database? For that, you'll need a tool like Sqoop.

Moving Data with Sqoop

You can use the Sqoop tool to move data into and out of relational databases, including Oracle, MySQL, PostgreSQL, and HSQLDB. Sqoop can place the data onto HDFS and from there move it into Hive, the Hadoop data warehouse. It provides the ability to incrementally load data and it supports many data formats—for example, CSV and Avro. It is integrated with such Hadoop-based tools as Hive, HBase, Oozie, Map Reduce, and HDFS. Sqoop is the popular default Hadoop-based tool of choice for moving this type of data.

In this section I demonstrate how to use Sqoop to import data from a MySQL database. Initially, I load the data onto HDFS, and then I add the step to load the data directly into a Hive table. But note: Before you start to work with Sqoop and Hadoop, always make sure that your database (which in my example is MySQL) is configured correctly on your Hadoop cluster. If it is not, you might get unexpected errors from the Sqoop job and low-level errors in your database might be masked.

▓ **Note**   For details on the specifics of Sqoop, refer to the Apache Sqoop website at `sqoop.apache.org`. Make sure you choose the documentation that matches the version you are using. (The examples in this chapter use version 1.4.3.)

Check the Database

To use Sqoop for this example, I have installed MySQL onto the Linux server hc1nn and MySQL clients onto each of the data nodes. That means that MySQL will be accessible on all servers in the cluster. Basically, I create a database called "sqoop" with a user called "sqoop" and a table called "rawdata." I place some data in the rawdata table, but the content is not important because whatever is there will be copied. Because I am concentrating on Sqoop, there's no need to describe the MySQL configuration or the data any further; there are plenty of sources on the web to describe the MySQL configuration, if you need more information.

Each data node that will access part of the Map Reduce-based Scoop job may try to access the MySQL database. So, before running the Sqoop task, I must be sure that MySQL (or whichever database you use) is accessible on each cluster node and that the test table can be accessed using the test user. To do this, I perform a test on each data node to ensure the data node can access the remote MySQL database.

For example, I begin the test with the line:

```
mysql --host=hc1nn --user=sqoop --password=xxxxxxxxxx
```

I have obscured the password above, but the command gives MySQL on the local node the username and proper password, as well as the hostname of the remote server. Once the mysql> prompt is available, I can try to get a row count from the rawdata table:

```
mysql> select count(*) from sqoop.rawdata;
+-----------+
| count(*)  |
+-----------+
|   20031   |
+-----------+
```

The results show that the rawdata table exists and is accessible by the sqoop database account, and that the table contains 20,031 rows of data for the import test. The data is just textual data contained in a table with a single column called "rawline," which contains the textual row data. I'm good to go.

░ **Note**   Any relational database access problems must be fixed before you proceed, given that MySQL is being used in this example. Further information can be found at the MySQL website at http://dev.mysql.com/doc/.

Install Sqoop

Given that the Cloudera stack was installed in Chapter 2, you can simply install Sqoop as the root user account on the server hc1nn as follows, using yum:

```
[root@hc1nn ~]# yum install sqoop
```

To check that the Sqoop installation was successful, you use the version option:

```
[hadoop@hc1nn conf]$ sqoop version

Warning: /usr/lib/hcatalog does not exist! HCatalog jobs will fail.
Please set $HCAT_HOME to the root of your HCatalog installation.
Warning: /usr/lib/sqoop/../accumulo does not exist! Accumulo imports will fail.
Please set $ACCUMULO_HOME to the root of your Accumulo installation.
14/07/17 18:41:17 INFO sqoop.Sqoop: Running Sqoop version: 1.4.3-cdh4.7.0
Sqoop 1.4.3-cdh4.7.0
git commit id 8e266e052e423af592871e2dfe09d54c03f6a0e8
Compiled by jenkins on Wed May 28 11:36:29 PDT 2014
```

Note that this check for my example yielded a couple of warnings: HCatalog and Accumulo (the database) are not installed. But as they are not used in the example, these warnings can be ignored.

In order to use MySQL, however, you must download and install a connector library for MySQL, as follows:

```
[root@hc1nn ~]# wget http://dev.mysql.com/get/Downloads/Connector-J/mysql-connector-java-5.1.22.tar.gz
```

The wget command downloads the tarred and compressed connector library file from the web address http://dev.mysql.com/get/Downloads/Connector-J/. As soon as the file is downloaded, you unzip and untar it, and then move it to the correct location so that it can be used by Sqoop:

```
[root@hc1nn ~]# ls -l mysql-connector-java-5.1.22.tar.gz
-rw-r--r--. 1 root root 4028047 Sep  6  2012 mysql-connector-java-5.1.22.tar.gz
```

This command shows the downloaded connector library, while the next commands show the file being unzipped using the gunzip command and unpacked using the tar command with the expand (x) and file (f) options:

```
[root@hc1nn ~]# gunzip mysql-connector-java-5.1.22.tar.gz
[root@hc1nn ~]# tar xf mysql-connector-java-5.1.22.tar

[root@hc1nn ~]# ls -lrt
total 9604
drwxr-xr-x. 4 root root    4096 Sep  6  2012 mysql-connector-java-5.1.22
-rw-r--r--. 1 root root 9809920 Sep  6  2012 mysql-connector-java-5.1.22.tar
```

Now, you copy the connector library to the /usr/lib/sqoop/lib directory so that it is available to Sqoop when it attempts to connect to a MySQL database:

```
[root@hc1nn ~]# cp mysql-connector-java-5.1.22/mysql-connector-java-5.1.22-bin.jar /usr/lib/sqoop/lib/
```

For this example installation, I use the Linux hadoop account. In that user's $HOME/.bashrc Bash shell configuration file, I have defined some Hadoop and Map Reduce variables, as follows:

```
#######################################################
# Set up Sqoop variables

# For each user who will be submitting MapReduce jobs using MapReduce v2 (YARN), or running
# Pig, Hive, or Sqoop in a YARN installation, set the HADOOP_MAPRED_HOME

export HADOOP_CONF_DIR=/etc/hadoop/conf
export HADOOP_COMMON_HOME=/usr/lib/hadoop
export HADOOP_HDFS_HOME=/usr/lib/hadoop-hdfs
export HADOOP_MAPRED_HOME=/usr/lib/hadoop-mapreduce
export YARN_HOME=/usr/lib/hadoop-yarn/
```

Use Sqoop to Import Data to HDFS

To import data from a database, you use the Sqoop import statement. For my MySQL database example, I use an options file containing the connection and access information. Because these details are held in a single file, this method requires less typing each time the task is repeated. The file that will be used to write table data to HDFS contains nine lines.

The import line tells Sqoop that data will be imported from the database to HDFS. The -- connect option with a connect string of jdbc:mysql://hc1nn/sqoop tells Sqoop that JDBC will be used to connect to a MySQL database on server hc1nn called "sqoop." I use the Linux cat command to show the contents of the Sqoop options file.

```
[hadoop@hc1nn sqoop]$ cat import.txt

1 import
2 --connect
3 jdbc:mysql://hc1nn/sqoop
4 --username
5 sqoop
6 --password
7 xxxxxxxxxx
8 --table
9 rawdata
```

The username and password options in the Sqoop options file provide account access to MySQL, while the table option shows that the table to be accessed is called "rawdata." Given that the Sqoop options file describes how to connect to the relational database that has been set up, the database itself should now be checked.

For MySQL on the server hc1nn to be accessed from all data nodes, the access must be granted within MySQL—otherwise, you will see errors like this:

```
14/07/19 20:20:28 ERROR manager.SqlManager: Error executing statement:
com.mysql.jdbc.exceptions.jdbc4.MySQLSyntaxErrorException: Access denied for user
  'sqoop'@'localhost' to database 'sqoop.rawdata'
```

To grant that access, you first log in as the root user. On the MySQL instance on the server hc1nn that contains the example database "sqoop," for instance, you log into the database as the root user with the line:

```
mysql  -u root -p
```

Then you grant access to the database access user (in this case, called "sqoop") on all servers, as follows:

```
GRANT ALL PRIVILEGES ON sqoop.rawdata to 'sqoop'@'hc1r1m1';
GRANT ALL PRIVILEGES ON sqoop.rawdata to 'sqoop'@'hc1r1m2';
GRANT ALL PRIVILEGES ON sqoop.rawdata to 'sqoop'@'hc1r1m3';
```

Also, you set password access for all the remote database users. (The actual passwords here have been crossed out, but the syntax to use is shown.)

```
SET PASSWORD FOR 'sqoop'@'hc1nn'   = PASSWORD('Xxxxxxxxxxxx');
SET PASSWORD FOR 'sqoop'@'hc1r1m1' = PASSWORD('Xxxxxxxxxxxx');
SET PASSWORD FOR 'sqoop'@'hc1r1m2' = PASSWORD('Xxxxxxxxxxxx');
SET PASSWORD FOR 'sqoop'@'hc1r1m3' = PASSWORD('Xxxxxxxxxxxx');
```

Finally, you flush the privileges in MySQL in order to make the changes take effect:

```
FLUSH PRIVILEGES;
```

Now that the options file has been created, and the MySQL database access has been checked, it is time to attempt to use Sqoop. The Sqoop command that executes the import by using the options file is as follows:

```
sqoop --options-file ./import.txt --table sqoop.rawdata -m 1
```

Make sure that you specify the -m option (as shown above) to perform a sequential import, or you will encounter an error like the following. If such an error occurs, just correct your sqoop command and try again.

```
14/07/19 20:29:46 ERROR tool.ImportTool: Error during import: No primary key could be
found for table rawdata. Please specify one with --split-by or perform a sequential
import with '-m 1'.
```

Another common error you might see is this one:

```
14/07/19 20:31:19 INFO mapreduce.Job: Task Id : attempt_1405724116293_0001_m_000000_0,
Status: FAILED
Error: java.lang.RuntimeException: java.lang.RuntimeException: com.mysql.jdbc.exceptions.jdbc4.
CommunicationsException: Communications link failure
```

This error may mean that MySQL access is not working. Check that you can log in to MySQL on each node and that the database on the test node (in this case, hc1nn) can be accessed as was tested earlier, in the section "Check the Database."

By default, the Sqoop import will attempt to install the data in the directory /user/hadoop/rawdata on HDFS. Before running the import command, though, make sure that the directory does not exist. This is done by using the HDFS file system -rm option with the -r recursive switch:

```
[hadoop@hc1nn sqoop]$  hdfs dfs  -rm -r /user/hadoop/rawdata
Moved: 'hdfs://hc1nn/user/hadoop/rawdata' to trash at: hdfs://hc1nn/user/hadoop/.Trash/Current
```

If the directory already exists, you will see an error like this:

```
14/07/20 11:33:51 ERROR tool.ImportTool: Encountered IOException running import job:
org.apache.hadoop.mapred.FileAlreadyExistsException: Output directory
hdfs://hc1nn/user/hadoop/rawdata already exists
```

So, to run the Sqoop import job, you use the following command:

```
[hadoop@hc1nn sqoop]$ sqoop --options-file ./import.txt --table sqoop.rawdata -m 1
```

The output will then look like this:

```
Please set $HCAT_HOME to the root of your HCatalog installation.
Please set $ACCUMULO_HOME to the root of your Accumulo installation.
14/07/20 11:35:28 INFO sqoop.Sqoop: Running Sqoop version: 1.4.3-cdh4.7.0
14/07/20 11:35:28 INFO manager.MySQLManager: Preparing to use a MySQL streaming resultset.
14/07/20 11:35:28 INFO tool.CodeGenTool: Beginning code generation
14/07/20 11:35:29 INFO manager.SqlManager: Executing SQL statement: SELECT t.* FROM `rawdata`
AS t LIMIT 1
14/07/20 11:35:29 INFO manager.SqlManager: Executing SQL statement: SELECT t.* FROM `rawdata`
AS t LIMIT 1
14/07/20 11:35:29 INFO orm.CompilationManager: HADOOP_MAPRED_HOME is /usr/lib/hadoop-mapreduce
14/07/20 11:35:31 INFO orm.CompilationManager: Writing jar file: /tmp/sqoop-hadoop/compile/
647e8646a006f6e95b0582fca9ccf4ca/rawdata.jar
14/07/20 11:35:31 WARN manager.MySQLManager: It looks like you are importing from mysql.
14/07/20 11:35:31 WARN manager.MySQLManager: This transfer can be faster! Use the --direct
14/07/20 11:35:31 WARN manager.MySQLManager: option to exercise a MySQL-specific fast path.
14/07/20 11:35:31 INFO manager.MySQLManager: Setting zero DATETIME behavior to convertToNull (mysql)
```

```
14/07/20 11:35:31 INFO mapreduce.ImportJobBase: Beginning import of rawdata
14/07/20 11:35:33 INFO service.AbstractService: Service:org.apache.hadoop.yarn.client.
YarnClientImpl is inited.
14/07/20 11:35:33 INFO service.AbstractService: Service:org.apache.hadoop.yarn.client.
YarnClientImpl is started.
14/07/20 11:35:37 INFO mapreduce.JobSubmitter: number of splits:1
14/07/20 11:35:37 INFO mapreduce.JobSubmitter: Submitting tokens for job: job_1405804878984_0001
14/07/20 11:35:38 INFO client.YarnClientImpl: Submitted application application_1405804878984_0001
to ResourceManager at hc1nn/192.168.1.107:8032
14/07/20 11:35:38 INFO mapreduce.Job: The url to track the job: http://hc1nn:8088/proxy/
application_1405804878984_0001/
14/07/20 11:35:38 INFO mapreduce.Job: Running job: job_1405804878984_0001
14/07/20 11:35:55 INFO mapreduce.Job: Job job_1405804878984_0001 running in uber mode : false
14/07/20 11:35:55 INFO mapreduce.Job:  map 0% reduce 0%
14/07/20 11:36:14 INFO mapreduce.Job:  map 100% reduce 0%
14/07/20 11:36:14 INFO mapreduce.Job: Job job_1405804878984_0001 completed successfully
14/07/20 11:36:14 INFO mapreduce.Job: Counters: 27
        File System Counters
                FILE: Number of bytes read=0
                FILE: Number of bytes written=92714
                FILE: Number of read operations=0
                FILE: Number of large read operations=0
                FILE: Number of write operations=0
                HDFS: Number of bytes read=87
                HDFS: Number of bytes written=1427076
                HDFS: Number of read operations=4
                HDFS: Number of large read operations=0
                HDFS: Number of write operations=2
        Job Counters
                Launched map tasks=1
                Other local map tasks=1
                Total time spent by all maps in occupied slots (ms)=16797
                Total time spent by all reduces in occupied slots (ms)=0
        Map-Reduce Framework
                Map input records=20031
                Map output records=20031
                Input split bytes=87
                Spilled Records=0
                Failed Shuffles=0
                Merged Map outputs=0
                GC time elapsed (ms)=100
                CPU time spent (ms)=3380
                Physical memory (bytes) snapshot=104140800
                Virtual memory (bytes) snapshot=823398400
                Total committed heap usage (bytes)=43712512
        File Input Format Counters
                Bytes Read=0
        File Output Format Counters
                Bytes Written=1427076
14/07/20 11:36:14 INFO mapreduce.ImportJobBase: Transferred 1.361 MB in 41.6577 seconds (33.4543 KB/sec)
14/07/20 11:36:14 INFO mapreduce.ImportJobBase: Retrieved 20031 records.
```

In this case, the output indicates that the Sqoop import was completed successfully. You check the HDFS data directory by using the HDFS file system `ls` command and see the results of the job:

```
[hadoop@hc1nn sqoop]$  hdfs dfs  -ls /user/hadoop/rawdata
Found 2 items
-rw-r--r--   2 hadoop hadoop          0 2014-07-20 11:36 /user/hadoop/rawdata/_SUCCESS
-rw-r--r--   2 hadoop hadoop    1427076 2014-07-20 11:36 /user/hadoop/rawdata/part-m-00000
```

These results show a _SUCCESS file and a part data file. You can dump the contents of the part file by using the HDFS file system `cat` command. You can then pipe the contents to the `wc` (word count) Linux command | `wc -l` by using the `-l` switch to give a file line count:

```
[hadoop@hc1nn sqoop]$  hdfs dfs  -cat /user/hadoop/rawdata/part-m-00000 | wc -l
20031
```

The output shows that there were 20,031 lines imported from MySQL to HDFS, which matches the data volume from MySQL. You can double-check the MySQL volume easily:

```
mysql --host=hc1nn --user=sqoop --password=xxxxxxxxxxxxx

mysql> select count(*) from sqoop.rawdata;
+-----------+
| count(*)  |
+-----------+
|   20031   |
+-----------+
```

Good; logging into MySQL as the user sqoop and getting a row count from the database table sqoop.rawdata by using count (*) gives you a row count of 20,031.

This is a good import and thus a good test of Sqoop. Although this simple example shows an import, you could also export data to a database. For example, you can easily import data, modify or enrich it and export it to another database.

Use Sqoop to Import Data to Hive

As you saw, Sqoop can move data to HDFS, but what if you need to move the data into the Hive data warehouse? Although you could use a Pig Latin or Hive script, Sqoop can directly import to Hive as well.

As for HDFS, you need to remember that Hive must be working on each data node before you attempt the Sqoop import. Testing before making the import is far better than getting strange errors later. On each node, you run a simple Hive `show tables` command, as follows:

```
[hadoop@hc1nn ~]$ hive
Logging initialized using configuration in file:/etc/hive/conf.dist/hive-log4j.properties
Hive history file=/tmp/hadoop/hive_job_log_ac529ba0-df48-4c65-9440-dbddf48f87b5_42666910.txt
hive>
    > show tables;
OK
Time taken: 2.089 seconds
```

For the Hive import, you need to add an extra line to the option file: --hive-import (line 10). Once the data is loaded onto HDFS, the new line will cause it to be loaded into Hive, this time to a table named to match its source:

```
1    import
2    --connect
3    jdbc:mysql://hc1nn/sqoop
4    --username
5    sqoop
6    --password
7    xxxxxxxxxxxx
8    --table
9    rawdata
10   --hive-import
```

The Sqoop command to import remains the same—only the contents of the options file change. The data will be loaded into Hive, and the table in Hive will be named the same as its source table in MySQL, as follows:

```
sqoop --options-file ./hive-import.txt --table sqoop.rawdata -m 1
```

Also, before you run the Sqoop command, you should be aware of some potential errors that can occur. For example, if the Hive Metastore server (the server that manages metadata for Hive) is not running, you will receive the following error:

```
14/07/20 11:45:44 INFO hive.HiveImport: org.apache.hadoop.hive.ql.metadata.HiveException:
java.lang.RuntimeException:
Unable to instantiate org.apache.hadoop.hive.metastore.HiveMetaStoreClient
```

As the Linux root user, you can check the state of the Hive Metastore server by using the following command:

```
[root@hc1nn conf]# service hive-metastore status
Hive Metastore is dead and pid file exists [FAILED]
```

Errors also can occur when the server cannot access the Derby database for read/write, as this error from the /var/log/hive/ hive-metastore.log shows:

```
2014-07-20 09:20:58,148 ERROR Datastore.Schema (Log4JLogger.java:error(125)) - Failed initialising
database.
Cannot get a connection, pool error Could not create a validated object, cause: A read-only user
or a user in a read-only database is not permitted to disable read-only mode on a connection.
org.datanucleus.exceptions.NucleusDataStoreException: Cannot get a connection, pool error Could
not create a validated object, cause: A read-only user or a user in a read-only database is not
permitted to disable read-only mode on a connection.
```

If you encounter read/write errors to the Hive Derby database, you can fix them by updating each Hive instance's hive-site.xml file under /etc/hive/conf to add the following:

```
<property>
  <name>hive.metastore.uris</name>
  <value>thrift://hc1nn:9083</value>
  <description>
    IP address (or fully-qualified domain name) and port of the metastore host
  </description>
</property>
```

These additions tell Hive the server and port of its Metastore server, with the server name matching the physical host on which it is installed (in this case, hc1nn).

Once all the errors are out of the way, you can import data to Hive. Here is the output of a successful job:

```
Please set $HCAT_HOME to the root of your HCatalog installation.
Please set $ACCUMULO_HOME to the root of your Accumulo installation.
14/07/21 15:53:07 INFO sqoop.Sqoop: Running Sqoop version: 1.4.3-cdh4.7.0
14/07/21 15:53:07 INFO tool.BaseSqoopTool: Using Hive-specific delimiters for output. You can
override
14/07/21 15:53:07 INFO tool.BaseSqoopTool: delimiters with --fields-terminated-by, etc.
14/07/21 15:53:07 INFO manager.MySQLManager: Preparing to use a MySQL streaming resultset.
14/07/21 15:53:07 INFO tool.CodeGenTool: Beginning code generation
14/07/21 15:53:08 INFO manager.SqlManager: Executing SQL statement: SELECT t.* FROM `rawdata`
AS t LIMIT 1
14/07/21 15:53:08 INFO manager.SqlManager: Executing SQL statement: SELECT t.* FROM `rawdata`
AS t LIMIT 1
14/07/21 15:53:08 INFO orm.CompilationManager: HADOOP_MAPRED_HOME is /usr/lib/hadoop-mapreduce
14/07/21 15:53:11 INFO orm.CompilationManager: Writing jar file: /tmp/sqoop-hadoop/compile/
6cdb761542523f9fe68bb9d0ffca26c3/rawdata.jar
14/07/21 15:53:11 WARN manager.MySQLManager: It looks like you are importing from mysql.
14/07/21 15:53:11 WARN manager.MySQLManager: This transfer can be faster! Use the --direct
14/07/21 15:53:11 WARN manager.MySQLManager: option to exercise a MySQL-specific fast path.
14/07/21 15:53:11 INFO manager.MySQLManager: Setting zero DATETIME behavior to convertToNull (mysql)
14/07/21 15:53:11 INFO mapreduce.ImportJobBase: Beginning import of rawdata
14/07/21 15:53:12 INFO service.AbstractService: Service:org.apache.hadoop.yarn.client.
YarnClientImpl is inited.
14/07/21 15:53:12 INFO service.AbstractService: Service:org.apache.hadoop.yarn.client.
YarnClientImpl is started.
14/07/21 15:53:16 INFO mapreduce.JobSubmitter: number of splits:1
14/07/21 15:53:16 INFO mapreduce.JobSubmitter: Submitting tokens for job: job_1405907667472_0001
14/07/21 15:53:17 INFO client.YarnClientImpl: Submitted application application_1405907667472_0001
to ResourceManager at hc1nn/192.168.1.107:8032
14/07/21 15:53:17 INFO mapreduce.Job: The url to track the job: http://hc1nn:8088/proxy/
application_1405907667472_0001/
14/07/21 15:53:17 INFO mapreduce.Job: Running job: job_1405907667472_0001
14/07/21 15:53:32 INFO mapreduce.Job: Job job_1405907667472_0001 running in uber mode : false
14/07/21 15:53:32 INFO mapreduce.Job:  map 0% reduce 0%
14/07/21 15:53:52 INFO mapreduce.Job:  map 100% reduce 0%
14/07/21 15:53:52 INFO mapreduce.Job: Job job_1405907667472_0001 completed successfully
14/07/21 15:53:52 INFO mapreduce.Job: Counters: 27
        File System Counters
                FILE: Number of bytes read=0
                FILE: Number of bytes written=92712
                FILE: Number of read operations=0
                FILE: Number of large read operations=0
                FILE: Number of write operations=0
                HDFS: Number of bytes read=87
                HDFS: Number of bytes written=1427076
                HDFS: Number of read operations=4
                HDFS: Number of large read operations=0
                HDFS: Number of write operations=2
```

```
        Job Counters
                Launched map tasks=1
                Other local map tasks=1
                Total time spent by all maps in occupied slots (ms)=17353
                Total time spent by all reduces in occupied slots (ms)=0
        Map-Reduce Framework
                Map input records=20031
                Map output records=20031
                Input split bytes=87
                Spilled Records=0
                Failed Shuffles=0
                Merged Map outputs=0
                GC time elapsed (ms)=97
                CPU time spent (ms)=3230
                Physical memory (bytes) snapshot=103755776
                Virtual memory (bytes) snapshot=823398400
                Total committed heap usage (bytes)=43712512
        File Input Format Counters
                Bytes Read=0
        File Output Format Counters
                Bytes Written=1427076
14/07/21 15:53:52 INFO mapreduce.ImportJobBase: Transferred 1.361 MB in 39.9986 seconds
(34.8419 KB/sec)
14/07/21 15:53:52 INFO mapreduce.ImportJobBase: Retrieved 20031 records.
14/07/21 15:53:52 INFO manager.SqlManager: Executing SQL statement: SELECT t.* FROM `rawdata`
AS t LIMIT 1
14/07/21 15:53:52 INFO hive.HiveImport: Loading uploaded data into Hive
14/07/21 15:53:56 INFO hive.HiveImport: Logging initialized using configuration in file:/etc/hive/
conf.dist/hive-log4j.properties
14/07/21 15:53:56 INFO hive.HiveImport: Hive history file=/tmp/hadoop/hive_job_log_ca16a3e3-8ece-
4b54-8c5b-e4da6779e121_58562623.txt
14/07/21 15:53:57 INFO hive.HiveImport: OK
14/07/21 15:53:57 INFO hive.HiveImport: Time taken: 0.958 seconds
14/07/21 15:53:59 INFO hive.HiveImport: Loading data to table default.rawdata
14/07/21 15:54:00 INFO hive.HiveImport: Table default.rawdata stats: [num_partitions: 0,
num_files: 2, num_rows: 0, total_size: 1427076, raw_data_size: 0]
14/07/21 15:54:00 INFO hive.HiveImport: OK
14/07/21 15:54:00 INFO hive.HiveImport: Time taken: 2.565 seconds
14/07/21 15:54:00 INFO hive.HiveImport: Hive import complete.
14/07/21 15:54:00 INFO hive.HiveImport: Export directory is empty, removing it.
```

As you can see, the data was imported to HDFS and then successfully imported to Hive. The warnings in this output indicate that you could have used a --direct flag in the import that would employ MySQL-specific functionality for a faster import. (I didn't use the flag so as to demonstrate a simple data import; I wasn't worried about performance.) To check the table's contents, use the following commands:

```
hive>
    > show tables;
OK
rawdata
Time taken: 1.344 seconds
```

You can see that the rawdata table was created, but how much data does it contain? To check the row count of the Hive rawdata table, you use the following:

```
hive> select count(*) from rawdata;
Total MapReduce jobs = 1
Launching Job 1 out of 1
......
Total MapReduce CPU Time Spent: 2 seconds 700 msec
OK
20031
Time taken: 25.098 seconds
```

Success is confirmed: the table in Hive contains 20,031 rows, which matches the MySQL table row count.

As you can see from this brief introduction, Sqoop is a powerful relational database data import/export tool for Hadoop. You can even use Sqoop in an Oozie workflow and schedule complex ETL flows using Sqoop, Pig, and Hive scripts. Also, you can carry out incremental loads with such Hive options as --incremental and --check-column. This would be useful if you were receiving periodic data-feed updates from a relational database. Check the sqoop.apache.org website to learn more.

So with Sqoop, you have seen an example of moving data between a relational database and Hadoop. But what if the data that you wish to move is in another type of data source? Well, that is where the Apache Flume tool comes into play. The next section describes it and provides an example of its use.

Moving Data with Flume

Apache Flume (flume.apache.org) is an Apache Software Foundation system for moving large volumes of log-based data. The Flume data model is defined in terms of agents, where an agent has an event source, a channel, and an event sink. Agents are defined in Flume configuration files. The source describes the data source. The channel receives event data from the source and stores it. The sink takes event data from the channel. Figure 6-1 provides a simple example of a Flume agent; it is the building block of the Flume architecture.

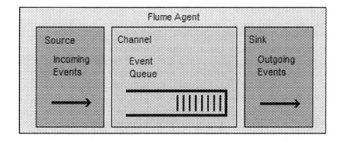

Figure 6-1. *The Flume agent*

The sink might pass data to the source of another agent or write the data to a store like HDFS, as I'll demonstrate in the sections that follow. You can build complex topologies to process log or event data, with multiple agents passing data to a single agent or to agents processing data in parallel. The following two examples show simple architectures for Flume. Figure 6-2 shows a hierarchical arrangement, with Flume agents on the left of the diagram passing data on to subagents that act as collectors for the data and then store the data to HDFS.

Agents

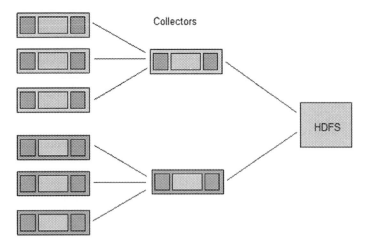

Figure 6-2. *Hierarchical example of Flume architecture*

Figure 6-3 shows a linear arrangement whereby the output from one agent is passed to the next and the next after that, until it is finally stored on HDFS. No specific architecture is advised here; instead, you should recognize that Flume is flexible and that you can arrange your Flume architecture to meet your needs.

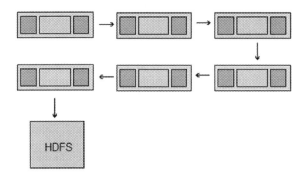

Figure 6-3. *Linear example of Flume architecture*

The Flume user guide shows some example topologies; see http://flume.apache.org/FlumeUserGuide.html.

Install Flume

Before you can move any log data, you must install Flume. Given that I am using the Cloudera CDH4.x stack here, installation is simple, using the Linux yum commands. You install the Flume server first, as follows:

```
yum install flume-ng
```

Then you install the component that allows Flume to start up at server boot time:

```
yum install flume-ng-agent
```

Finally, you install Flume documentation:

```
yum install flume-ng-doc
```

Next, you set up the basic configuration:

```
[root@hc1nn etc]# cd /etc/flume-ng/conf
[root@hc1nn conf]# ls
flume.conf  flume-conf.properties.template  flume-env.sh.template  log4j.properties

cp flume-conf.properties.template   flume.conf
```

You won't customize the configuration for the agent now; an agent-based configuration file will be defined shortly.

As with many other Cloudera stack Apache components, you can find the Flume configuration and logs in the standard places. For instance, logs are located under /var/log/flume-ng, the configuration files are under /etc/flume-ng/conf, and the flume-ng executable file is /usr/bin/flume-ng.

A Simple Agent

As an example of how Flume works, I build a single Flume agent to asynchronously take data from a Centos Linux-based message file called /var/log/messages. The message file acts as the data source and is stored in a single Linux-based channel called channel1. The data sink is on HDFS in a directory called "flume/messages."

In the Linux hadoop account I have created a number of files to run this example of a Flume job, display the resulting data, and clean up after the job. These files make the job easier to run; there is minimal typing, and it is easier to rerun the job because the results have been removed from HDFS. The files will also display the results of the job that reside on HDFS. You can use scripts like these if you desire.

```
[hadoop@hc1nn ~]$ cd $HOME/flume
[hadoop@hc1nn flume]$ ls
agent1.cfg  flume_clean_hdfs.sh  flume_exec_hdfs.sh  flume_show_hdfs.sh
```

The file agent1.cfg is the Flume configuration file for the agent, while the Bash (.sh) files are for running the agent (flume_exec_hdfs.sh), showing the results on HDFS (flume_show_hdfs.sh), and cleaning up the data on HDFS (flume_clean_hdfs.sh). Examining each of these files in turn, we see that the *show* script just executes a Hadoop file system ls command against the directory /flume/messages, where the agent will write the data.

```
[hadoop@hc1nn flume]$ cat flume_show_hdfs.sh

#!/bin/bash

hdfs dfs -ls /flume/messages
```

The *clean* script executes a Hadoop file system remove command with a recursive switch:

```
[hadoop@hc1nn flume]$ cat flume_clean_hdfs.sh

#!/bin/bash

hdfs dfs -rm -r /flume/messages
```

The *execution* script, flume_execute_hdfs.sh, runs the Flume agent and needs nine lines:

```
[hadoop@hc1nn flume]$ cat flume_exec_hdfs.sh

 1 #!/bin/bash
 2
 3 # run the bash agent
 4
 5 flume-ng agent \
 6          --conf /etc/flume-ng/conf \
 7          --conf-file agent1.cfg \
 8          -Dflume.root.logger=DEBUG,INFO,console \
 9          -name agent1
```

This execution script runs the Flume agent within a Linux Bash shell and is easily repeatable because a single script has been run, rather than retyping these options each time you want to move log file content. Line 5 actually runs the agent, while lines 6 and 7 specify the configuration directory and agent configuration file. Line 8 specifies the log4j log configuration via a -D command line option to show DEBUG, INFO, and console messages. Finally, line 9 specifies the Flume agent name agent1.

The Flume agent configuration file (agent1.cfg, in this case) must contain the agent's source, sink, and channel. Consider the contents of this example file:

```
[hadoop@hc1nn flume]$ cat agent1.cfg

 1 # ---------------------------------------------------------------------
 2 # define agent src, channel and sink
 3 # ---------------------------------------------------------------------
 4
 5 agent1.sources  = source1
 6 agent1.channels = channel1
 7 agent1.sinks = sink1
 8
 9 # ---------------------------------------------------------------------
10 # define agent channel
11 # ---------------------------------------------------------------------
12
13 agent1.channels.channel1.type = FILE
14 agent1.channels.channel1.capacity = 2000000
15 agent1.channels.channel1.checkpointInterval = 60000
16 agent1.channels.channel1.maxFileSize = 10737418240
17
```

```
18 # ----------------------------------------------------------------------
19 # define agent source
20 # ----------------------------------------------------------------------
21
22 agent1.sources.source1.type = exec
23 agent1.sources.source1.command = tail -F /var/log/messages
24 agent1.sources.source1.channels = channel1
25
26 # ----------------------------------------------------------------------
27 # define agent sink
28 # ----------------------------------------------------------------------
29
30 agent1.sinks.sink1.type = hdfs
31 agent1.sinks.sink1.hdfs.path = hdfs://hc1nn/flume/messages
32 agent1.sinks.sink1.hdfs.rollInterval = 0
33 agent1.sinks.sink1.hdfs.rollSize = 1000000
34 agent1.sinks.sink1.hdfs.batchSize = 100
35 agent1.sinks.sink1.channel = channel1
```

As already defined in the agent execution script, the Flume agent name in this example is agent1. Lines 5 to 7 define the names of the source, channel, and sink.

```
5 agent1.sources  = source1
6 agent1.channels = channel1
7 agent1.sinks    = sink1
```

The channel (channel1) is described between lines 13 and 16. Line 13 specifies that the channel type will be a file. Line 14 indicates that the maximum capacity of the channel will be 2 million events. Line 15, in milliseconds, indicates the time between checkpoints. Line 16 specifies the maximum channel file size in bytes.

```
13 agent1.channels.channel1.type = FILE
14 agent1.channels.channel1.capacity = 2000000
15 agent1.channels.channel1.checkpointInterval = 60000
16 agent1.channels.channel1.maxFileSize = 10737418240
```

The configuration file lines (22 to 24) show how the Flume data source source1 is defined.

```
22 agent1.sources.source1.type = exec
23 agent1.sources.source1.command = tail -F /var/log/messages
24 agent1.sources.source1.channels = channel1
```

In this example, I may need to ensure that the Linux account I am using to run this Flume job has access to read the log file /var/log/messages. Therefore, I grant access using the root account as follows: I use the Linux su (switch user) command to change the user ID to root. Then I use the Linux chmod command to grant global read privileges while maintaining current access. The two Linux ls command listings show that extra access has been granted:

```
su -
ls -l /var/log/messages
-rw------- 1 root root 410520 Nov 22 09:20 /var/log/messages
chmod 644 /var/log/messages
ls -l /var/log/messages
-rw-r--r-- 1 root root 410520 Nov 22 09:25 /var/log/messages
exit
```

The source type is defined as "exec" in line 22, but Flume also supports sources of Avro, Thrift, Syslog, jms, spooldir, twittersource, seq, http, and Netcat. You also could write custom sources to consume your own data types; see the Flume user guide at flume.apache.org for more information.

The executable command is specified at line 23 as tail -F /var/log/messages. This command causes new messages in the file to be received by the agent. Line 24 connects the source to the Flume agent channel, channel1. Finally, lines 30 through 35 define the HDFS data sink:

```
30 agent1.sinks.sink1.type = hdfs
31 agent1.sinks.sink1.hdfs.path = hdfs://hc1nn/flume/messages
32 agent1.sinks.sink1.hdfs.rollInterval = 0
33 agent1.sinks.sink1.hdfs.rollSize = 1000000
34 agent1.sinks.sink1.hdfs.batchSize = 100
35 agent1.sinks.sink1.channels = channel1
```

In this example, the sink type is specified at line 30 to be HDFS, but it could also be a value like logger, avro, irc, hbase, or a custom sink (see the Flume user guide at flume.apache.org for futher alternatives). Line 31 specifies the HDFS location as a URI, saving the data to /flume/messages.

Line 32 indicates that the logs will not be rolled by time, owing to the value of 0, while the value at line 33 indicates that the sink will be rolled based on size. Line 34 specifies a sink batch size of 100 for writing to HDFS, and line 35 connects the channel to the sink.

For this example, I encountered the following error owing to a misconfiguration of the channel name:

```
2014-07-26 14:45:10,177 (conf-file-poller-0) [WARN - org.apache.flume.conf.FlumeConfiguration
$AgentConfiguration.
validateSources(FlumeConfiguration.java:589)] Could not configure source  source1 due to: Failed to
configure component!
```

This error message indicated a configuration error—in this case, it was caused by putting an "s" on the end of the channels configuration item at line 24. When corrected, the line reads as follows:

```
24 agent1.sources.source1.channel = channel1
```

Running the Agent

To run your Flume agent, you simply run your Bash script. In my example, to run the Flume agent agent1, I run the Centos Linux Bash script flume_exec_hdfs.sh, as follows:

```
[hadoop@hc1nn flume]$ cd $HOME/flume
[hadoop@hc1nn flume]$ ./flume_execute_hdfs.sh
```

This writes the voluminous log output to the session window and to the logs under /var/log/flume-ng. For my example, I don't provide the full output listing here, but I identify the important parts. Flume validates the agent configuration and so displays the source, channel, and sink as defined:

```
2014-07-26 17:50:01,377 (conf-file-poller-0) [DEBUG - org.apache.flume.conf.FlumeConfiguration$Agent
Configuration.isValid(FlumeConfiguration.java:313)] Starting validation of configuration for agent:
agent1, initial-configuration: AgentConfiguration[agent1]

SOURCES: {source1={ parameters:{command=tail -F /var/log/messages, channels=channel1, type=exec} }}
```

```
CHANNELS: {channel1={ parameters:{checkpointInterval=60000, capacity=2000000,
maxFileSize=10737418240, type=FILE} }}

SINKS: {sink1={ parameters:{hdfs.path=hdfs://hc1nn/flume/messages, hdfs.batchSize=100,
hdfs.rollInterval=0, hdfs.rollSize=1000000, type=hdfs, channel=channel1} }}
```

Flume then sets up the file-based channel:

```
2014-07-26 17:50:02,858 (lifecycleSupervisor-1-0) [INFO - org.apache.flume.channel.file.FileChannel.
start(FileChannel.java:254)] Starting FileChannel channel1 { dataDirs: [/home/hadoop/.flume/file-
channel/data] }...
```

The channel on the Linux file system contains checkpoint and log data:

```
[hadoop@hc1nn flume]$ ls -l $HOME/.flume/file-channel/*
/home/hadoop/.flume/file-channel/checkpoint:
total 15652
-rw-rw-r--. 1 hadoop hadoop 16008232 Jul 26 17:51 checkpoint
-rw-rw-r--. 1 hadoop hadoop       25 Jul 26 17:51 checkpoint.meta
-rw-rw-r--. 1 hadoop hadoop       32 Jul 26 17:51 inflightputs
-rw-rw-r--. 1 hadoop hadoop       32 Jul 26 17:51 inflighttakes
drwxrwxr-x. 2 hadoop hadoop     4096 Jul 26 17:50 queueset

/home/hadoop/.flume/file-channel/data:
total 2060
-rw-rw-r--. 1 hadoop hadoop        0 Jul 26 15:44 log-6
-rw-rw-r--. 1 hadoop hadoop       47 Jul 26 15:44 log-6.meta
-rw-rw-r--. 1 hadoop hadoop  1048576 Jul 26 15:55 log-7
-rw-rw-r--. 1 hadoop hadoop       47 Jul 26 15:56 log-7.meta
-rw-rw-r--. 1 hadoop hadoop  1048576 Jul 26 17:50 log-8
-rw-rw-r--. 1 hadoop hadoop       47 Jul 26 17:51 log-8.meta
```

The Flume agent sets up the data sink by creating a single empty file on HDFS. The log message indicating this is as follows:

```
2014-07-26 17:50:10,532 (SinkRunner-PollingRunner-DefaultSinkProcessor) [INFO - org.apache.
flume.sink.hdfs.BucketWriter.open(BucketWriter.java:220)] Creating hdfs://hc1nn/flume/messages/
FlumeData.1406353810397.tmp
```

The script flume_show_hdfs.sh can be run as follows, using the Linux hadoop account to show the Flume data sink file on HDFS:

```
[hadoop@hc1nn flume]$ ./flume_show_hdfs.sh
Found 1 items
-rw-r--r--   2 hadoop hadoop          0 2014-07-26 17:50 /flume/messages/FlumeData.1406353810397.tmp
```

The script reveals that the file is empty, with a zero in the fifth column. When the number of new messages in the messages file reaches 100 (as defined by batchSize at line 34 of the agent configuration file), the data is written to HDFS from the channel:

```
34 agent1.sinks.sink1.hdfs.batchSize = 100
```

I can see this behavior by running the Flume show script again:

```
[hadoop@hc1nn flume]$ ./flume_show_hdfs.sh
Found 11 items
-rw-r--r--   2 hadoop hadoop       1281 2014-07-26 17:50 /flume/messages/FlumeData.1406353810397
-rw-r--r--   2 hadoop hadoop       1057 2014-07-26 17:50 /flume/messages/FlumeData.1406353810398
-rw-r--r--   2 hadoop hadoop        926 2014-07-26 17:50 /flume/messages/FlumeData.1406353810399
-rw-r--r--   2 hadoop hadoop       1528 2014-07-26 17:50 /flume/messages/FlumeData.1406353810400
-rw-r--r--   2 hadoop hadoop       1281 2014-07-26 17:50 /flume/messages/FlumeData.1406353810401
-rw-r--r--   2 hadoop hadoop       1214 2014-07-26 17:50 /flume/messages/FlumeData.1406353810402
-rw-r--r--   2 hadoop hadoop       1190 2014-07-26 17:50 /flume/messages/FlumeData.1406353810403
-rw-r--r--   2 hadoop hadoop       1276 2014-07-26 17:50 /flume/messages/FlumeData.1406353810404
-rw-r--r--   2 hadoop hadoop       1387 2014-07-26 17:50 /flume/messages/FlumeData.1406353810405
-rw-r--r--   2 hadoop hadoop       1107 2014-07-26 17:50 /flume/messages/FlumeData.1406353810406
-rw-r--r--   2 hadoop hadoop       1281 2014-07-26 17:51 /flume/messages/FlumeData.1406353810407
```

As you can see in this example, those 100 messages have been written to HDFS and the data is now available for further processing in an ETL chain by one of the other processing languages. This has further possibilities. For instance, you could use Apache Pig native to strip information from these files and employ an Oozie workflow to organize that processing into an ETL chain.

This simple example uses a simple agent with a single source and sink. You could also organize agents to act as sources or sinks for later agents in the chain so the feeds can fan in and out. You could build complex agent processing topologies with many different types, depending upon your needs. Check the Apache Flume website at flume.apache.org for further configuration examples.

You've now seen how to process relational database data with Sqoop and log-based data with Flume, but what about streamed data? How is it possible to process an endless stream of data from a system like Twitter? The data would not stop—it would just keep coming. The answer is that systems like Storm allow processing on data streams. For instance, by using this tool, you can carry out trend analysis continuously on current data in the stream. The next section examines some uses of Storm.

Moving Data with Storm

Apache Storm (storm.incubator.apache.org) from the Apache Software Foundation is an Apache incubator project for processing unbounded data streams in real time. (The term "incubator" means that this is a new Apache project; it is not yet mature. It needs to follow the Apache process before it can "graduate," and this might mean that its release process or documentation is not complete.) The best way to understand the significance of Storm is with a comparison. On Hadoop, a Map Reduce job will start, process its data set, and exit; however, a topology (a Storm job architecture) will run forever because its data feed is unlimited.

Consider a feed of events from the website Twitter; they just keep coming. When Storm processes a feed from such a source, it processes the data it receives in real time. So, at any point, what Storm presents is a window on a stream of data at the current time. Because of this, it also presents current trends in the data. In terms of Twitter, that might indicate what many people are talking about right now. But also, because the data set is a stream that never ends, Storm needs to be manually stopped.

A topology is a Storm job architecture. It is described in terms of spouts, steams, and bolts. *Streams* are streams of data created from a sequence of data records called *tuples*.

Figure 6-4 shows a simple Storm data record, or tuple; a sequence or pipe of these data records forms a stream, which is shown in Figure 6-5.

Figure 6-4. *A Storm tuple*

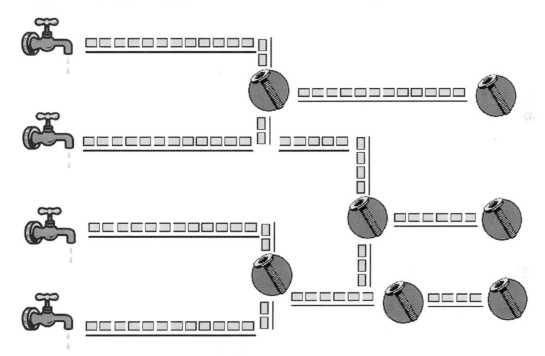

Figure 6-5. *A Storm stream*

Data sources in Storm are call spouts, while the joints between the streams are called bolts. The input to a stream might be a spout or a bolt. Bolts can connect stream outputs to inputs from or to other streams. Figure 6-6 shows a simple Storm topology. Multiple spouts and bolts have been used to merge stream data.

Figure 6-6. *An simple Storm topology*

Storm has a master process called a Nimbus and slave processes called supervisors. Configuration is managed via ZooKeeper servers. The Nimbus master handles the monitoring and distribution of code and tasks to the slaves. Hadoop runs potentially long-running batch jobs that will eventually end. Storm provides real-time trend processing of endless stream-based data that will run until it is manually stopped.

As an incubator project, Storm has not yet matured to the level of a full Apache project. It demands a little more work than is required of more mature systems to source and build the components. Storm depends on ZeroMQ (a messaging system) and JZMQ (Java Bindings for ZeroMQ), so you need to install these before you install Storm itself. In the next sections, I show how to install ZeroMQ, JXMQ, and Storm. Remember, though, that each of the following installations should be carried out on each server on which Storm will run. You will also need ZooKeeper, the installation for which was described in Chapter 2. Once you check that everything is operating without error, you'll be able to follow my demonstration of the Storm interface and an example of the code samples that Storm provides on the Storm cluster that I build.

Install ZeroMQ

Install ZeroMQ on the server hc1nn as the Linux user hadoop into a working directory that you create for the task called $HOME/storm/zeromq. Download version 2.1.7 of ZeroMQ from the ZeroMQ website as a zipped tar file, as follows:

```
[hadoop@hc1nn zeromq]$ pwd
/home/hadoop/storm/zeromq

[hadoop@hc1nn zeromq]$ wget http://download.zeromq.org/zeromq-2.1.7.tar.gz
[hadoop@hc1nn zeromq]$ ls -l
total 1836
-rw-rw-r--. 1 hadoop hadoop 1877380 May 12  2011 zeromq-2.1.7.tar.gz
```

Next, unpack the compressed tar file using tar -xzf. The z option accepts a file with a gzipped file extension and unzips it. The x option extracts the tar contents, and the f option allows the tar file name to be specified:

```
[hadoop@hc1nn zeromq]$ tar -xzf zeromq-2.1.7.tar.gz
```

This shows the contents of the ZeroMQ release:

```
[hadoop@hc1nn zeromq]$ cd zeromq-2.1.7

[hadoop@hc1nn zeromq-2.1.7]$ ls
acinclude.m4  builds     configure.in    foreign       Makefile.am  README      zeromq.spec
aclocal.m4    ChangeLog  COPYING         include       Makefile.in  src
AUTHORS       config     COPYING.LESSER  INSTALL       NEWS         tests
autogen.sh    configure  doc             MAINTAINERS   perf         version.sh
```

The configure script is used to prepare the release for a build. Note: When I initially used this script, I encountered this error:

```
checking for uuid_generate in -luuid... no
configure: error: cannot link with -luuid, install uuid-dev.
```

The error alerted me that a dependency was missing that ZeroMQ required. To fix this, I installed the following components as root, using the yum command:

```
[root@hc1nn ~]#  yum install  libuuid-devel   gcc-c++.x86_64   libtool
```

Then I set up the build configuration as follows:

```
[hadoop@hc1nn zeromq]$ ./configure

.............
config.status: creating builds/redhat/zeromq.spec
config.status: creating src/platform.hpp
config.status: executing depfiles commands
config.status: executing libtool commands
```

I have truncated the output I received next; if you get these last few lines and see no errors, then you know that the command has completed successfully.

Now, you run a make command to build ZeroMQ:

```
[hadoop@hc1nn zeromq-2.1.7]$ make

..............
make[1]: Leaving directory `/home/hadoop/storm/zeromq/zeromq-2.1.7/tests'
make[1]: Entering directory `/home/hadoop/storm/zeromq/zeromq-2.1.7'
make[1]: Nothing to be done for `all-am'.
make[1]: Leaving directory `/home/hadoop/storm/zeromq/zeromq-2.1.7'
```

Again, the preceding is only a portion of the output to give you the general idea. Having built the release, you are ready to install it, but you will need to do this as root:

```
[root@hc1nn zeromq-2.1.7]# make install

make[1]: Entering directory `/home/hadoop/storm/zeromq/zeromq-2.1.7'
make[2]: Entering directory `/home/hadoop/storm/zeromq/zeromq-2.1.7'
make[2]: Nothing to be done for `install-exec-am'.
make[2]: Nothing to be done for `install-data-am'.
make[2]: Leaving directory `/home/hadoop/storm/zeromq/zeromq-2.1.7'
make[1]: Leaving directory `/home/hadoop/storm/zeromq/zeromq-2.1.7'
```

With ZeroMQ, Storm's messaging component successfully installed, you can move on to installing JZMQ, the Java binding component.

Install JZMQ

Create a working directory for this installation at $HOME/storm/jzmq, from which you will carry out the installation. Use the git command to download a JZMQ release. Use the Linux yum command as root in the first three lines that follow to install the git command, which enables software downloads:

```
[hadoop@hc1nn jzmq]$ su -
[root@hc1nn ~]$ yum install git
[root@hc1nn ~]$ exit

[hadoop@hc1nn jzmq]$ git clone https://github.com/nathanmarz/jzmq.git

Initialized empty Git repository in /home/hadoop/storm/jzmq/jzmq/.git/
remote: Counting objects: 611, done.
remote: Compressing objects: 100% (257/257), done.
remote: Total 611 (delta 239), reused 611 (delta 239)
Receiving objects: 100% (611/611), 348.62 KiB | 216 KiB/s, done.
Resolving deltas: 100% (239/239), done.
```

Move it into the release directory and list the contents:

```
[hadoop@hc1nn jzmq]$ cd     jzmq
[hadoop@hc1nn jzmq]$ ls

AUTHORS      ChangeLog     COPYING.LESSER  Makefile.am  pom.xml      src
autogen.sh   configure.in  debian          NEWS         README       test
builds       COPYING       jzmq.spec       perf         README-PERF
```

Now run the autogen.sh script to prepare this release for a build:

```
[hadoop@hc1nn jzmq]$ ./autogen.sh

................
configure.in:14: installing `config/install-sh'
configure.in:14: installing `config/missing'
src/Makefile.am: installing `config/depcomp'
Makefile.am: installing `./INSTALL'
autoreconf: Leaving directory `.'
```

(Again, the output is cropped.) You use the configure script to set up build Makefiles:

```
[hadoop@hc1nn jzmq]$ ./configure
.........................
configure: creating ./config.status
config.status: creating Makefile
config.status: creating src/Makefile
config.status: creating perf/Makefile
config.status: creating src/config.hpp
config.status: executing depfiles commands
config.status: executing libtool commands
```

Now, build the release by executing the make command:

```
[hadoop@hc1nn jzmq]$ make

................
echo timestamp > classdist_noinst.stamp
/usr/bin/jar cf  zmq-perf.jar *.class
make[2]: Leaving directory `/home/hadoop/storm/jzmq/jzmq/perf'
make[1]: Leaving directory `/home/hadoop/storm/jzmq/jzmq/perf'
make[1]: Entering directory `/home/hadoop/storm/jzmq/jzmq'
make[1]: Nothing to be done for `all-am'.
make[1]: Leaving directory `/home/hadoop/storm/jzmq/jzmq'
```

Having built the release successfully, you install it as the Linux root account:

```
[root@hc1nn jzmq]# pwd
/home/hadoop/storm/jzmq/jzmq
[root@hc1nn jzmq]# make install
```

```
..........................
make[1]: Entering directory `/home/hadoop/storm/jzmq/jzmq'
make[2]: Entering directory `/home/hadoop/storm/jzmq/jzmq'
make[2]: Nothing to be done for `install-exec-am'.
make[2]: Nothing to be done for `install-data-am'.
make[2]: Leaving directory `/home/hadoop/storm/jzmq/jzmq'
make[1]: Leaving directory `/home/hadoop/storm/jzmq/jzmq'
```

That completes the installations for the Storm dependencies; now it's time to install Storm itself.

Install Storm

Create a working directory $HOME/storm/storm to carry out this installation. Use wget to download version 0.9.2 of Storm as a gzipped tar file.

```
[hadoop@hc1nn storm]$ pwd
/home/hadoop/storm/storm

[hadoop@hc1nn storm]$ wget http://supergsego.com/apache/incubator/storm/apache-storm-0.9.2-
incubating/apache-storm-0.9.2-incubating.tar.gz

[hadoop@hc1nn storm]$ ls -l
total 19608
-rw-rw-r--. 1 hadoop hadoop 20077564 Jun 25 02:49 apache-storm-0.9.2-incubating.tar.gz
```

Using the tar command, unpack the zipped tar file; x means extract, f specifies the archive file, and z decompresses the file:

```
[hadoop@hc1nn storm]$  tar -xzf apache-storm-0.9.2-incubating.tar.gz
```

Using the Linux root account, move the Storm release to the /usr/local directory:

```
[root@hc1nn ~]# cd /home/hadoop/storm/storm
[root@hc1nn storm]# mv apache-storm-0.9.2-incubating /usr/local
[root@hc1nn storm]# cd /usr/local
```

Create a symbolic link for the release under /usr/local and name the link "storm." Using this link to refer to the release will simplify the environment and make the Storm configuration release version independent:

```
[root@hc1nn local]# ln -s apache-storm-0.9.2-incubating storm
[root@hc1nn local]# ls -ld *storm*
drwxrwxr-x. 9 hadoop hadoop 4096 Jul 27 11:28 apache-storm-0.9.2-incubating
lrwxrwxrwx. 1 root    root     29 Jul 27 11:30 storm -> apache-storm-0.9.2-incubating
```

Now, create a Storm-related environment variable in the $HOME/.bashrc shell file for the Linux hadoop account:

```
#######################################################
# Set up Storm variables

export STORM_HOME=/usr/local/storm
export PATH=$PATH:$STORM_HOME/bin
```

This action creates a STORM_HOME environment variable that points to the installation. Also, the Storm installation binary directory has been added to the path so that the Storm executable can be located. Now, you create a Storm working directory; I place mine under /app/storm:

```
[root@hc1nn app]# cd /app
[root@hc1nn app]# mkdir storm
[root@hc1nn app]# chown hadoop:hadoop storm

[root@hc1nn app]# ls -l
total 8
drwxr-xr-x. 3 hadoop hadoop 4096 Mar 15 15:38 hadoop
drwxr-xr-x. 2 hadoop hadoop  4096 Jul 27 11:38 storm
```

I am using my Linux hadoop user account to run Storm, but you might want to use a dedicated Storm account. You set up the Storm configuration files under $STORM_HOME/conf. Set the following values in the file storm.yaml:

```
storm.zookeeper.servers:
    - "hc1r1m1"
    - "hc1r1m2"
    - "hc1r1m3"

nimbus.host: "hc1nn"

nimbus.childopts: "-Xmx1024m -Djava.net.preferIPv4Stack=true"

ui.childopts: "-Xmx768m -Djava.net.preferIPv4Stack=true"

supervisor.childopts: "-Djava.net.preferIPv4Stack=true"

worker.childopts: "-Xmx768m -Djava.net.preferIPv4Stack=true"

storm.local.dir: "/app/storm"
```

These settings specify three ZooKeeper servers (hc1r1m1, hc1r1m2, hc1r1m3) to be the nodes on which the slave Storm processes will run. The master (Nimbus) Storm process is set to run on the server hc1nn. Also, the Storm local directory is defined as " /app/storm."

Now, it is time to check that ZooKeeper is running before you attempt to run Storm.

Start and Check Zookeeper

As the Linux root account, you start the ZooKeeper server on each slave node; root is used because these servers are Linux-based services:

```
service zookeeper-server start
```

The configuration for the ZooKeeper server is stored under /etc/zookeeper/conf/zoo.cfg, while the logs can be found under /var/log/zookeeper. Check the logs for errors and ensure that each ZooKeeper server is running.

Now, you use a four-letter acronym RUOK ("Are you OK?") with each server to check that it is running correctly. The response that you can expect is IMOK ("I am OK").

The ZooKeeper installation and use was already described in Chapter. Here are the successful outputs from the ZooKeeper checks:

```
nc  hc1r1m1 2181
ruok
imok

nc  hc1r1m2 2181
ruok
imok

nc  hc1r1m3 2181
ruok
imok
```

Run Storm

With the prep work finished, you're now ready to run the Storm servers. Start the supervisor processes on the slave nodes first, and then the Nimbus process on the master server. Also, start the Storm user interface process so that Storm can be monitored via a web page. On each of the slave nodes (hc1r1m1, hc1r1m2, hc1r1m3), you run the Storm supervisor slave process (via the Linux hadoop account) as a background process ("&"). This will free up the terminal session if it is required:

```
storm supervisor &
```

Run the Storm Nimbus master process on the master server hc1nn, and run the user interface instance on the same server:

```
storm nimbus &
storm ui &
```

The Storm cluster is now running, but you need to check the logs under /usr/local/storm/logs for errors on each server. Shown in Figure 6-7, the Storm user interface is available at `http://hc1nn:8080/`.

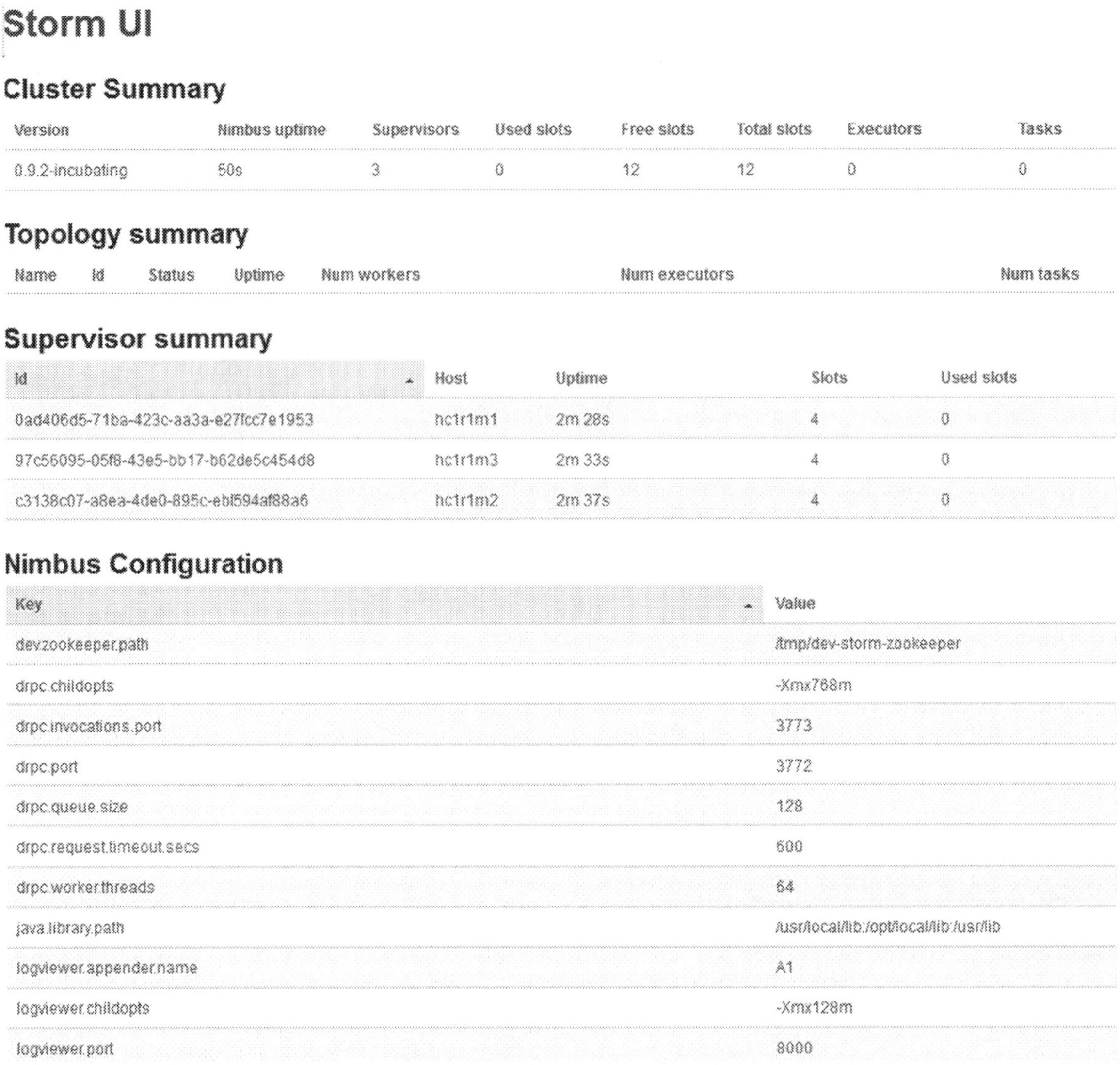

Storm UI

Cluster Summary

Version	Nimbus uptime	Supervisors	Used slots	Free slots	Total slots	Executors	Tasks
0.9.2-incubating	50s	3	0	12	12	0	0

Topology summary

Name	Id	Status	Uptime	Num workers	Num executors	Num tasks

Supervisor summary

Id	Host	Uptime	Slots	Used slots
0ad406d5-71ba-423c-aa3a-e27fcc7e1953	hc1r1m1	2m 28s	4	0
97c56095-05f8-43e5-bb17-b62de5c454d8	hc1r1m3	2m 33s	4	0
c3138c07-a8ea-4de0-895c-ebf594af88a6	hc1r1m2	2m 37s	4	0

Nimbus Configuration

Key	Value
dev.zookeeper.path	/tmp/dev-storm-zookeeper
drpc.childopts	-Xmx768m
drpc.invocations.port	3773
drpc.port	3772
drpc.queue.size	128
drpc.request.timeout.secs	600
drpc.worker.threads	64
java.library.path	/usr/local/lib:/opt/local/lib:/usr/lib
logviewer.appender.name	A1
logviewer.childopts	-Xmx128m
logviewer.port	8000

Figure 6-7. *The basic Storm user interface*

The Storm user interface shows the supervisor processes running, the Nimbus master, the topologies that are running, and a cluster summary. In the Figure 6-7 window, no topology can be seen running at the moment (but I add one in the next section). The Storm release comes with some example topologies in a storm-starter subdirectory. Next, I build these examples into a jar file and run one of them on the Storm cluster; this will demonstrate the process and the available tools.

An Example of Storm Topology

On the master server, I build the storm-starter topology code under $STORM_HOME/examples/storm-starter.
The build creates a jar file that can be run against the cluster under a directory called "target" at this level:

```
[hadoop@hc1nn ~]$ cd $STORM_HOME/examples/storm-starter
[hadoop@hc1nn storm-starter]$ ls -l
total 2888
drwxr-xr-x. 3 hadoop hadoop    4096 May  6 07:13 multilang
-rw-r--r--. 1 hadoop hadoop    5191 Jun 14 08:35 pom.xml
-rw-r--r--. 1 hadoop hadoop    4825 May 29 04:24 README.markdown
drwxr-xr-x. 4 hadoop hadoop    4096 May  6 07:13 src
-rw-r--r--. 1 hadoop hadoop 2927299 Jun 14 08:57 storm-starter-topologies-0.9.2-incubating.jar
drwxr-xr-x. 3 hadoop hadoop    4096 May  6 07:13 test
```

I then use Apache Maven version 3.2.1 to build this package:

```
[hadoop@hc1nn storm-starter]$ mvn package


....................
[INFO] ------------------------------------------------------------------------
[INFO] BUILD SUCCESS
[INFO] ------------------------------------------------------------------------
[INFO] Total time: 04:16 min
[INFO] Finished at: 2014-07-27T13:57:49+12:00
[INFO] Final Memory: 25M/59M
[INFO] ------------------------------------------------------------------------
```

I have truncated the build output for purposes of this example, but if you get to this success line, then all is good.
So, I check to see that the built library exists:

```
[hadoop@hc1nn storm-starter]$ ls -lrt  target/storm-starter-*-incubating-jar-with-dependencies.jar

-rw-rw-r--. 1 hadoop hadoop 2927301 Jul 27 13:57 target/storm-starter-0.9.2-incubating-jar-with-
dependencies.jar
```

▓ **Note** If you want to know which topologies are available for use, you can look at the Java source code under
$STORM_HOME/examples/storm-starter/src/jvm/storm/starter. It might be useful to have a look at this code to familiarize
yourself with how it works. Also, check the Apache Storm website (storm.incubator.apache.org) documentation for
topology coding examples.

```
[hadoop@hc1nn storm-starter]$ cd $STORM_HOME/examples/storm-starter/src/jvm/storm/starter
[hadoop@hc1nn starter]$ ls *.java
BasicDRPCTopology.java    ReachTopology.java              TransactionalWords.java
ExclamationTopology.java  RollingTopWords.java            WordCountTopology.java
ManualDRPC.java           SingleJoinExample.java
PrintSampleStream.java    TransactionalGlobalCount.java
```

Now that the storm-start jar file is built, I can run a topology from its contents. Remember that this topology will run forever, processing a simulated data feed created by the example code. So, I run the exclamation topology, which just randomly adds exclamation marks to the incoming data. (This may seem like a simple process, but the aim here is to show how to use and run topologies on Storm. Later, you can investigate building and running your own.) The storm command line tool launches the topology onto the cluster and has four parameters. The first is the `jar` parameter, which is followed by the jar file name that was just built. The third and fourth are the class name and topology name:

```
storm jar target/storm-starter-*-incubating-jar-with-dependencies.jar \
        storm.starter.ExclamationTopology \
        exclamation-topology
```

I now check the status of the topology on the cluster by using the Storm `list` option. The output that follows shows that the topology is active, provides timing information, and shows the number of tasks:

```
[hadoop@hc1nn starter]$  storm list
```

Topology_name	Status	Num_tasks	Num_workers	Uptime_secs
exclamation-topology	ACTIVE	18	3	62

As mentioned, this topology will run forever, providing a series of snapshots of data processed within given data windows. If I check the Storm user interface (Figure 6-8), I see that there is an active topology called "exclamation-topology" shown in the Topology Summary section.

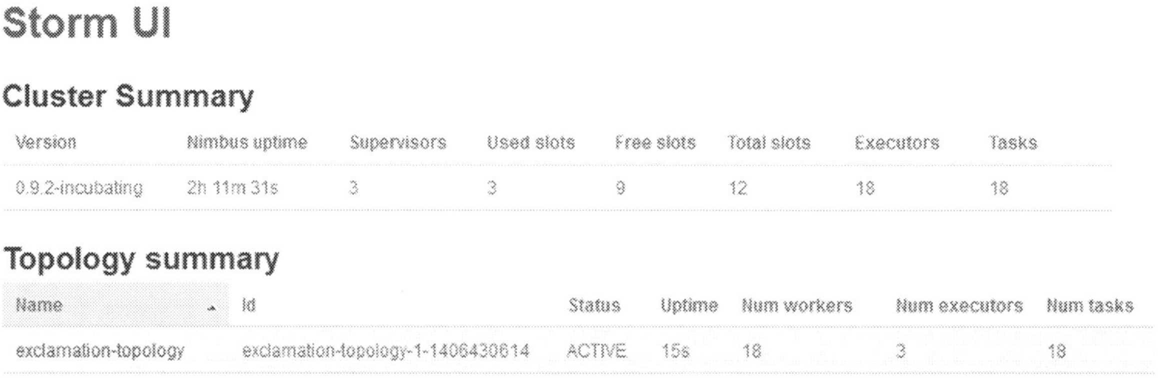

Storm UI

Cluster Summary

Version	Nimbus uptime	Supervisors	Used slots	Free slots	Total slots	Executors	Tasks
0.9.2-incubating	2h 11m 31s	3	3	9	12	18	18

Topology summary

Name	Id	Status	Uptime	Num workers	Num executors	Num tasks
exclamation-topology	exclamation-topology-1-1406430614	ACTIVE	15s	18	3	18

Figure 6-8. *Storm user interface with running topology*

By clicking on the topology name in the user interface, I can drill down into the topology to get more information. The detailed topology view in Figure 6-9 gives information about the spouts and bolts within the topology, for example. Remember that spouts provide data sources while bolts process the streams of data. My example includes a single spout data source, called "word," whose data is being passed to two bolts, called "exclaim1" and "exclaim2." The detailed topology view also lists the volume of data processed and the number of tasks involved.

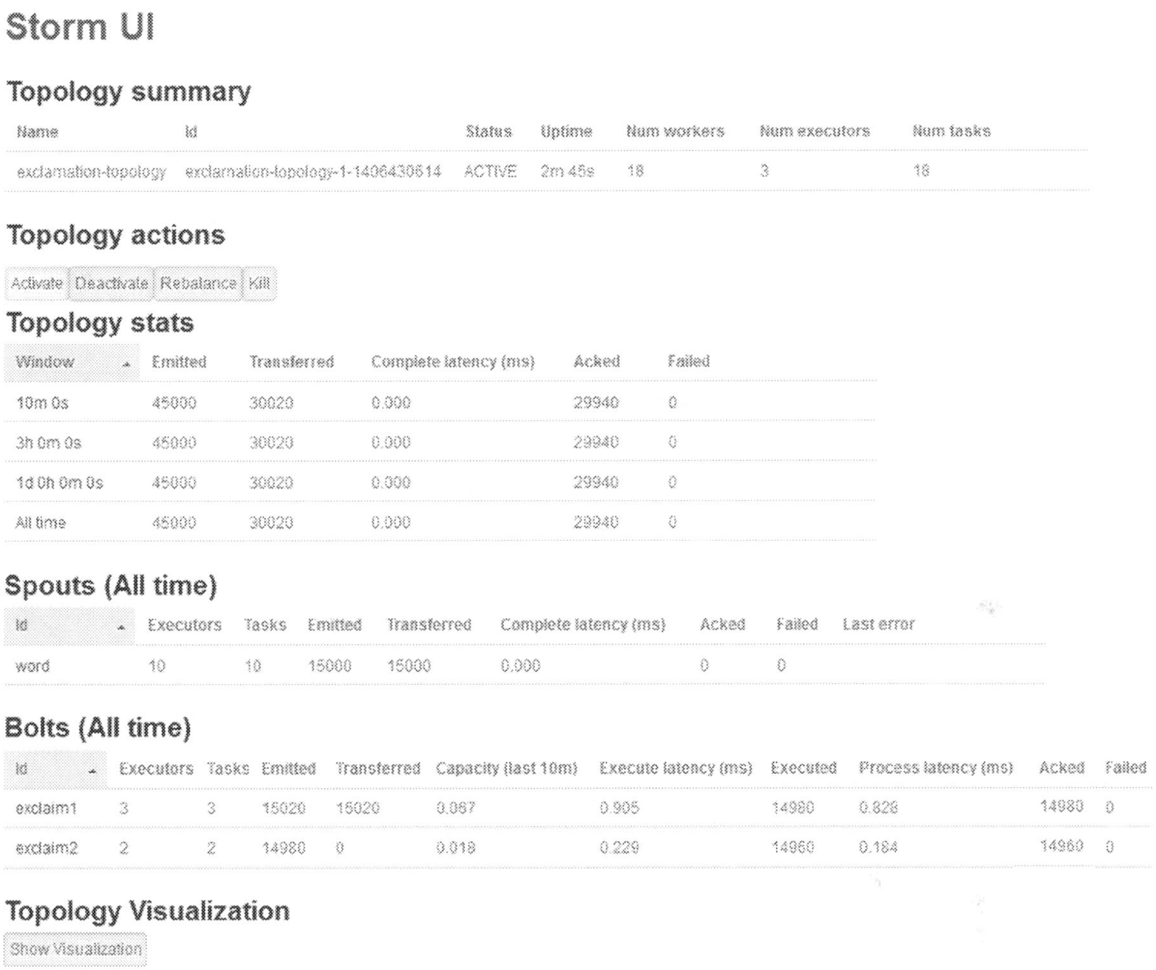

Storm UI

Topology summary

Name	Id	Status	Uptime	Num workers	Num executors	Num tasks
exclamation-topology	exclamation-topology-1-1406430514	ACTIVE	2m 45s	18	3	18

Topology actions

Activate | Deactivate | Rebalance | Kill

Topology stats

Window	Emitted	Transferred	Complete latency (ms)	Acked	Failed
10m 0s	45000	30020	0.000	29940	0
3h 0m 0s	45000	30020	0.000	29940	0
1d 0h 0m 0s	45000	30020	0.000	29940	0
All time	45000	30020	0.000	29940	0

Spouts (All time)

Id	Executors	Tasks	Emitted	Transferred	Complete latency (ms)	Acked	Failed	Last error
word	10	10	15000	15000	0.000	0	0	

Bolts (All time)

Id	Executors	Tasks	Emitted	Transferred	Capacity (last 10m)	Execute latency (ms)	Executed	Process latency (ms)	Acked	Failed
exclaim1	3	3	15020	15020	0.067	0.905	14980	0.828	14980	0
exclaim2	2	2	14980	0	0.018	0.229	14960	0.184	14960	0

Topology Visualization

Show Visualization

Figure 6-9. *Detailed view of Storm user interface topology*

I also can obtain a visual view of my example's topology structure by clicking the Show Visualization button. Figure 6-10 provides an enlarged image of the resulting topology. As you can see, the spout passes data to the bolt exclaim1, which then passes it on to exclaim2.

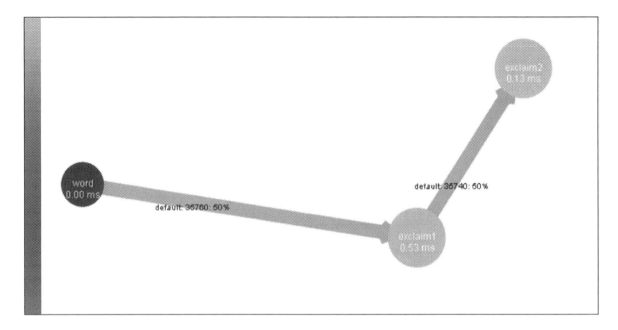

Figure 6-10. *Storm user interface showing topology structure*

The green circles in Figure 6-10 are bolts and the blue circle is a spout. The gray arrows are data flows, and the timing information is shown in each object. This is a simple topology. To do anything meaningful, you would build a larger structure. If you refresh the user interface (press F5), you will see the statistics update. You will always see a current window of data showing the current data trend.

The data stream that this topology processes is boundless, meaning that if you want to stop it, you have to manually kill it. You can do this by using the Storm kill command with the topology name, as I have done for my example:

```
[hadoop@hc1nn starter]$  storm kill exclamation-topology
2530 [main] INFO  backtype.storm.thrift - Connecting to Nimbus at hc1nn:6627
2689 [main] INFO  backtype.storm.command.kill-topology - Killed topology: exclamation-topology
```

Now, when I list the running Storm topologies, I can see that there are none running:

```
[hadoop@hc1nn starter]$  storm list

3326 [main] INFO  backtype.storm.thrift - Connecting to Nimbus at hc1nn:6627
No topologies running.
```

To shut down Storm, you first stop the supervisor processes on each slave node. To find out which processes are running, you use the jps command to see a list of processes and their process numbers. In my example, I can see that the supervisor process on the host hc1r1m2 has a process ID of 17617. I can use this number with the Linux kill command to kill the process, as follows:

```
[hadoop@hc1r1m2 logs]$ jps
17617 supervisor
18134 Jps
```

```
[hadoop@hc1r1m2 logs]$ kill -9 17617
[hadoop@hc1r1m2 logs]$
[1]+ Killed                 storm supervisor
```

Next, I kill the Nimbus and the user interface processes on the master server hc1nn. Again, I use the jps command to show the running processes. The Storm user interface shows the name "Core" instead of a meaningful name. Remember, though, that Storm is an incubator project and so problems like this will be resolved in future releases. However, when I kill the associated process numbers by using the Linux kill command, I can see that the Nimbus and UI have stopped:

```
[hadoop@hc1nn starter]$ jps
24718 core
27661 Jps
24667 nimbus

[hadoop@hc1nn starter]$ kill -9 24718 24667
[hadoop@hc1nn starter]$
[1]- Killed                 storm nimbus  (wd: /usr/local/storm/conf)
(wd now: /usr/local/storm/examples/storm-starter/src/jvm/storm/starter)
[2]+ Killed                 storm ui  (wd: /usr/local/storm/conf)
(wd now: /usr/local/storm/examples/storm-starter/src/jvm/storm/starter)
```

If this very short introduction to Storm leaves you curious for more information, take a look at the other example topologies and examine the code. Try running some of these other topologies and see what they do. You can read the Apache Storm website, but be aware that because Storm is an incubator project, the documentation is a little thin.

Summary

This chapter has highlighted some, but not all, of the many tools and alternatives for moving data. For instance, the Sqoop2 tool was just recently released. Remember that although most of the examples in this chapter have processed data in to Hadoop, these same tools can be used to send data out of Hadoop as well. Also, each of the tools examined, especially Sqoop and Flume, can process multiple types of data. You can also embed your Sqoop data-processing scripts into Oozie workflows for management and scheduling. This chapter has examined only a small portion of the functionality that is offered by Sqoop, Flume, and Storm for processing data. You could also examine a tool called Apache Chukwa (chukwa.apache.org), which has features similar to Flume. Note also that Chapter 10 examines tools like Pentaho and Talend, with which you can move data using visual building blocks.

The next chapter surveys monitoring systems like Hue to provide a visual view of Hadoop cluster processing. Hue provides a single, integrated web-based interface by which scripting and monitoring functionality can be accessed. Examples here and in earlier chapters have used Sqoop, Hive, Pig, and Oozie; next, you'll be accessing these tools within Hue.

## CHAPTER 7

Monitoring Data

No matter how carefully you set up your big data system, you need to continually monitor HDFS, as well as the Hadoop jobs and workflows running on it, to ensure the system is running as efficiently as possible. This chapter examines the Hadoop and third-party tools available for monitoring a big data system, including tools for monitoring the system-level resources on each node in the cluster and determining how processing is spread across the cluster.

For example, user interface systems such as Hue ease both the use and the monitoring of Hadoop by centralizing access to Hadoop-based functionality via a single well-designed interface. Systems like Ganglia and Nagios provide rich open-source, resource-level monitoring and alerting. In the sections that follow, I will provide working examples for sourcing these systems, installing them, and putting them into use. Because I will use the Cloudera CDH4 stack, this chapter's examples will be based upon Hadoop V2.

In the first section I examine the Hadoop Hue browser, sourcing the software, installing and configuring it, and then I present the user interface in operation.

The Hue Browser

An Apache Software Foundation top-level project released under an Apache 2 license, the Hue browser offers a web-based user interface on top of Hadoop, including user interfaces for tools like Oozie, Pig, Impala, HBase, and Hive. It also has a file browser for HDFS and interactive scripting for Pig, Sqoop, Hive, and HBase. There is also a job browser, a job designer, and an Oozie editor and dashboard.

The Hue browser provides a single location for accessing multiple Hadoop-based tools. For instance, suppose you were developing an ETL (extract, transform, and load) chain that might pull data from a remote relational database, run a Pig script on the data in HDFS, and then move the data to Hive. You could develop, test, and run the ETL components from within Hue. You could even run an Oozie job that would group and schedule the linked ETL tasks from Hue. For the simple convenience alone, that must be worth considering.

As I use the Cloudera stack version 4, I show how to install and use Hue 2.5.0, the version that comes with Cloudera CDH4. Later versions are available, but this installation should be a good introduction. Along the way, I also point out some solutions to common errors that you might encounter while installing and working with Hue. For the latest news and details on Hue, visit its official website at gethue.com.

Installing Hue

To install Hue, you need to ensure that the components it will integrate with are properly installed. For this example, that means I have to ensure that the connections to HDFS, YARN, HBase, Oozie, Sqoop, and Sqoop2 are working correctly before I move on to use Hue itself. It's advised that you follow each section completely before moving on to the Hue interface, thereby avoiding unnecessary errors.

By example, I install Hue on the CentOS 6 server hc1nn, using the Linux-based yum command as the root user, as follows:

```
[root@hc1nn ~]#  yum install  hue
[root@hc1nn ~]#  yum install  hue-server
```

Next, I add some configuration items to Hadoop for Hue. Specifically, under /etc/hadoop/conf, I add the following entry to the hdfs-site.xml file at the bottom of the configuration section:

```
<property>
  <name>dfs.webhdfs.enabled</name>
  <value>true</value>
</property>
```

This enables a webhdfs Rest API on the name node and data nodes. I repeat this addition on all Hadoop nodes in my cluster, then add the following changes to the Hadoop core-site.xml file in the same location:

```
<!-- Hue WebHDFS proxy user setting -->

<property>
  <name>hadoop.proxyuser.hue.hosts</name>
  <value>*</value>
</property>

<property>
  <name>hadoop.proxyuser.hue.groups</name>
  <value>hadoop</value>
</property>

<property>
  <name>hadoop.proxyuser.hcat.hosts</name>
  <value>*</value>
</property>

<property>
  <name>hadoop.proxyuser.hcat.groups</name>
  <value>hadoop</value>
</property>

<!-- set up hdfs trash collection advised by hue -->

<property>
  <name>fs.trash.interval</name>
  <value>10060</value>
</property>
```

By defining Hue's Hadoop proxy user settings, these first four entries define the host and group access for hue and hcat. The final entry specifies the Hue file system trash interval.

After making these changes to Hadoop, I restart the Hadoop servers to pick up the changes. Next, I set up the Hue configuration file under /etc/hue/conf called hue.ini. To begin, I set the secret key to a suitable alpha numeric value for session hashing. This secret key string should be between 30 and 60 characters long, and it should be random. It is used for Internet browser cookie session security:

```
secret_key=kdntwdfjgmxnsprngpwekspfnsmdpwtyiubkdn
```

I then define the web host and port with the following:

```
http_host=hc1nn
http_port=8888
```

I set my web services URL, which I will use later to access Hue:

```
webhdfs_url=http://hc1nn:50070/webhdfs/v1/
```

I leave the Hadoop paths as the default values; they are correct for Cloudera Hadoop CDH4:

```
hadoop_hdfs_home=/usr/lib/hadoop-hdfs
hadoop_bin=/usr/bin/hadoop
hadoop_conf_dir=/etc/hadoop/conf
```

I now define my YARN configuration as follows:

```
resourcemanager_host=hc1nn
resourcemanager_port=8032
submit_to=True
resourcemanager_api_url=http://localhost:8088
proxy_api_url=http://localhost:8088
history_server_api_url=http://localhost:19888
node_manager_api_url=http://localhost:8042
```

My liboozie section to enable the Hue Oozie browser is as follows. (If you remember, this is the Oozie URL that was used to connect to the Oozie web browser in Chapter 5.) I connect Hue to the Oozie functionality:

```
oozie_url=http://localhost:11000/oozie
```

Sqoop2 Server Setup for Hue

Next, I install and set up the Sqoop2 server. Sqoop2 is a server-based version of Sqoop that, at the time of this writing, does not yet have a full complement of functionals. For instance, right now Sqoop2 cannot transfer data from a relational database to the Hadoop HBase database. Given that Hue integrates with Sqoop2, I install it so that I can demonstrate its features via Hue.

It should also be noted at this point that Sqoop2 and Sqoop should not be installed on the same server. I have installed Sqoop on hc1nn, while Sqoop2 is installed on hc1r1m1. Sqoop2 is installed as the Linux root user, as follows:

```
[root@hc1r1m1 ~]# yum install sqoop2-server
[root@hc1r1m1 ~]# yum install sqoop2-client
```

The mysql driver library mysql-connector-java-5.1.22-bin.jar in the directory /usr/lib/sqoop/lib that is installed for Sqoop on hc1nn is copied to the Sqoop2 server hc1r1m1, as follows:

```
[root@hc1nn lib]# pwd
/usr/lib/sqoop/lib
[root@hc1nn lib]# ls -l mysql-connector-java-5.1.22-bin.jar
-rw-r--r--. 1 root root 832960 Jul 17 18:50 mysql-connector-java-5.1.22-bin.jar
```

It is copied via ftp to the following directory on hc1r1m1:

```
/usr/lib/sqoop2/webapps/sqoop/WEB-INF/lib
```

This MySQL driver gives Sqoop2 the ability to access the MySQL databases from the server hc1r1m1. Having made these changes, I restart the Sqoop2 server using the Linux server restart command as root:

```
[root@hc1r1m1 ~]# service sqoop2-server restart
```

I then update the section in the Hue configuration file hue.ini for Sqoop2 to reflect these changes. The port number comes from the value of the SQOOP_HTTP_PORT variable in the setenv.sh under /etc/sqoop2/conf on the sqoop2 install server hc1r1m1:

```
# Sqoop server URL
  server_url=http://hc1r1m1:12000/sqoop
```

HBase Cluster Setup for Hue

My section for HBase in the hue.ini file has the following entry (which will be used when the HBase cluster is set up). I determined the HBase port number by looking for the HBase Thrift server port number in the HBase logs. I looked in the directory /var/log/hbase and searched the log files there for the term "TBoundedThreadPoolServer," and the related log message, then provide the port number. The term "Cluster" represents the fact that this is a clustered version of HBase running on many servers:

```
hbase_clusters=(Cluster|hc1r1m1:9090)
```

There are some new HBase servers to introduce in this section, so before I start installing them, let's review their purposes. The *HBase Region* server manages the HBase regions comprising the storages files and blocks. The *HBase Thrift* server provides a thrift API for HBase; it means that HBase clients can be developed in multiple languages and can be used to access HBase. The *Hbase Rest* server uses an HTTP-based method to access HBase, and access is achieved and data passed via a web address. The *HBase Master* process is the main server that manages the other servers.

Before you can use the HBase browser in Hue, you must set up HBase to run as a cluster because Hue attempts to connect to the HBase Thrift server. For this example, I will install and run HBase on the three nodes where my ZooKeeper servers are running (hc1r1m1, hc1r1m2, hc1r1m3). HBase needs a master node in the cluster, so the first step for me is to install the HBase Master server (hc1r1m1, for the example):

```
yum install hbase-master
```

Next, I install the HBase Thrift server on the HBase master node (hc1r1m1) and I install the Rest and Region servers on all HBase cluster nodes:

```
yum install hbase-thrift
yum install hbase-rest
yum install hbase-regionserver
```

Next, I modify the HBase configuration files on all HBase cluster nodes under /etc/hbase/conf. In the regionservers file used in the Linux Cat command, I insert the name of the HBase master node:

```
[root@hc1r1m1 conf]# cat regionservers
hc1r1m1
```

In the HBase environment file hbase-env.sh, I make sure that logging is set up and I ensure that HBase is configured to not manage ZooKeeper. This is because the ZooKeeper servers are already running and are being used by multiple Hadoop components. HBase can use them, but does not need to manage them, thereby avoiding any impact on their availability for other components:

```
export HBASE_LOG_DIR=/var/log/hbase
export HBASE_MANAGES_ZK=false
```

I create a link in the HBase configuration directory to the Hadoop core configuration file so that HBase has access to the Hadoop configuration:

```
[root@hc1r1m1 conf]# ln -s /etc/hadoop/conf/core-site.xml   core-site.xml
[root@hc1r1m1 conf]# ls -l
total 28
lrwxrwxrwx. 1 root root   30 Aug  9 12:07 core-site.xml -> /etc/hadoop/conf/core-site.xml
```

I next set up the hbase-site.xml file by adding the following property entries between the file's configuration tags:

```
<property>
  <name>hbase.zookeeper.quorum</name>
  <value>hc1r1m1,hc1r1m2,hc1r1m3</value>
  <description>
    Comma separated list of Zookeeper servers (match to what is specified
    in zoo.cfg but without portnumbers)
  </description>
</property>

<property>
  <name>hbase.cluster.distributed</name>
  <value>true</value>
</property>

<property>
  <name>hbase.master.wait.on.regionservers.mintostart</name>
  <value>1</value>
</property>

<property>
  <name>hbase.rootdir</name>
  <value>hdfs://hc1nn:8020/hbase</value>
</property>

<property>
  <name>hbase.rest.port</name>
  <value>60050</value>
</property>
```

```
<property>
  <name>hbase.master</name>
  <value>hc1r1m1:60000</value>
</property>
```

The hbase.zookeeper.quorum parameter sets the ZooKeeper quorum servers to be the three listed machines on which ZooKeeper is installed. Setting the hbase.cluster.distributed parameter to True tells HBase that it is set up as a cluster. Finally, the HBase parameter hbase.rootdir tells HBase where to access HDFS, while the HBase hbase.master parameter tells HBase which node is the master node.

With all the pieces in place, I can start the HBase servers. To do so, I use the following script as the Linux root user on each of the HBase cluster nodes:

```
for x in `cd /etc/init.d ; ls hbase-*` ; do service $x start ; done
```

This example installation was intended as a guide for completing your own installation. If all has gone well, you have installed the components that Hue depends on and you are ready to start using Hue.

Starting Hue

Once the Hue service is started, the Hue error logs can be checked for errors. Some possible errors are described in the section that follows. Also, you can try to access the Hue interface described in the section "Running Hue." Any configuration or operational errors will quickly become apparent. If you encounter a problem, recheck your configuration steps for a mistake.

You now start the Hue service as the Linux root user using the service command. The Hue service is called "hue":

```
[root@hc1nn init.d]# service hue start
Starting hue:   [  OK  ]
```

Once the Hue service is running, you can find the Hue logs under the Linux file system directory /var/log/hue. Look through the logs for any error messages; if you find any, then check the section on errors. If you don't find a solution there, check the Hue website at gethue.com:

```
[hadoop@hc1nn ~]$ cd /var/log/hue
[hadoop@hc1nn hue]$ ls -lrt
total 80
-rw-r--r--. 1 hue hue     0 Jul 17 18:57 access.log
-rw-r--r--. 1 hue hue     0 Jul 17 18:57 shell_output.log
-rw-r--r--. 1 hue hue     0 Jul 17 18:57 shell_input.log
-rw-r--r--. 1 hue hue 15943 Jul 17 19:00 syncdb.log
-rw-r--r--. 1 hue hue     0 Jul 17 19:38 supervisor.out
-rw-r--r--. 1 hue hue     0 Jul 17 19:38 kt_renewer.out
-rw-r--r--. 1 hue hue   526 Jul 17 19:41 error.log
-rw-r--r--. 1 hue hue   310 Jul 19 10:26 kt_renewer.log
-rw-r--r--. 1 hue hue  1784 Jul 19 10:26 beeswax_server.log
-rw-r--r--. 1 hue hue  2753 Jul 19 10:26 supervisor.log
-rw-r--r--. 1 hue hue   278 Jul 19 10:26 runcpserver.out
-rw-r--r--. 1 hue hue  1254 Jul 19 10:26 runcpserver.log
-rw-r--r--. 1 hue hue 40208 Jul 19 10:26 beeswax_server.out
```

Potential Errors

Here are some of the errors that can occur while you are configuring and installing Hue. The details of the errors are provided along with their causes and solutions. You may encounter errors that are not mentioned here; if so, consult the Hue website at gethue.com.

For instance, this error occurred in the beeswax_server.out log file:

```
Booting Derby version The Apache Software Foundation - Apache Derby - 10.4.2.0 - (689064): instance
a816c00e-0147-4341-f063-0000008eef18
on database directory /var/lib/hive/metastore/metastore_db in READ ONLY mode

Database Class Loader started - derby.database.classpath=''
14/07/17 00:38:21 ERROR Datastore.Schema: Failed initialising database.
Cannot get a connection, pool error Could not create a validated object, cause: A read-only user or
a user in a read-only database is not permitted to disable read-only mode on a connection.
```

The Hue user account needs to be able to access the Hive Metastore server directory /var/lib/hive/metastore/metastore_db, but this error message indicates that it currently lacks the correct permissions. To rectify this, add a Hue Linux user account (hue2) that you can use for Hue access to the Linux hadoop group.

If you see this error:

```
could not create home directory
```

it implies that there is a configuration error in the file core-site.xml. Check your proxy user settings that were set up earlier in that file.

If you see an error like the secret key issue shown here:

```
Secret key should be configured as a random string
```

it means that the secret key has not been defined in the hue.ini file. See the secret key example configuration previously used in the "Installing Hue" section.

When attempting to use the Sqoop functionality in Hue, you may see an error like this one:

```
shell.shelltypes      Command '/usr/bin/sqoop2' for entry 'sqoop2' in Shell app configuration
cannot be found on the path.
```

It means that the Sqoop2 server that Hue depends on has not been installed. See the earlier section "Sqoop2 Server Setup for Hue."

If a Linux-based account has an associated account in the Hue browser but does not have a home directory, any of these errors may occur:

```
Failed to access filesystem root
hadoop.mapred_clusters.default          Failed to contact JobTracker plugin at localhost:9290.
Failed to determine superuser of WebHdfs at
Failed to obtain user group information:
User: hue is not allowed to impersonate hue (error 401)
```

Just as your Linux hadoop account must have a home directory under /home/hadoop, and an associated .bashrc file, so too must your Hue user account. The account that I use in these examples is Hue2; it has a home directory on Linux (/home/hue2) and a home directory in HDFS (/user/hue2). It also has the same .bashrc contents as the Linux hadoop user to set up its environment and a Linux user ID number greater than 500. If your account's user ID is not greater than 500, you may encounter an error when trying to create scripts.

To test webhdfs Hue access, use the Hue LISTSTATUS function. You can test access to the webhdfs Rest interface using a general URL of the form:

```
curl -i http://<server>:<port>/webhdfs/v1/tmp?op=LISTSTATUS
```

This is where the server is the Name Node server name and the port is the port number of the name node. The command is passed using the curl command, which retrieves information from the URL. The -I option causes header information to be included in the output. So the full command is:

```
[hadoop@hc1nn ~]$ curl -i http://hc1nn:50070/webhdfs/v1/tmp?op=LISTSTATUS
```

And the trimmed output looks like this:

```
HTTP/1.1 200 OK
Cache-Control: no-cache
Expires: Tue, 05 Aug 2014 07:03:42 GMT
Date: Tue, 05 Aug 2014 07:03:42 GMT
Pragma: no-cache
Expires: Tue, 05 Aug 2014 07:03:42 GMT
Date: Tue, 05 Aug 2014 07:03:42 GMT
Pragma: no-cache
Content-Type: application/json
Content-Length: 1167
Server: Jetty(6.1.26.cloudera.2)

{"FileStatuses":{"FileStatus":[
{"accessTime":0,"blockSize":0,"group":"hadoop","length":0,"modificationTime":1406363236301,"owner":"
hadoop","pathSuffix":"flume","permission":"755","replication":0,"type":"DIRECTORY"},
................................
```

The output is the same as an HDFS ls command on the /tmp directory. If you get this type of output, then you know that your Hue webhdfs Rest interface is working and so Hue should work for you.

Running Hue

It should be possible to connect to the Hue web-based user interface at this point. Any log-based errors encountered in your installation should have been resolved. Given that, in my example, Hue was installed on the Centos Linux server hc1nn, I can access the Hue web interface via the URL, http://hc1nn:8888/. For your installation, you simply substitute your own server name.

On your first login, you will be prompted to create an account and guided through further steps by the Quick Start wizard. The wizard displays any configuration problems it encounters, as shown in Figure 7-1.

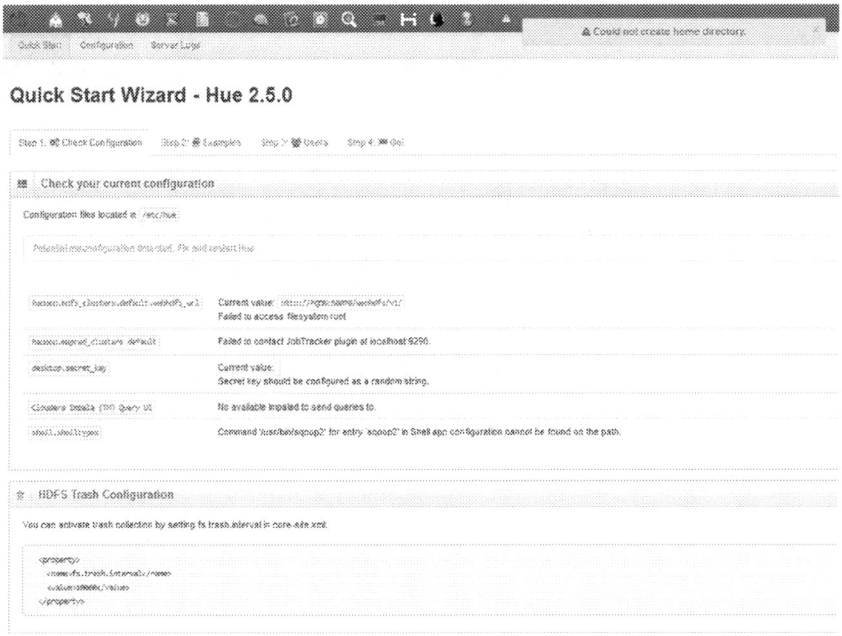

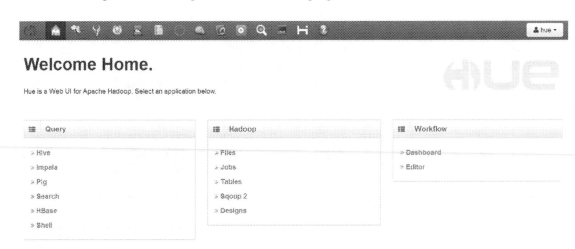

Figure 7-1. *The Hue Quick Start wizard*

Most of the problems listed in Figure 7-1 relate to missing components (such as Impala not having been installed) or a home directory not existing for the login account on Linux. Make sure that the home directories exist and that there is a suitable environment shell file, such as .bashrc, set up for the account. Also, as stated earlier, make sure that the account has a home directory on HDFS under /user. Finally, make sure that the Linux user ID number of the Linux account is greater than 500 to avoid scripting errors in Hue.

If there are no errors, you will see the Home page for Hue (Figure 7-2), which is divided into three sections of functions: those for Query, Hadoop, and Workflow. The same functions are displayed with icons across the top of the Hue screen. In Figure 7-3, for example, the Home icon is highlighted.

Figure 7-2. *The Hue home page*

Figure 7-3. *The Hue icons, with Home highlighted*

Next to the Home icon in Figure 7-3 you'll see the Bee icon, which provides access to the Hive user interface, shown in Figure 7-4. This page allows you to load and save Hive queries in Hive QL. It enables you to create queries and choose which database to run them against. There is also an Explain option, which is similar to an Explain plan in a relational database system, helping to describe how your query will run.

Figure 7-4. *The user interface for Hue Hive*

The next icon is for access to Impala. (You'll learn about using Impala via Hue in Chapter 9.) To the right of the Impala icon is the Pig icon. Click it to see the Pig interface shown in Figure 7-5. In this color-coded Pig file editor you can load and save Pig scripts, as well as examine logs and run Pig scripts. In addition, it includes a dashboard feature to examine your running jobs.

Figure 7-5. *The Hue Pig editor*

The next icon opens the HDSF browser (Figure 7-6). Here, you can upload and download files from the browser, as well as rename, move, and delete those files. You can also change file permissions and navigate the file system via clickable directory links. Clicking on a file link allows you to edit that file's contents, as illustrated in Figure 7-7.

File Browser

| | | Search for file name | A Rename | ⤢ Move | Copy | Change permissions | Download | Move to trash | ▾ | Upload ▾ | New ▾ |

| ⌂ Home | / flume / **messages** | ✎ | | | | | | 🗑 View trash |

	Type	Name	Size	User	Group	Permissions	Date
	📁			hadoop	hadoop	drwxr-xr-x	July 25, 2014 10:51 PM
	📁	.		hadoop	hadoop	drwxr-xr-x	July 25, 2014 10:50 PM
	🗋	FlumeData.1406353810397	1.3 KB	hadoop	hadoop	-rw-r--r--	July 25, 2014 10:50 PM
	🗋	FlumeData.1406353810398	1.0 KB	hadoop	hadoop	-rw-r--r--	July 25, 2014 10:50 PM
	🗋	FlumeData.1406353810399	926 bytes	hadoop	hadoop	-rw-r--r--	July 25, 2014 10:50 PM
	🗋	FlumeData.1406353810400	1.5 KB	hadoop	hadoop	-rw-r--r--	July 25, 2014 10:50 PM
	🗋	FlumeData.1406353810401	1.3 KB	hadoop	hadoop	-rw-r--r--	July 25, 2014 10:50 PM
	🗋	FlumeData.1406353810402	1.2 KB	hadoop	hadoop	-rw-r--r--	July 25, 2014 10:50 PM
	🗋	FlumeData.1406353810403	1.2 KB	hadoop	hadoop	-rw-r--r--	July 25, 2014 10:50 PM
	🗋	FlumeData.1406353810404	1.2 KB	hadoop	hadoop	-rw-r--r--	July 25, 2014 10:50 PM
	🗋	FlumeData.1406353810405	1.4 KB	hadoop	hadoop	-rw-r--r--	July 25, 2014 10:50 PM
	🗋	FlumeData.1406353810406	1.1 KB	hadoop	hadoop	-rw-r--r--	July 25, 2014 10:50 PM
	🗋	FlumeData.1406353810407	1.3 KB	hadoop	hadoop	-rw-r--r--	July 25, 2014 10:51 PM

Figure 7-6. *The Hue HDFS browser*

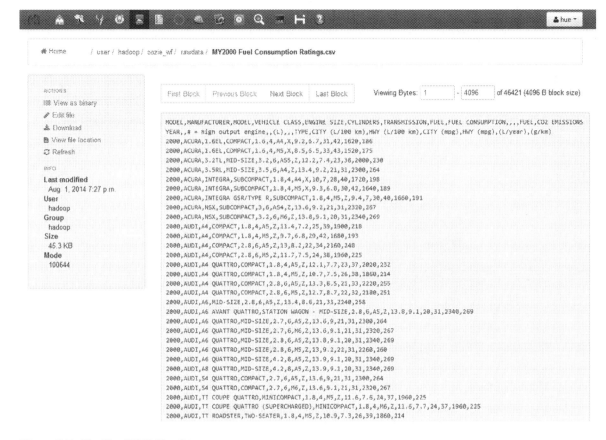

Figure 7-7. *The Hue HDFS file editor*

You can view file contents block by block, or you can edit the files by clicking the Edit option on the left. The Actions list at the left also contains options to view the contents as binary and to download the file.

Following the HDFS Browser icon in the bar is the icon for the Metastore manager (Figure 7-8), which enables you to navigate among your databases. Here, you can manage the databases and tables, as well as import data into those tables. You can also sample table data to check on the content.

Table feed1

Databases	staging / **feed1**

ACTIONS

Import Data
Browse Data
Drop Table
View File Location

Columns	Sample

⏶ Name	Type	Comment
code	string	
cost	double	
description	string	
id	int	
name	string	
value	float	

Figure 7-8. *The Hue Metastore manager*

To the right once more is the icon for the Sqoop user interface (Figure 7-9), with which you can specify and run Sqoop import and export jobs from Hue.

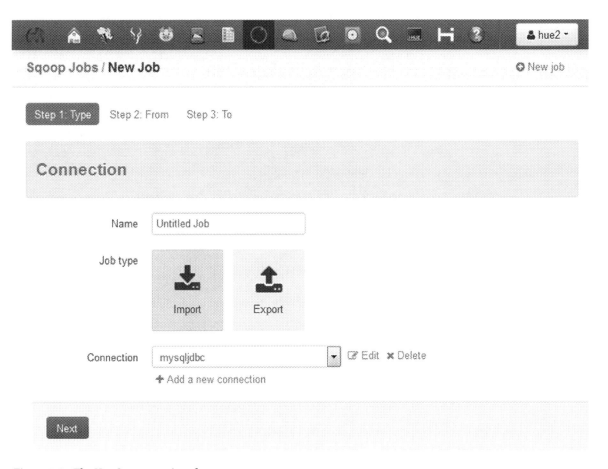

Figure 7-9. *The Hue Sqoop user interface*

An import job is specified in terms of its type, followed by where the data is coming from; in the case of Figure 7-9, it's mysql), as well as where the data is going to (HDFS, as shown in Figure 7-10).

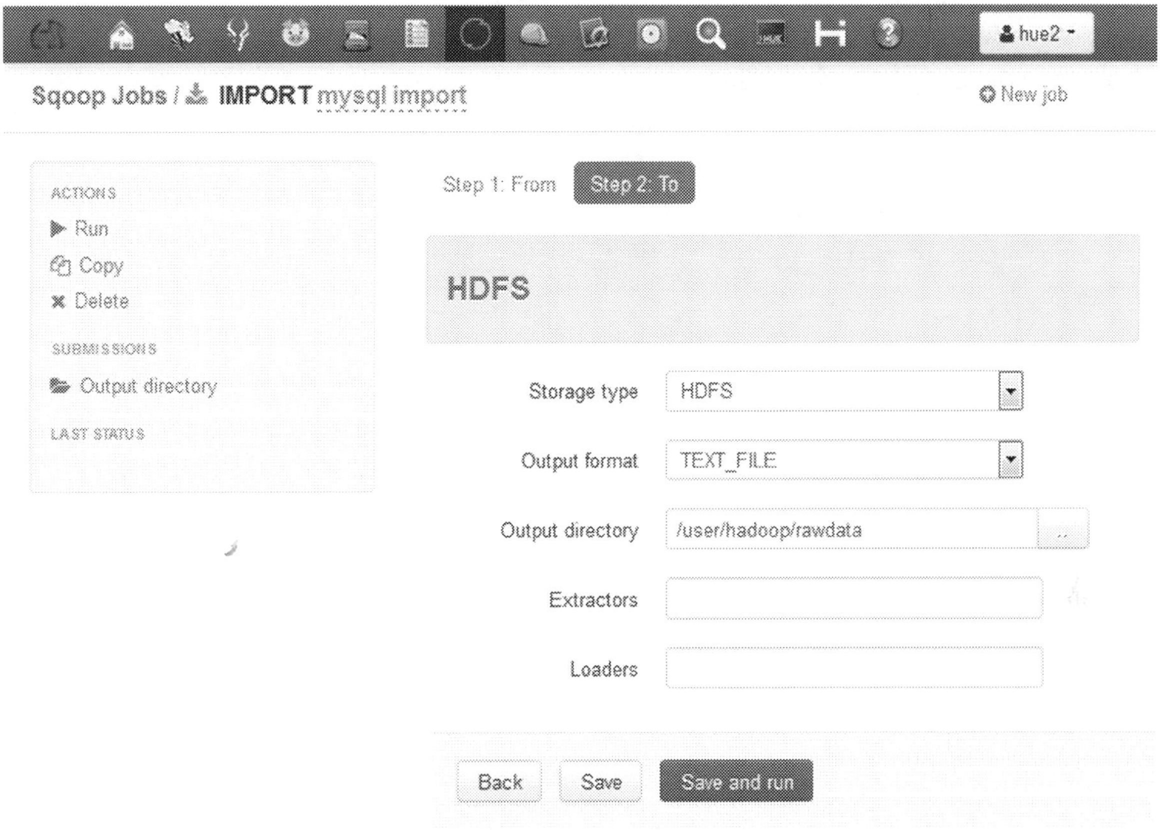

Sqoop Jobs / ⚓ **IMPORT** mysql import   ○ New job

ACTIONS
▶ Run
📋 Copy
✖ Delete

SUBMISSIONS
📁 Output directory

LAST STATUS

Step 1: From   **Step 2: To**

HDFS

Storage type   HDFS   ▾

Output format   TEXT_FILE   ▾

Output directory   /user/hadoop/rawdata   ..

Extractors   [   ]

Loaders   [   ]

Back   Save   Save and run

Figure 7-10. *The Hue Sqoop user interface: output to HDFS*

The next icon in the top bar, the Job Designer, allows you to create job actions of many different types, including but not limited to Map Reduce, Pig, Hive, and Sqoop. You can design a job by adding properties and parameters and then submit that job from the same interface. For example, Figure 7-11 shows a Sqoop action.

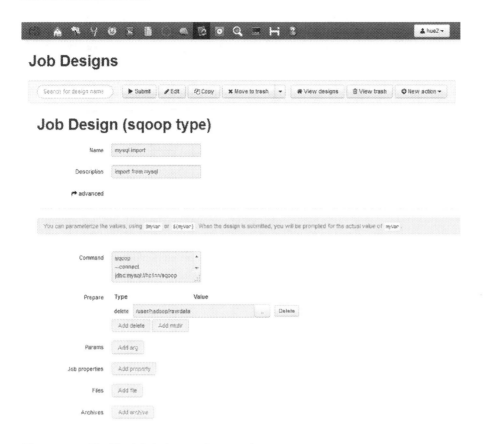

Figure 7-11. *The Hue job designer: a Sqoop action*

Continuing to the right in the top bar, the Oozie icon opens the Oozie workflow interface, which I feel is an improvement on the default Oozie user interface. From here, you can filter the job list by job type and drill down into workflows to see further details (Figure 7-12).

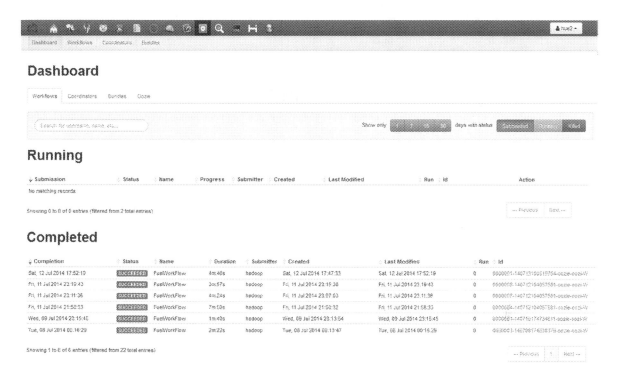

Figure 7-12. *The Hue Oozie interface*

If you click the next icon, for Hue Shell, you get the option of performing a Pig, Sqoop2, or HBase shell that will execute an adhoc script (Figure 7-13).

Figure 7-13. *The Hue shell: Pig Grunt*

The next-to-last icon, which looks like an "H," is for the HBase browser. Using this browser, you can examine HBase tables, determine their structure, and manipulate their data. In addition, you can add rows or execute a bulk upload of data (Figure 7-14).

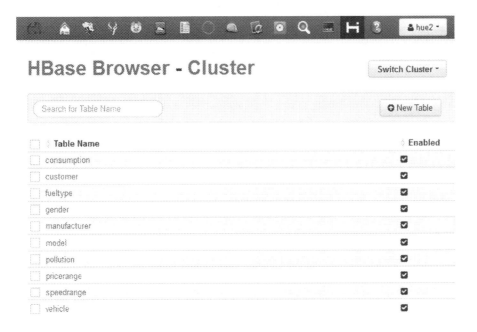

Figure 7-14. *The Hue HBase user interface*

The right-most icon (the question mark) is the Hue help option, which is comprehensive. It has a series of sub-help sections that include helpful topics on all of the Hadoop tools that have been integrated into Hue. If using Hue or its components becomes unclear at any point, the Hue help pages are a good place to find extra information.

That completes a very quick tour of the Hue browser. It packs a lot of Hadoop functionality into a single user interface, thereby enabling you to visually create and monitor Hadoop-based jobs and workflows from one interface. What it doesn't offer, however, is any low-level monitoring of system-level resources. For that, you need Ganglia and Nagios. The next sections present these monitoring tools, which can be used with a pre-existing Hadoop cluster.

Ganglia

An open-source monitoring system released under a BSD license, Ganglia is designed for monitoring on distributed high-performance systems. With Ganglia installed you can monitor a Hadoop-based cluster. The official Ganglia website is ganglia.sourceforge.net. Ganglia has been integrated into the Ambari Hadoop cluster manager, which you will examine in Chapter 8.

While Hue provided a single, web-based user interface for accessing Hadoop-based components, Ganglia offers true monitoring functionality. For example, it is possible to set up graph-based dashboards in Ganglia that show the state of Hadoop cluster resources. At a glance, it is possible to determine if there is a cluster problem at the present time or there has been one during the lifetime of each graph. Also, it is possible to add graphs for multiple types of resources on a single dashboard. (Ganglia will again be discussed in regard to the Ambari Cluster manager, in the next chapter.)

Here, I first tell how to source and install Ganglia. I also discuss some of the errors that might be encountered. Then, I discuss its user interface.

Installing Ganglia

To install Ganglia on Centos Linux, you first install the Epel repository so that you have a safe location for the Linux yum command to source the software from. Be sure to execute the Epel repository steps on all Hadoop cluster nodes to support the Ganglia install. For my example installation, I enable the Epel repository for Centos 6.x (on each node) working in /tmp/epel:

```
[root@hc1nn ~]# cd /tmp ; mkdir epel ; cd epel
[root@hc1nn epel]# wget http://dl.fedoraproject.org/pub/epel/6/x86_64/epel-release-6-8.noarch.rpm
[root@hc1nn epel]# wget http://rpms.famillecollet.com/enterprise/remi-release-6.rpm
[root@hc1nn epel]# rpm -Uvh remi-release-6*.rpm epel-release-6*.rpm
```

When I check the downloaded files in /etc/yum.repos.d/, I see that four files have been sourced:

```
[root@hc1nn epel]# ls -1 /etc/yum.repos.d/epel* /etc/yum.repos.d/remi.repo
/etc/yum.repos.d/epel-apache-maven.repo
/etc/yum.repos.d/epel.repo
/etc/yum.repos.d/epel-testing.repo
/etc/yum.repos.d/remi.repo
```

Next, I enable the remi repository by editing the remi.repo file and setting Enabled to 1 in the [remi] section:

```
[root@hc1nn epel]# vi /etc/yum.repos.d/remi.repo

[remi]
name=Les RPM de remi pour Enterprise Linux 6 - $basearch
#baseurl=http://rpms.famillecollet.com/enterprise/6/remi/$basearch/
mirrorlist=http://rpms.famillecollet.com/enterprise/6/remi/mirror
enabled=1
```

Now, I am ready to install the Ganglia software on the Hadoop name node (in my example, this is hc1nn) and all of the data nodes. I install the following packages on the Name Node server hc1nn, using the Linux yum command as the root user:

```
yum install ganglia
yum install ganglia-gmetad
yum install ganglia-web
yum install ganglia-gmond
```

On the data nodes (hc1r1m1, hc1r1m2, hc1r1m3), I install the following components:

```
yum install ganglia
yum install ganglia-gmond
```

The Ganglia gmond processes will collect data and pass it to the gmetad process on hc1nn. I can then view the data via the Ganglia web component. I must, however, tell Ganglia the frequency at which to collect the data. On hc1nn, I specify the Ganglia data-collection frequency in the file gmetad.conf; in my example, I have set it to be two minutes (120 seconds):

```
vi /etc/ganglia/gmetad.conf

data_source "my cluster" 120 hc1nn
```

In the same line, I named the cluster "my cluster" for purposes of Ganglia data collection. On the collector server hc1nn, I set up the gmond.conf file using this name and the other parameters that follow, like this:

```
[root@hc1nn etc]# vi /etc/ganglia/gmond.conf

cluster {
  name = "my cluster"
  owner = "unspecified"
  latlong = "unspecified"
  url = "unspecified"
}

udp_send_channel {
  mcast_join = hc1nn
  port = 8649
}

udp_recv_channel {
  port = 8649
}

tcp_accept_channel {
  port = 8649
}
```

I then restart the collector and node daemons on the collector server hc1nn by using the Linux `service` command:

```
service gmetad restart
service gmond  restart
```

On each data node to be monitored (hc1r1m1, hc1r1m2, hc1r1m3), I set up the configuration of the Ganglia gmond process by editing the gmond.conf file under /etc/ganglia:

```
[root@hc1r1m1 etc]# vi /etc/ganglia/gmond.conf

cluster {
  name = "my cluster"
  owner = "unspecified"
  latlong = "unspecified"
  url = "unspecified"
}

udp_send_channel {
  mcast_join = hc1nn
  port = 8649
}
```

Then I restart the Ganglia gmond server by using the Linux service command:

```
service gmond restart
```

I make sure that the httpd daemon is running on hc1nn; this supports the Ganglia web interface:

```
service httpd   restart
```

Also, I make sure that selinux is disabled on all nodes by checking that the SELINUX value is set to the file /etc/sysconfig/selinux. If this is not the case, then I will need to make the following change and restart my Linux servers:

```
[root@hc1nn ganglia]# vi /etc/sysconfig/selinux
```

```
SELINUX=disabled
```

Now I wait for the minimum monitoring interval of 120 seconds and attempt to access my Ganglia web interface by using the URL http://<collector server>/ganglia/. For example, the name of my collector server is hc1nn, with an IP address of 192.168.1.107, making the URL http://hc1nn/ganglia/.

At this point, I can enable Hadoop metrics collection on all Hadoop servers by editing the hadoop-metrics.properties file in /etc/hadoop/conf.

```
[root@hc1r1m1 conf]# cd /etc/hadoop/conf
```

```
vi hadoop-metrics.properties
```

```
dfs.class=org.apache.hadoop.metrics.ganglia.GangliaContext
dfs.period=10
dfs.servers=localhost:8649
```

```
mapred.class=org.apache.hadoop.metrics.ganglia.GangliaContext
mapred.period=10
mapred.servers=localhost:8649
```

```
jvm.class=org.apache.hadoop.metrics.ganglia.GangliaContext
jvm.period=10
jvm.servers=localhost:8649
```

```
rpc.class=org.apache.hadoop.metrics.ganglia.GangliaContext
rpc.period=10
rpc.servers=localhost:8649
```

I restart the Hadoop cluster to make the changes take effect.

Potential Errors

Potential Ganglia errors can relate to the web server configuration, permissions, access, and Ganglia servers that might be down. Here are some common error messages and some possible solutions.

For example, if you encounter an error like this on the Ganglia web page:

```
Not Found

The requested URL /ganglia was not found on this server.
Apache/2.2.15 (CentOS) Server at 192.168.1.107 Port 80
```

the solution is to edit the Ganglia web configuration file /etc/httpd/conf.d/ganglia.conf and change the Deny line to an Allow line. The # character comments out a line so that the httpd web server will ignore that line:

```
#Deny from all
Allow from all
```

Then you can restart the httpd service on the server hc1nn by using the Linux service command as the Linux root user:

```
service httpd restart
```

Displayed on the Ganglia web user interface, an error like the following implies that the SELINUX option needs to be disabled on the cluster:

```
There was an error collecting ganglia data (127.0.0.1:8652): fsockopen error: Permission denied
```

This means that Ganglia is not compatable with the option enabled. You edit the file /etc/sysconfig/selinux and set the SELINUX option to the value disabled. You will need to do this as the Linux root user account and restart each server afterwards. If you're in doubt, consult a systems administrator; otherwise, use the following command:

```
[root@hc1nn ganglia]# vi /etc/sysconfig/selinux
SELINUX=disabled
```

If an error like this occurs in the /var/log/messages file:

```
Aug 16 17:12:52 hc1nn /usr/sbin/gmetad[4575]: Please make sure that /var/lib/ganglia/rrds is owned
by ganglia
```

then the directory mentioned, /var/lib/ganglia/rrds, is not owned by the Linux ganglia user account. Check the ownership of the directory by using the Linux ls long listing and it will show that it is owned by "nobody." You can reset ownership recursively using the Linux chmod -R command:

```
[root@hc1nn log]# ls -ld /var/lib/ganglia/rrds
drwxrwxrwx. 4 nobody root 4096 Aug 16 15:06 /var/lib/ganglia/rrds
[root@hc1nn log]# chown -R ganglia /var/lib/ganglia/rrds
```

Next, you restart the Ganglia servers on the data collector server hc1nn:

```
service gmetad restart
service gmond  restart
```

If you see an error similar to the following in the /var/log/messages file:

```
Aug 16 17:28:55 hc1nn /usr/sbin/gmetad[7861]: data_thread() got no answer from any [my cluster]
datasource
```

check on each server that the Ganglia gmond processes are running. If necessary, restart them using the Linux `service` command as the root user:

```
service gmond restart
```

The Ganglia Interface

The Ganglia cluster web interface offers an overview of the cluster and a series of clickable icons that represent each server in the cluster that has been configured to collect data. Figure 7-15 shows the default CPU-based display of the cluster.

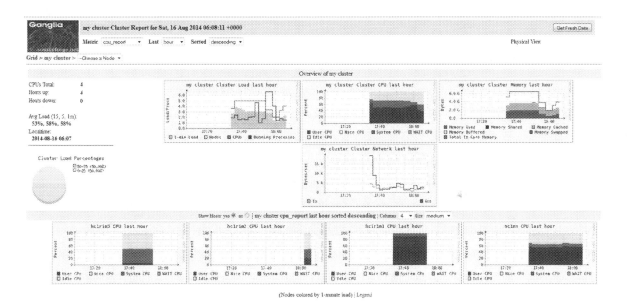

Figure 7-15. *Ganglia overview display*

In Figure 7-15, notice the Get Fresh Data button (top right) for refreshing your data and the Metric drop-down menu (top left), from which you can choose to display network, packets, memory, and load information. The visual display of server states enables you to drill down into those servers that appear to be taking the greatest load. For example, click on the Name Node display (hc1nn), and you'll get a detailed display for that server. As you can see in Figure 7-16, the display provides information about the Ganglia processes, such as when they started and how long they have been running. It also provides information about the server's memory and CPU configuration. Click the images on the right of the screen to reveal detailed graphs for the server's load, memory, CPU, network, and packet usage.

Figure 7-16. *Ganglia node display*

The second half of the screen, shown in Figure 7-17, contains a large number of sections and resource graphs. Because there are so many available, I have selected just a few to display here. However, each is composed of CPU, disk, load, memory, network, and process graphs. Each category of function being monitored has a number of aspects: for instance, CPU shows system user and input/output characteristics.

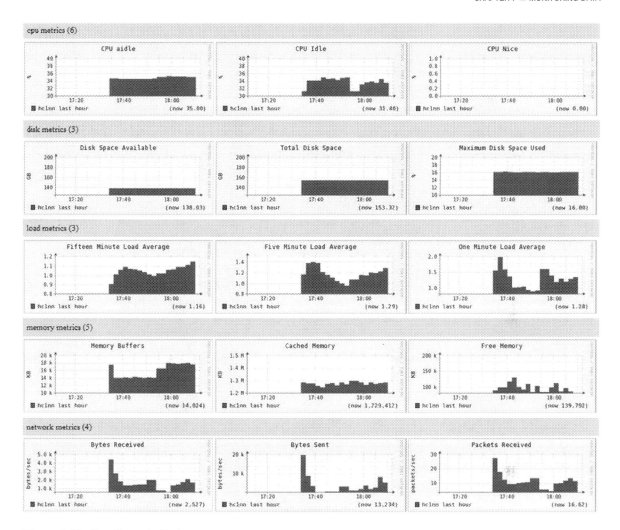

Figure 7-17. *Ganglia node display*

While this multitude of resource graphs may be bewildering at first, it does present a wide range of possible features that Ganglia can be used to monitor. It should be obvious that Ganglia provides suitable functionality to monitor a Hadoop cluster. Check websites like ganglia.sourceforge.net for further information. Also, try installing Ganglia yourself and compare the graphs that it produces by default against your own Hadoop cluster.

This chapter also presents the Nagios monitoring system. Nagios extends the possibilities for system monitoring by providing the ability to create alerts based on the problems that it finds.

Nagios

Nagios is an open-source cluster monitoring system that is available from Nagios Enterprises (www.nagios.org). While systems like Ganglia offer a wide range of monitoring graph options, the Nagios monitoring system provides the ability to monitor cluster server attributes and create alerts when problems occur. Nagios has a similar structure to Ganglia, in that it has a Nagios Master server and client (nrpe) programs. In this section, I show how it can be sourced, installed, configured, and used to monitor alerts.

It is important to reiterate Nagios's alerting capability because not only can you determine that there is a problem but you can also create an automatic alert and raise awareness of the issue. So, the combination of Ganglia and Nagios for system monitoring is a highly functional duo. Nagios alerts you to a problem while Ganglia provides a color-coded graph to display the problem. A typical problem might be that Hadoop has run out of memory or disk space. Now, you can be alerted to the problem and assess the potential solutions.

Installing Nagios

To demonstrate how to install Nagios, I place the Nagios server on the name node, hc1nn, and the Nagios clients on all of the servers so that the whole cluster will be monitored and the Nagios map (more on this in a moment) will contain the entire cluster. So, on hc1nn, I install the following servers via the Linux yum command as root:

```
[root@hc1nn ~]#  yum install nagios
[root@hc1nn ~]#  yum install nagios-plugins-all
[root@hc1nn ~]#  yum install nagios-plugins-nrpe
[root@hc1nn ~]#  yum install nrpe
```

In your installation, you will also need the php and httpd components, which you probably already have installed. When you execute the following commands, you will receive a "Nothing to do" message if the components are already installed:

```
[root@hc1nn ~]#  yum install php
[root@hc1nn ~]#  yum install httpd
```

I use the chkconfig command to configure these servers (httpd, nrpe, and Nagios) to start when the server hc1nn is rebooted. The Nagios server uses the nrpe components on each machine, which in turn use the Nagios plug-ins to monitor different features, like hosts, devices, and services:

```
[root@hc1nn ~]#  chkconfig httpd on
[root@hc1nn ~]#  chkconfig nagios on
[root@hc1nn ~]#  chkconfig nrpe on
```

Now, I install the the following Nagios components on the data nodes (hc1r1m1, hc1r1m2, hc1r1m3) by using the Linux yum command as root:

```
[root@hc1nn ~]#  yum install nagios
[root@hc1nn ~]#  yum install nagios-plugins-all
[root@hc1nn ~]#  yum install nrpe

[root@hc1nn ~]#  chkconfig nrpe on
```

On the name node, I set the password for the Nagios administration account, nagiosadmin. I will use this account and password combination to access the Nagios web browser:

```
[root@hc1nn ~]#  htpasswd -c /etc/nagios/passwd nagiosadmin
```

Thus, I can access the web browser via a URL in this form: http://hc1nn/nagios.

My example uses the name of the name node in my cluster, but it won't be accesssible until the Nagios server is started after the configuration has been finalized. Therefore, on the Name Node server hc1nn, I edit the file /etc/nagios/nagios.cfg, which is the configuration file for the central Nagios server component. I comment out the line that defines the cfg_file value for localhost (I will later define each of the cluster servers explicitly.)

```
#cfg_file=/etc/nagios/objects/localhost.cfg
```

In the same file on hc1nn, I use the cfg_dir attribute to specify the directory location where Nagios will look for the configuration files:

```
[root@hc1nn nagios]# grep "^cfg_dir" /etc/nagios/nagios.cfg
cfg_dir=/etc/nagios/conf.d
```

I go to this location and create empty configuration files for each of the servers in the cluster by using the Linux touch command:

```
[root@hc1nn nagios]# cd /etc/nagios/conf.d

[root@hc1nn conf.d]# touch hc1r1m1.cfg
[root@hc1nn conf.d]# touch hc1r1m2.cfg
[root@hc1nn conf.d]# touch hc1r1m3.cfg
[root@hc1nn conf.d]# touch hc1nn.cfg
```

The contents of each of the server configuration files mirrors the following listing (to which I added line numbers for easy reference). The important parts of a single server configuration file—in this case, hc1nn.cfg—are described below. The file sections have been listed along with some text to describe their function. The sum of the numbered configuration file parts forms a whole that Nagios uses to monitor and alert on a server.

The first entry defines the server details, while the later ones define such services as ping, ssh, and load. For example, the host definition defines the hostname and its alias, plus its IP address:

```
1     define host {
2             use                  linux-server
3             host_name            hc1nn
4             alias                     hc1nn
5             address              192.168.1.107
6             }
```

Each of the file's remaining entries define the services available to Nagios on the server. Notice that they take advantage of the terms "generic-service" and "local-service." These terms are templates specified in the file /etc/nagios/objects/templates.cfg, and they provide a way of using pre-defined attributes for the service. For example, the check_ping service (lines 8 to 13) defines the hostname to ping, hc1nn. In addition, it specifies an average roundtrip time of 100 milliseconds and a packet loss of 20 percent will produce a warning message. A roundtrip time of 500 milliseconds and a packet loss of 60 percent will trigger a critical error. The commands referred to in the check_command line (i.e., check_ping) are further defined in the file /etc/nagios/objects/commands.cfg.

```
8     define service {
9             use                          generic-service
10            host_name                    hc1nn
11            service_description          PING
12            check_command                check_ping!100.0,20%!500.0,60%
13            }
```

The rest of the configuration file contains a list of services defined for the server—in this case, hc1nn. Although space prevents me from explaining every service definition, you can find details at http://nagios.sourceforge.net/docs.

```
15      define service {
16              use                             generic-service
17              host_name                       hc1nn
18              service_description             SSH
19              check_command                   check_ssh
20              notifications_enabled           0
21              }
22
23      define service {
24              use                             generic-service
25              host_name                       hc1nn
26              service_description             Current Load
27              check_command                   check_local_load!5.0,4.0,3.0!10.0,6.0,4.0
28              }
29
30      ##### extra checks
31
32      define service{
33              use                             local-service
34              host_name                       hc1nn
35              service_description             Root Partition
36              check_command                   check_local_disk!20%!10%!/
37              }
38
39      define service{
40              use                             local-service
41              host_name                       hc1nn
42              service_description             Current Users
43              check_command                   check_local_users!20!50
44              }
45
46      define service{
47              use                             local-service
48              host_name                       hc1nn
49              service_description             Total Processes
50              check_command                   check_local_procs!250!400!RSZDT
51              }
```

I used the file /etc/nagios/objects/localhost.cfg as a template for each server configuration file, simply by changing the server name and IP address for each to match the corresponding server. These files give Nagios a map of servers that it needs to do the monitoring.

With the service configuration files finished, I configure the nrpe.cfg file on each server under /etc/nagios. The nrpe (Nagios remote plugin executor) process on each host executes the Nagios plug-ins:

```
vi /etc/nagios/nrpe.cfg
```

In the file, I find the `allowed_hosts` line and change the server name to indicate which of my servers are allowed to contact nrpe; for example, I use the following:

```
allowed_hosts=hc1nn
```

Also, I add the following command lines to each nrpe.cfg file to define which service commands will be used to monitor each server—specifically, which check will be carried out on users, loads, disks (/ and /home), and processes:

```
command[check_users]=/usr/lib/nagios/plugins/check_users -w 5 -c 10
command[check_load]=/usr/lib/nagios/plugins/check_load -w 15,10,5 -c 30,25,20
command[check_root]=/usr/lib64/nagios/plugins/check_disk -w 20% -c 10% -p /
command[check_home]=/usr/lib64/nagios/plugins/check_disk -w 20% -c 10% -p /home
command[check_zombie_procs]=/usr/lib/nagios/plugins/check_procs -w 5 -c 10 -s Z
command[check_total_procs]=/usr/lib/nagios/plugins/check_procs -w 150 -c 200
```

Now I am ready to start the Nagios servers. On each machine, I start the nrpe server as the Linux root user using the service command :

```
[root@hc1nn ~]# service nrpe start
```

On the name node (hc1nn), I start the httpd and Nagios servers:

```
[root@hc1nn ~]# service httpd start
[root@hc1nn ~]# service nagios start
```

Potential Errors

Error messages may be generated while you are installing and trying to run Nagios. For instance, you might see the following:

```
error: Starting nagios:CONFIG ERROR!  Start aborted.  Check your Nagios configuration.
```

This error may be due to a "hostgroup" entry in the Nagios configuration file. In this case, Nagios won't start and there are no log files.

First determine what the error is, then use the Linux which command to determine where the Nagios executable resides. Run that Nagios executable under /usr/sbin/nagios with a −v parameter and the full path to the Nagios configuration file. The −v option verifies the configuration file before starting, and so provides extra logged output, as follows:

```
which nagios
/usr/sbin/nagios
```

```
/usr/sbin/nagios -v /etc/nagios/nagios.cfg
```

This provides details of a configuration error message, which states that a hostgroup in the configuration file is incorrect. You comment out this section (with # characters at line position 0) and restart Nagios.

The Nagios Interface

With your servers up and running, the Nagios web interface is available at `http://hc1nn/nagios/`. The home page is shown in Figure 7-18.

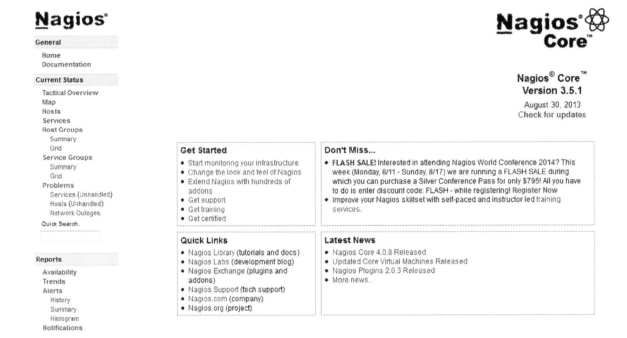

Figure 7-18. *Nagios home page*

For my example, I can click the Map option on the left of the home page to see the Nagios cluster map showing the server objects that were created under /etc/nagios/conf.d on hc1nn (bottom right of Figure 7-19). I select a server icon (such as for hc1nn) in the Nagios map to display the details for that server, as shown on the left of Figure 7-19.

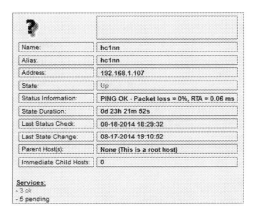

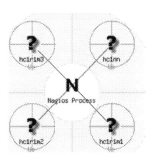

Figure 7-19. *Nagios map display*

If I click one of the servers in the map to drill down to determine server status, I get the results shown in Figure 7-20.

Current Network Status
Last Updated: Mon Aug 18 18:49:43 NZST 2014
Updated every 90 seconds
Nagios® Core™ 3.5.1 - www.nagios.org
Logged in as nagiosadmin

View History For This Host
View Notifications For This Host
View Service Status Detail For All Hosts

Host Status Totals			
Up	Down	Unreachable	Pending
1	0	0	0
All Problems	All Types		
0	1		

Service Status Totals				
Ok	Warning	Unknown	Critical	Pending
7	0	0	0	0
All Problems	All Types			
0	7			

Service Status Details For Host 'hc1nn'

Limit Results: 100

Host	Service	Status	Last Check	Duration	Attempt	Status Information
hc1nn	Current Load	OK	08-18-2014 18:48:02	0d 23h 38m 51s	1/4	OK - load average: 0.92, 0.93, 0.83
	Current Users	OK	08-18-2014 18:47:54	0d 0h 18m 49s	1/4	USERS OK - 2 users currently logged in
	PING	OK	08-18-2014 18:48:02	0d 23h 36m 51s	1/3	PING OK - Packet loss = 0%, RTA = 0.08 ms
	Root Partition	OK	08-18-2014 18:49:18	0d 0h 15m 25s	1/4	DISK OK - free space: / 42209 MB (84% inode=95%).
	SSH	OK	08-18-2014 18:45:00	0d 23h 34m 51s	1/4	SSH OK - OpenSSH_5.3 (protocol 2.0)
	Swap Usage	OK	08-18-2014 18:48:42	0d 0h 14m 1s	1/4	SWAP OK - 100% free (4031 MB out of 4031 MB)
	Total Processes	OK	08-18-2014 18:46:25	0d 0h 13m 18s	1/4	PROCS OK: 175 processes with STATE = RSZDT

Figure 7-20. *Nagios server details*

The Nagios display follows a traffic light system: green is good, warnings are yellow, and errors are red. Clicking the root partition service, as shown in Figure 7-20, allows you to drill down further to determine service details, as shown in Figure 7-21.

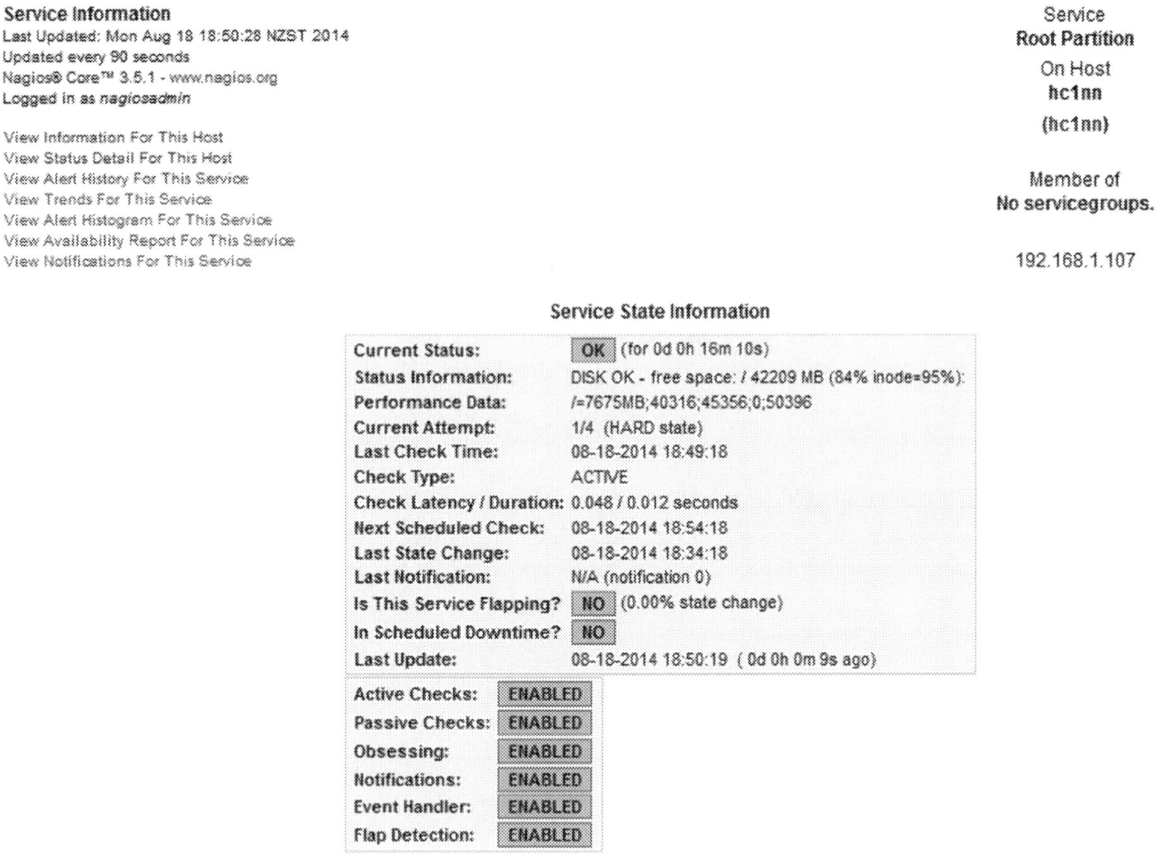

Service Information
Last Updated: Mon Aug 18 18:50:28 NZST 2014
Updated every 90 seconds
Nagios® Core™ 3.5.1 - www.nagios.org
Logged in as *nagiosadmin*

View Information For This Host
View Status Detail For This Host
View Alert History For This Service
View Trends For This Service
View Alert Histogram For This Service
View Availability Report For This Service
View Notifications For This Service

Service
Root Partition
On Host
hc1nn
(hc1nn)

Member of
No servicegroups.

192.168.1.107

Service State Information

Current Status:	OK (for 0d 0h 16m 10s)
Status Information:	DISK OK - free space: / 42209 MB (84% inode=95%):
Performance Data:	/=7675MB;40316;45356;0;50396
Current Attempt:	1/4 (HARD state)
Last Check Time:	08-18-2014 18:49:18
Check Type:	ACTIVE
Check Latency / Duration:	0.048 / 0.012 seconds
Next Scheduled Check:	08-18-2014 18:54:18
Last State Change:	08-18-2014 18:34:18
Last Notification:	N/A (notification 0)
Is This Service Flapping?	NO (0.00% state change)
In Scheduled Downtime?	NO
Last Update:	08-18-2014 18:50:19 (0d 0h 0m 9s ago)

Active Checks:	ENABLED
Passive Checks:	ENABLED
Obsessing:	ENABLED
Notifications:	ENABLED
Event Handler:	ENABLED
Flap Detection:	ENABLED

Figure 7-21. *Details of the Nagios root partition*

To demonstrate what a warning alert looks like, I changed the hc1nn.cfg configuration file on the server hc1nn, defining the root parition check_command line to provide a warning if free space reaches 90 percent. (A silly measure, I know, but it will show what a warning alert looks like.)

```
define service{
        use                     local-service
        host_name               hc1nn
        service_description     Root Partition
        check_command           check_local_disk!90%!10%!/
        }
```

I then saved the configuration file and restarted the Nagios server to pick up the changes:

```
[root@hc1nn ~]#  service nagios restart
```

Not surprisingly, at the next scheduled check, the alert shown in Figure 7-22 was raised, stating that the root partition contained less than 90 percent free space.

Service State Information

Current Status:	WARNING (for 0d 2h 7m 39s)
Status Information:	DISK WARNING - free space: / 42183 MB (84% inode=95%):
Performance Data:	/=7701MB;5039;45356;0;50396
Current Attempt:	4/4 (HARD state)
Last Check Time:	08-18-2014 21:27:18
Check Type:	ACTIVE
Check Latency / Duration:	0.194 / 0.014 seconds
Next Scheduled Check:	08-18-2014 21:32:18
Last State Change:	08-18-2014 19:24:18
Last Notification:	08-18-2014 21:27:20 (notification 3)
Is This Service Flapping?	NO (0.00% state change)
In Scheduled Downtime?	NO
Last Update:	08-18-2014 21:31:50 (0d 0h 0m 7s ago)

Figure 7-22. *Nagios root partition alert*

This is an example of a monitored alert that could enable critical intervention when your Hadoop cluster resources become limited. Monitoring and alerting on Hadoop cluster resources should be a mandatory consideration when you are setting up a cluster yourself.

A combination of Naios and Ganglia could be used on the Hadoop cluster to provide a rich selection of historical graphs and alerts. Nagios could trigger an alert for Hadoop cluster resources as those resources run low; then Ganglia could examine the graphs that have recorded the condition of the resources over time.

Summary

This chapter discussed three major enhancements for Hadoop functionality. Hue provides a central location for scripting and Hadoop-based job monitoring. Through Hue, you can inspect Hive and HBase databases and can manipulate their data. Hue also offers the ability to visually browse the Hadoop file system HDFS.

Although Hue consolidates Hadoop fuctionality and access into one useful interface, it isn't a full-scale monitoring system. (For instance, it doesn't have the ETL or reporting functionality of Pentaho or Talend, which are covered in Chapters 10 and 11.) To supplement Hue, Ganglia and Nagios offer cluster monitoring. When used together, Nagios and Ganglia complement each other: Nagios alerts the user to potential problems, while Ganglia provides graph-based details to show what has happened, on which server and its type. This combined action will help in problem investigation before system failure.

CHAPTER 8

Cluster Management

From its inception, Hadoop has been progressing and evolving to help you more easily manage your big data needs. Compared to version 1, installation of Hadoop V2 via Cloudera's version 4.x stack was an advance; Hadoop tool binaries were configured as Linux services and Hadoop's tool-related logging and functionality were moved to logical places within the Linux file system. The progression continues with the move to cluster managers, which consolidate all of the tools examined thus far in this book into a single management user interface. Cluster managers automate much of the difficult task of Hadoop component installation—and their configuration, as well.

This chapter examines Apache Ambari and the Cloudera Cluster Manager, two of several Hadoop cluster managers that enable you to install the whole Hadoop stack in one go. Management systems like Ambari also use cluster monitoring tools like Ganglia and Nagios to provide a user interface for management and monitoring within a single system. In addition, in this chapter you'll learn about the Apache Bigtop tool, with which you can install the whole stack, as well as run smoke tests during the installation to test the stack operation.

Although the installation of these components will include the whole Hadoop stack, this chapter primarily demonstrates the ease of use and overall functionality of the installation systems themselves. (I would need an entire book to cover each piece of subfunctionality within the Hadoop server stack.) Consider this chapter a snapshot of the current systems and their functionality. Which system is best for your purposes is a question you can answer only after matching your needs to their capabilities.

Because systems like Ambari install the whole Hadoop cluster, they are not compatible with pre-existing Hadoop installs, and therefore they cannot use the same set of servers, as were discussed in earlier chapters (hc1nn for the name node and hc1r1m1 to hc1r1m3 for the data nodes). For this chapter's example, I install the cluster on a new set of 64-bit machines but I preserve the work to date on the old set of machines whose Name Node server was called hc1nn. The new Name Node server is called hc2nn, and the four data nodes are called hc2r1m1, hc2r1m2, hc2r1m3, and hc2r1m4. As for the original servers, the "h" in these server names stands for Hadoop, the "c" indicates the cluster number, the "r" represents the rack number in the cluster, and the "m" represents the machine number within the Hadoop cluster rack. So, hc2nn is the Name Node server for Hadoop cluster 2. The server hc2r1m4 is the number 4 machine in rack 1 for Hadoop cluster 2. Also, because the systems examined in this chapter are intended for fresh servers, I reinstall Centos 6 on each machine prior to sourcing each system.

Initially, I install the Ambari Hadoop cluster manager. You will note that I am sourcing Ambari from the Hortonworks site, so I use it to install the latest Hortonworks Hadoop stack. If you have attempted all of the Hadoop tool installations up to this point in the book, you will have discovered that Hadoop installations and the necessary configuration are time-consuming and can be difficult. You may encounter errors that take a lot of time to solve. You might also find that versions of the components will not work with one another. But at this point, all you really want to do is use the software.

This is where cluster managers become useful: they automate the installation and configure the Hadoop cluster and the Hadoop tool set. They provide wizards that advise you when you need to make changes. They have monitoring tools to automatically check the health of your Hadoop cluster. They also offer a means to continuously upgrade the Hadoop stack.

The Cloudera cluster manager is Cloudera's own Hadoop cluster manager release, while the Ambari tool is actually an Apache project. Hortonworks has used Ambari as the mechanism to both release and manage its Hadoop stack. Given that Cloudera and Hortonworks are two of the best-known Hadoop stack suppliers, it is their cluster managers that I have chosen to source and install here.

As a last thought before launching into the installation of Ambari, I mention that there are licensing fees associated with these releases of Hadoop cluster manager. However, the costs are easily offset by the savings in time and trouble when big problems can be avoided.

The Ambari Cluster Manager

Although Ambari can be used to install many different Hadoop stacks, and is a top-level Apache project in its own right, I install only the latest HDP stack to demonstrate its functionality. As in prior chapters, I use an example installation to show how it's accomplished. I start by sourcing the Ambari code and install the Ambari agents and server on the new cluster nodes: hc2nn (the Name Node server) and hc2r1m1 through hc2r1m4 (four data nodes).

░ **Note**   For the latest information on Ambari and up-to-date release documentation, see the official Apache Software Foundation Ambari website at `http://ambari.apache.org/`.

Ambari Installation

To install the latest Hortonworks Ambari release, I first install a Hortonworks Centos 6 repository to match the version of CentOS that is installed, by running the following commands as root:

```
[root@hc2nn /]# cd /etc/yum.repos.d/
```

```
[root@hc2nn yum.repos.d]# wget http://public-repo-1.hortonworks.com/ambari/centos6/1.x/
updates/1.6.1/ambari.repo
```

The wget command downloads the ambari.repo file to the directory /etc/yum/yum.repos.d. Now, I use the Linux yum command to install the Ambari server on the Name Node server hc2nn:

```
[root@hc2nn yum.repos.d]# yum install ambari-server
```

I run the Ambari server setup command as root to configure the Ambari server component:

```
[root@hc2nn yum.repos.d]#  ambari-server setup
```

```
Using python  /usr/bin/python2.6
Setup ambari-server
Checking SELinux...
SELinux status is 'enabled'
SELinux mode is 'enforcing'
Temporarily disabling SELinux
WARNING: SELinux is set to 'permissive' mode and temporarily disabled.
OK to continue [y/n] (y)? y
Customize user account for ambari-server daemon [y/n] (n)? y
Enter user account for ambari-server daemon (root):
Adjusting ambari-server permissions and ownership...
```

```
Checking iptables...
WARNING: iptables is running. Confirm the necessary Ambari ports are accessible. Refer to the Ambari
documentation for more details on ports.
OK to continue [y/n] (y)?
Checking JDK...
[1] - Oracle JDK 1.7
[2] - Oracle JDK 1.6
[3] - Custom JDK
```

At this point, I choose option 3 for the Java JDK because I have already installed the openJDK on these servers and I am familiar with its use.

```
Enter choice (1): 3
WARNING: JDK must be installed on all hosts and JAVA_HOME must be valid on all hosts.
WARNING: JCE Policy files are required for configuring Kerberos security. If you plan to use
Kerberos,please make sure JCE Unlimited Strength Jurisdiction Policy Files are valid on all hosts.
Path to JAVA_HOME: /usr/lib/jvm/java-1.6.0-openjdk-1.6.0.0.x86_64
Validating JDK on Ambari Server...done.
Completing setup...
Configuring database...
Enter advanced database configuration [y/n] (n)?
Default properties detected. Using built-in database.
Checking PostgreSQL...
Running initdb: This may take upto a minute.
Initializing database: [  OK  ]

About to start PostgreSQL
Configuring local database...
Connecting to local database...done.
Configuring PostgreSQL...
Restarting PostgreSQL
Ambari Server 'setup' completed successfully.
```

Whichever JDK option you choose, if the last line of the setup output indicates that the setup is successful, you can start the Ambari server as root. I do just that, as follows:

```
ambari-server start

Using python  /usr/bin/python2.6
Starting ambari-server
Ambari Server running with 'root' privileges.
Organizing resource files at /var/lib/ambari-server/resources...
Waiting for server start...
Server PID at: /var/run/ambari-server/ambari-server.pid
Server out at: /var/log/ambari-server/ambari-server.out
Server log at: /var/log/ambari-server/ambari-server.log
Ambari Server 'start' completed successfully.
```

I can now access the Ambari web-based user interface via the Name Node server name and port number 8080:

```
http://hc2nn:8080
```

The default login account is called "admin" with a password of "admin." When I log in, the Ambari installation wizard will start automatically and guide me through the installation.

From the Ambari installation wizard, I set the cluster name and select the installation stack. For my example, I name the cluster "cluster2" and I select HDP 2.1, the lastest installation stack. Next, I specify the server names on which to install, making sure to use FQDNs (fully qualified domain names). So, instead of specifying the server short name of "hc2nn," I use hc2nn.semtech-solutions.co.nz. When I check the hostnames, Ambari does the equivalent of the Linux hostname -f command to determine the FQDN; if I hadn't used FQDNs, this check would fail. My server list now looks like this:

```
hc2nn.semtech-solutions.co.nz
hc2r1m1.semtech-solutions.co.nz
hc2r1m2.semtech-solutions.co.nz
hc2r1m3.semtech-solutions.co.nz
hc2r1m4.semtech-solutions.co.nz
```

If at this point in your own installation you encounter the following permission-based error, you can install the Ambari agent manually. This error message is probably related to SELinux and the file labeling used by SSH (secure shell).

```
Permission denied (publickey,gssapi-keyex,gssapi-with-mic,password).
```

A manual installation of the Ambari agent on each node is a workaround for this problem. Then, you would just be manually installing a component that Ambari has not been able to install automatically. Once you have done this manually, Ambari will complete the cluster installation for you.

If this error is encountered follow these steps; make sure that the Ambari Centos 6 repository is installed on each machine:

```
cd /etc/yum.repos.d/
wget http://public-repo-1.hortonworks.com/ambari/centos6/1.x/updates/1.6.1/ambari.repo
```

Next, I use the Linux yum command as root to install the Ambari agent on each machine, as follows:

```
yum install ambari-agent
```

I edit the Ambari configuration file and set the server hostname to be the fully qualified name of the server:

```
vi  /etc/ambari-agent/conf/ambari-agent.ini
hostname=hc2nn.semtech-solutions.co.nz
```

To register the Ambari machines, I install the Linux time service ntpd on all servers I intend to use. I use the Linux yum command as root to install the ntp servers and documentation, and I set the server to start at boot time:

```
yum install ntp ntpdate ntp-doc
chkconfig ntpd on
```

Next, I initialize and start the ntpd server:

```
ntpdate pool.ntp.org
```

```
service ntpd start
```

Now, I restart the Ambari agent as the root user:

```
ambari-agent restart
```

When the agent starts, it registers with the server so that the server has a list of the agent hostnames. I can check this list by using the curl command and see all the fully qualified server names I have previously specified. For example, the following curl command accesses the Ambari server URL by using the username and password admin provided by the -u option. It requests a list of hosts from the server, which is displayed as shown here. For instance, I ran this command as the Linux root user from the cluster2 Name Node machine hc2nn:

```
[root@hc2nn ~]# curl -u admin:admin http://hc2nn:8080/api/v1/hosts

{
  "href" : "http://192.168.1.103:8080/api/v1/hosts",
  "items" : [
    {
      "href" : "http://192.168.1.103:8080/api/v1/hosts/hc2nn.semtech-solutions.co.nz",
      "Hosts" : {
        "host_name" : "hc2nn.semtech-solutions.co.nz"
      }
    },
    {
      "href" : "http://192.168.1.103:8080/api/v1/hosts/hc2r1m1.semtech-solutions.co.nz",
      "Hosts" : {
        "host_name" : "hc2r1m1.semtech-solutions.co.nz"
      }
    },
    {
      "href" : "http://192.168.1.103:8080/api/v1/hosts/hc2r1m2.semtech-solutions.co.nz",
      "Hosts" : {
        "host_name" : "hc2r1m2.semtech-solutions.co.nz"
      }
    },
    {
      "href" : "http://192.168.1.103:8080/api/v1/hosts/hc2r1m3.semtech-solutions.co.nz",
      "Hosts" : {
        "host_name" : "hc2r1m3.semtech-solutions.co.nz"
      }
    },
    {
      "href" : "http://192.168.1.103:8080/api/v1/hosts/hc2r1m4.semtech-solutions.co.nz",
      "Hosts" : {
        "host_name" : "hc2r1m4.semtech-solutions.co.nz"
      }
    }
  ]
}
```

If the full list of server names in this output looks correct, I return to the Ambari user interface and manually register the agent servers. I can see that all of the servers in cluster2 are listed in the output of the curl command just given. Figure 8-1 shows notification of a successful registration.

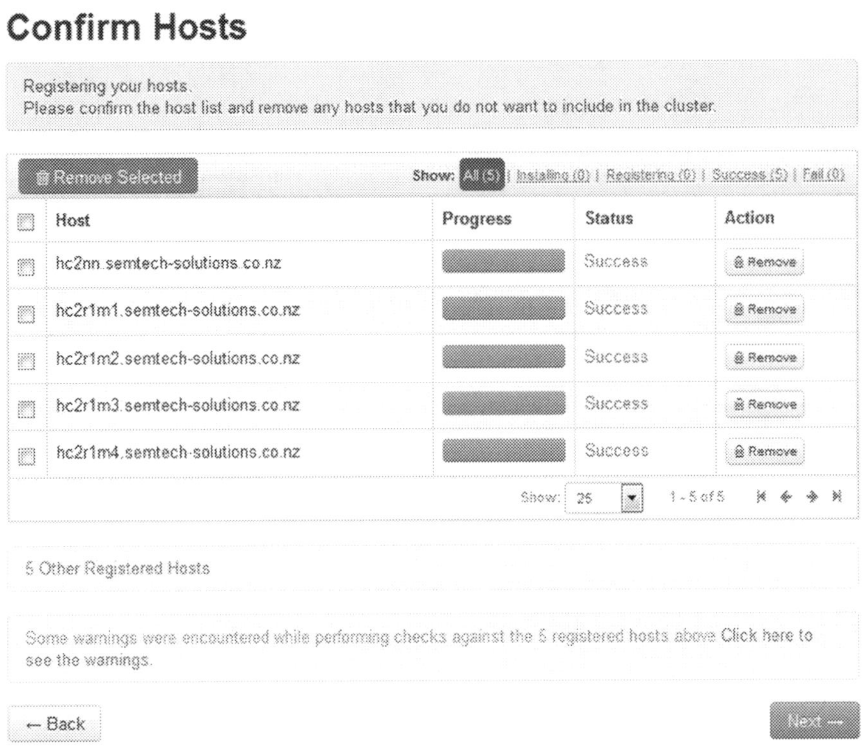

Confirm Hosts

Registering your hosts.
Please confirm the host list and remove any hosts that you do not want to include in the cluster.

🗑 Remove Selected		Show: All (5)	Installing (0)	Registering (0)	Success (5)	Fail (0)	
☐	**Host**	**Progress**	**Status**	**Action**
☐	hc2nn.semtech-solutions.co.nz	▭	Success	🔒 Remove
☐	hc2r1m1.semtech-solutions.co.nz	▭	Success	🔒 Remove
☐	hc2r1m2.semtech-solutions.co.nz	▭	Success	🔒 Remove
☐	hc2r1m3.semtech-solutions.co.nz	▭	Success	🔒 Remove
☐	hc2r1m4.semtech-solutions.co.nz	▭	Success	🔒 Remove

Show: 25 ▾   1 - 5 of 5 |◄ ← → ►|

5 Other Registered Hosts

Some warnings were encountered while performing checks against the 5 registered hosts above. Click here to see the warnings.

← Back   Next →

Figure 8-1. *Ambari server registration*

I click Next to access the Assign Master window. I keep the default settings here, because the Name Node server was set as hc2nn and components like Oozie are on a different server, which is hc2r1m1. I click Next again to set up the slaves and clients. As shown in Figure 8-2, this window defines which servers will be data nodes, node managers, and region servers, and so on. For my configuration, all of the non-Name Node (hc2nn) machines are set as data nodes and they all have clients.

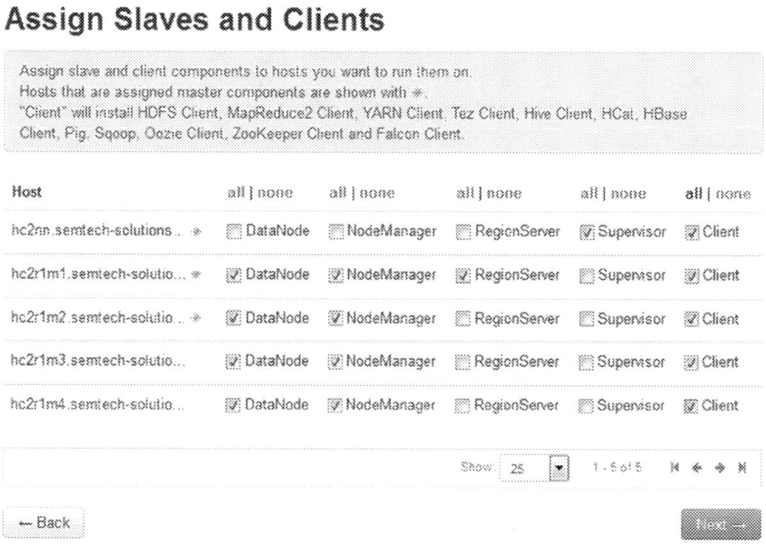

Assign Slaves and Clients

Assign slave and client components to hosts you want to run them on.
Hosts that are assigned master components are shown with ✳.
"Client" will install HDFS Client, MapReduce2 Client, YARN Client, Tez Client, Hive Client, HCat, HBase
Client, Pig, Sqoop, Oozie Client, ZooKeeper Client and Falcon Client.

Host	all | none	all | none	all | none	all | none	all | none
hc2nn.semtech-solutions .. ✳	☐ DataNode	☐ NodeManager	☐ RegionServer	☑ Supervisor	☑ Client
hc2r1m1.semtech-solutio.. ✳	☑ DataNode	☑ NodeManager	☑ RegionServer	☐ Supervisor	☑ Client
hc2r1m2.semtech-solutio... ✳	☑ DataNode	☑ NodeManager	☐ RegionServer	☐ Supervisor	☑ Client
hc2r1m3.semtech-solutio...	☑ DataNode	☑ NodeManager	☐ RegionServer	☐ Supervisor	☑ Client
hc2r1m4.semtech-solutio...	☑ DataNode	☑ NodeManager	☐ RegionServer	☐ Supervisor	☑ Client

Show: 25 ▼   1 - 5 of 5   ⏮ ◀ ▶ ⏭

← Back   Next →

Figure 8-2. *List of Ambari slaves and clients*

Clicking Next once more brings me to the Customize Services window, where I again use the defaults. Here also I can specify the passwords for the databases and monitoring. I specify passwords for Hive and Oozie, as well as a Nagios monitoring admin password and email address. I click Next to proceed to the Review Configuration window, as shown in Figure 8-3. At this stage, I am able to check the configuration that is going to be installed on my Hadoop cluster.

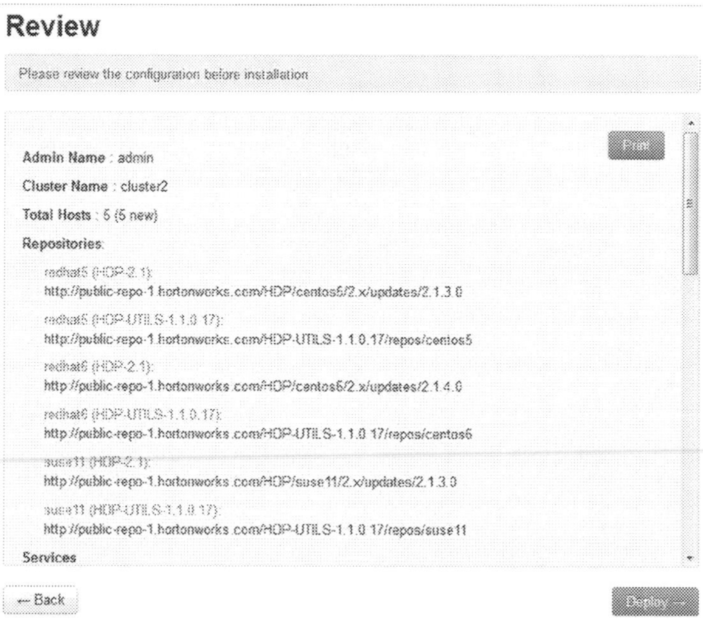

Review

Please review the configuration before installation

Print

Admin Name : admin

Cluster Name : cluster2

Total Hosts : 5 (5 new)

Repositories:

redhat5 (HDP-2.1):
http://public-repo-1.hortonworks.com/HDP/centos5/2.x/updates/2.1.3.0

redhat5 (HDP-UTILS-1.1.0.17):
http://public-repo-1.hortonworks.com/HDP-UTILS-1.1.0.17/repos/centos5

redhat6 (HDP-2.1):
http://public-repo-1.hortonworks.com/HDP/centos6/2.x/updates/2.1.4.0

redhat6 (HDP-UTILS-1.1.0.17):
http://public-repo-1.hortonworks.com/HDP-UTILS-1.1.0.17/repos/centos6

suse11 (HDP-2.1):
http://public-repo-1.hortonworks.com/HDP/suse11/2.x/updates/2.1.3.0

suse11 (HDP-UTILS-1.1.0.17):
http://public-repo-1.hortonworks.com/HDP-UTILS-1.1.0.17/repos/suse11

Services

← Back   Deploy →

Figure 8-3. *Ambari window for reviewing the configuration*

I take the time to go through the configuration list to be sure that I am happy both with what is going to be installed and with the servers on which everything will be installed.

In the Review Configuration window, I click Deploy to move to the next page. The next window gives me a list of components that will be installed. I select Next to start the installation, which takes around 30 minutes. Figure 8-4 shows the results of the successful installation.

Install, Start and Test

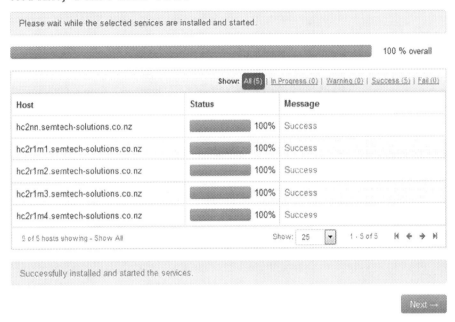

Figure 8-4. *Ambari announcement of installation success*

From the Install, Start and Test window I click Next to access the Ambari dashboard. The dashboard provides a visual overview of the Hadoop cluster, showing the services available and the state of resources in the cluster. Figure 8-5, for example, displays the dashboard's Metrics tab.

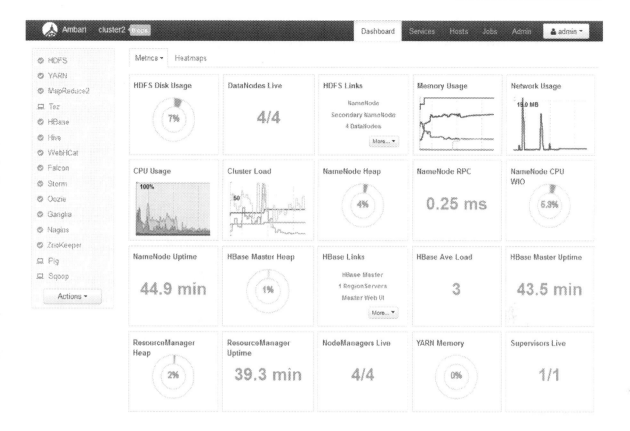

Figure 8-5. *Ambari dashboard*

The dashboard is the cluster manager home page, and the black menu bar at the top of the page allows me to select the Dashboard, Services, Hosts, Jobs, or Admin functions. The service list on the left side of the display allows me to access service-specific details—for example, HDFS. The dashboard display has two tabs—Metrics and Heatmaps—each of which I will examine shortly. For instance, the Metrics window basically shows the state of the cluster's resources.

Clicking one the resource icons provides me with a larger, more detailed display of that cluster resource type. For instance, I click CPU Usage and get the results shown in Figure 8-6. Specifically, what Figure 8-6 shows is CPU usage for the last hour. The display is color-coded, with a corresponding key; in this instance, user CPU usage exceeds that of system usage. It also shows that the cluster resources are being under-utilized, as CPU usage rarely exceeds 50 percent. Also, there is a table of minimum, average, and maximum CPU values for each category. I click OK to close that display.

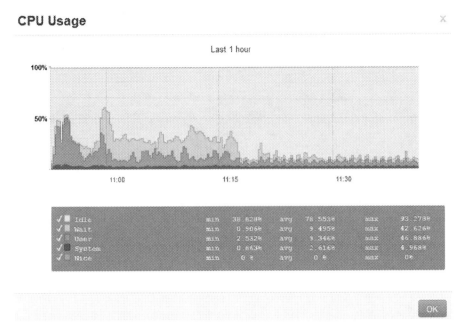

Figure 8-6. Ambari dashboard showing CPU usage

Clicking the Heatmaps tab at the top of the dashboard yields a color-coded display for the Hadoop cluster state for a given metric—in this instance, it is disk usage, as shown in Figure 8-7.

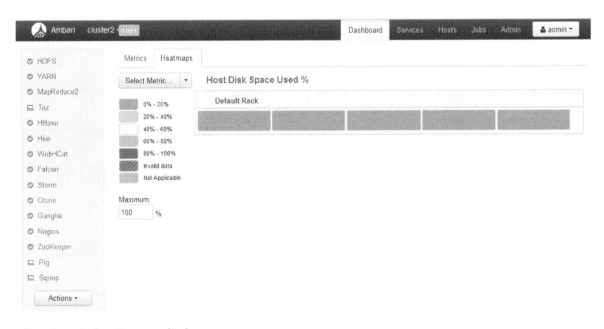

Figure 8-7. Ambari Heatmap display

The cluster that has been installed has a single server rack called "rack 1." In Figure 8-7, this is represented by the five green bars in "Host Disk Space Used %." Each bar represents a server in the rack. If there were more servers and more racks, there would be more rows of colored bars. Green represents a good state, while red would represent a problem.

The color-coded key on the left gives the meaning of the color state in the display. In this case, the key presents disk space usage and it warns when the disk space becomes low on each server.

There is a drop-down menu on the top left of the display so that different Hadoop cluster metrics can be examined—for instance, memory.

Clicking Services or Hosts in the top bar (next to Dashboard) enables you to see the cluster state from a service or server point of view. For example, by selecting an individual server, such as the Name Node, in the Hosts interface, I obtain a detailed view of the state of the Name Node server, as shown in Figure 8-8.

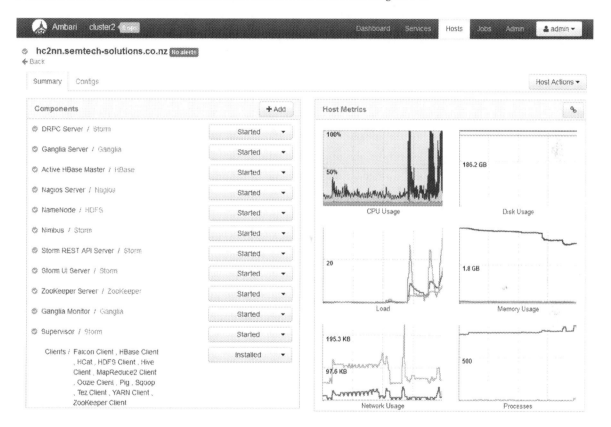

Figure 8-8. *State of Ambari server*

Figure 8-8 indicates that the system CPU (blue) on the top left graph is spiking to 100 percent. Also, it shows that memory usage for processes and disk usage seems to be high. I'm not worried about this, though, as I know that my cluster is a little under-powered, especially where memory is concerned. However, if you see graphs maxing out like this on your own cluster, you may need to invest in extra resources or examine the cluster loading.

Clicking Admin in the top menu bar will allow you to access the administration functions within Ambari, thus enabling you to examine users, high availability, the cluster, security, and access.

Finally, from the Dashboard you can also access user interfaces via links for such components as the Name Node, Resource Manager, HBase, Storm, Oozie, Ganglia, and Nagios by using the menu on the left of the screen. Many of these interfaces are familiar because I've covered them in earlier chapters of this book; it's just that Ambari brings them all together in one place. For example, Figure 8-9 illustrates how the Resource Manager user interface lists successful jobs that have completed.

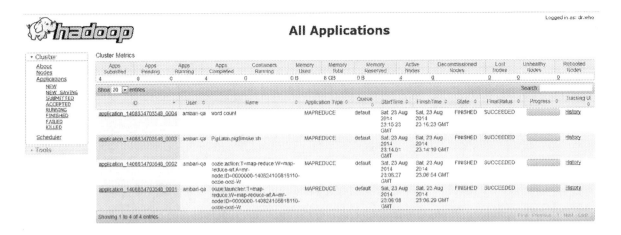

Figure 8-9. *Ambari Resource Manager user interface*

That completes this short introduction to Ambari. Given that Ganglia and Nagios monitoring systems were discusssed in Chapter 7, the Ambari displays should look familiar. That's because Ambari uses Nagios and Ganglia for its cluster manager graphs and alerts. In summary, Ambari can help you to install your cluster and provide automatic monitoring. It can also provide quick access to the web-based user interfaces of the Hadoop components.

Ambari is used for the Hortonworks Hadoop stack, but if you want to use the Cloudera stack, what does that cluster manager look like? Well, Cloudera has developed its own cluster manager application, which I demonstrate next. To avoid any conflict between installations, however, I need to reinstall CentOS Linux on all of my cluster2 servers so that they provide a clean base from which to install the next cluster version.

The Cloudera Cluster Manager

In this section I take a look at the Cloudera cluster manager for CDH5. As mentioned in this chapter's introduction, both Ambari and the Cloudera cluster managers automate the installation and management of the Hadoop stack. They also both provide a means for future software updates to be automated. So if you plan to install Hortonworks, choose Ambari; if you want Cloudera's releases, choose the Cloudera cluster manager.

Installing Cloudera Cluster Manager

For this example, I use the same 64-bit cluster of machines with 2 GB of memory, but I increase the Name Node machine's memory to 4 GB because the Cloudera cluster manager, especially the name node, needs more memory to avoid swapping. As with the installation of Ambari, I reinstall Centos 6 onto the servers so that the machines are fresh and free of conflict. Unless otherwise stated, the work is executed as the root user.

You can download the enterprise Cloudera manager binary installer from the Cloudera website at www.cloudera. com/content/support/en/downloads/cloudera_manager.html. For this example, I store the installer in /tmp on the Name Node machine hc2nn:

```
[root@hc2nn tmp]# pwd
/tmp
[root@hc2nn tmp]# ls -l cloudera-manager-installer.bin
-rw-r--r--. 1 root root 510569 Aug 30 19:11 cloudera-manager-installer.bin
```

I make the binary executable using the Linux chmod command:

```
[root@hc2nn tmp]# chmod 755 cloudera-manager-installer.bin
```

```
[root@hc2nn tmp]# ls -l cloudera-manager-installer.bin
-rwxr-xr-x. 1 root root 510569 Aug 30 19:11 cloudera-manager-installer.bin
```

Before running the installation, I disable SELinux on each server and reboot the server. I do this as the root user:

```
vi /etc/selinux/config
```

I set the SELINUX value to "disabled":

```
SELINUX=disabled
```

```
[root@hc2nn tmp]# reboot
```

Then I execute the Cloudera manager binary and follow the prompts until I can access the installation web browser at http://hc2nn:7180/. I log in as the user admin with a password of "admin" to see a list of the packages that I can install. I click Continue.

At the next page, I enter the fully qualified domain names (FQDN) hostnames on which I intend to install, in the form as follows:

```
192.168.1.103          hc2nn.semtech-solutions.co.nz          hc2nn
```

Then I click the Search button to allow the manager to obtain details about the servers. When the servers are located, I see "host ready" messages similar to those shown in Figure 8-10.

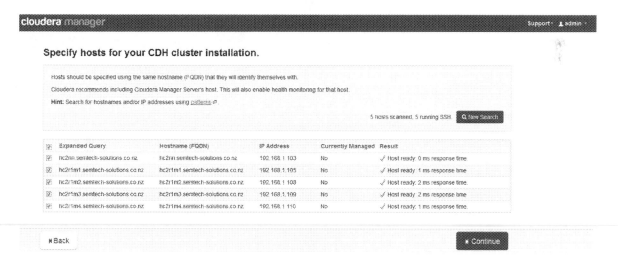

Figure 8-10. *List of Cloudera installation hosts*

I click the Continue button to move to the Select Repository page and choose the installation method, packages, or parcels (advised). Here, I can also specify the version of CDH to install and any extra parcels, such as Accumulo. Figure 8-11 shows this repository, with additional parcels like Accumulo and Sqoop to be chosen for installation, if needed.

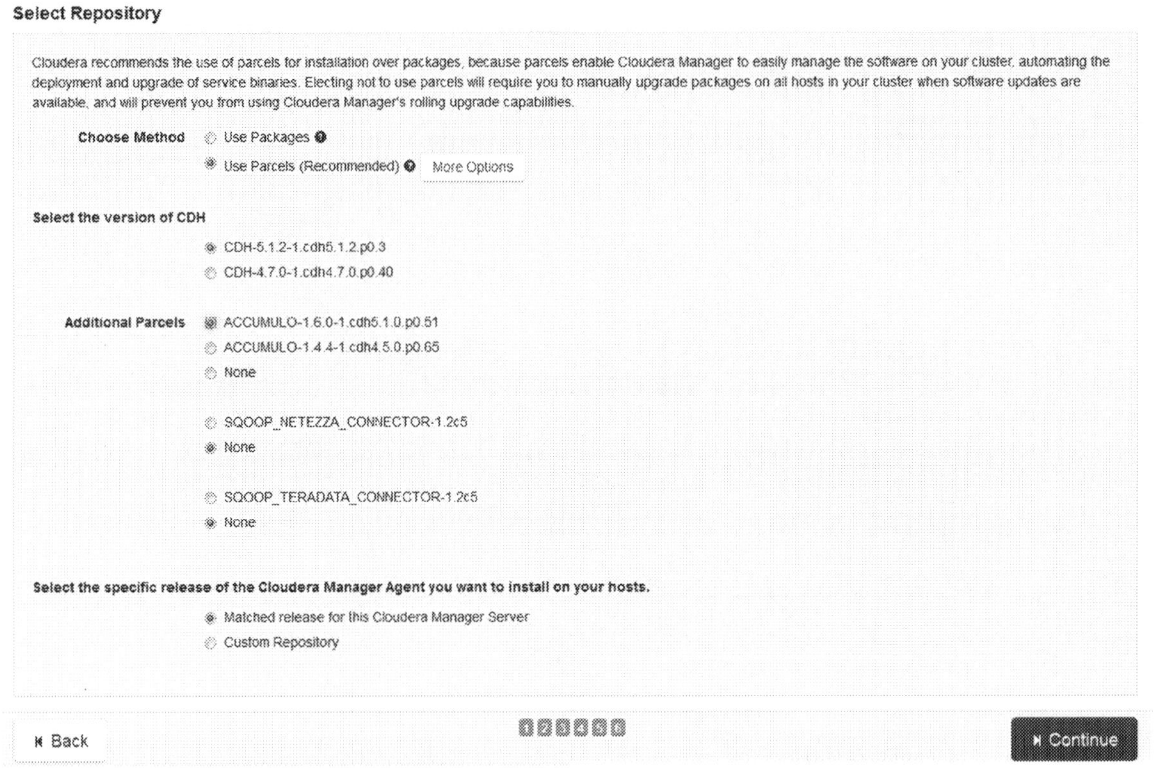

Select Repository

Cloudera recommends the use of parcels for installation over packages, because parcels enable Cloudera Manager to easily manage the software on your cluster, automating the deployment and upgrade of service binaries. Electing not to use parcels will require you to manually upgrade packages on all hosts in your cluster when software updates are available, and will prevent you from using Cloudera Manager's rolling upgrade capabilities.

Choose Method
○ Use Packages ❷
◉ Use Parcels (Recommended) ❷   More Options

Select the version of CDH
◉ CDH-5.1.2-1.cdh5.1.2.p0.3
○ CDH-4.7.0-1.cdh4.7.0.p0.40

Additional Parcels
◉ ACCUMULO-1.6.0-1.cdh5.1.0.p0.51
○ ACCUMULO-1.4.4-1.cdh4.5.0.p0.65
○ None

○ SQOOP_NETEZZA_CONNECTOR-1.2c5
◉ None

○ SQOOP_TERADATA_CONNECTOR-1.2c5
◉ None

Select the specific release of the Cloudera Manager Agent you want to install on your hosts.
◉ Matched release for this Cloudera Manager Server
○ Custom Repository

⬛ Back   🔒🔒🔒🔒🔒🔒   ⬛ Continue

Figure 8-11. *Cloudera selection repository*

I choose to install CDH 5.1.2, as shown in Figure 8-11. I had already installed CDH 4 in Chapter 2, so that installing CDH5 with the cluster manager is a progression from that previous work.

I click Continue once more, and consider enabling Java encryption. In this example, I leave my installation as the default option without encryption, simply by clicking Continue to move on.

At the next cluster installation page, which has the title "Provide SSH Login Credentials," I enter my installation login credentials. I choose to use the root user and enter the user's password. This time when I click Continue the actual installation of the Cloudera agent and server modules begins. Banners indicate both the overall progress and the individual progress per server. Figure 8-12 shows a successful installation. I click Continue again to move on to installing the selected parcels.

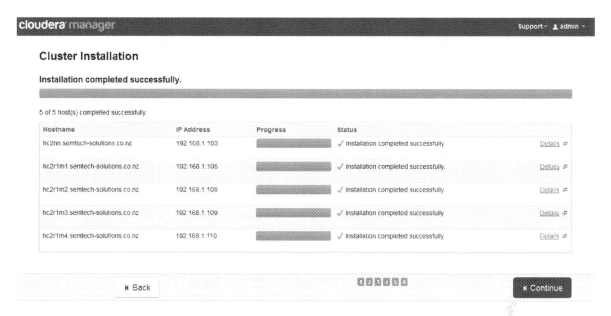

Figure 8-12. *Notification of successful Cloudera installation*

After each parcel is downloaded, distributed, and activated, my page looks like that shown in Figure 8-13.

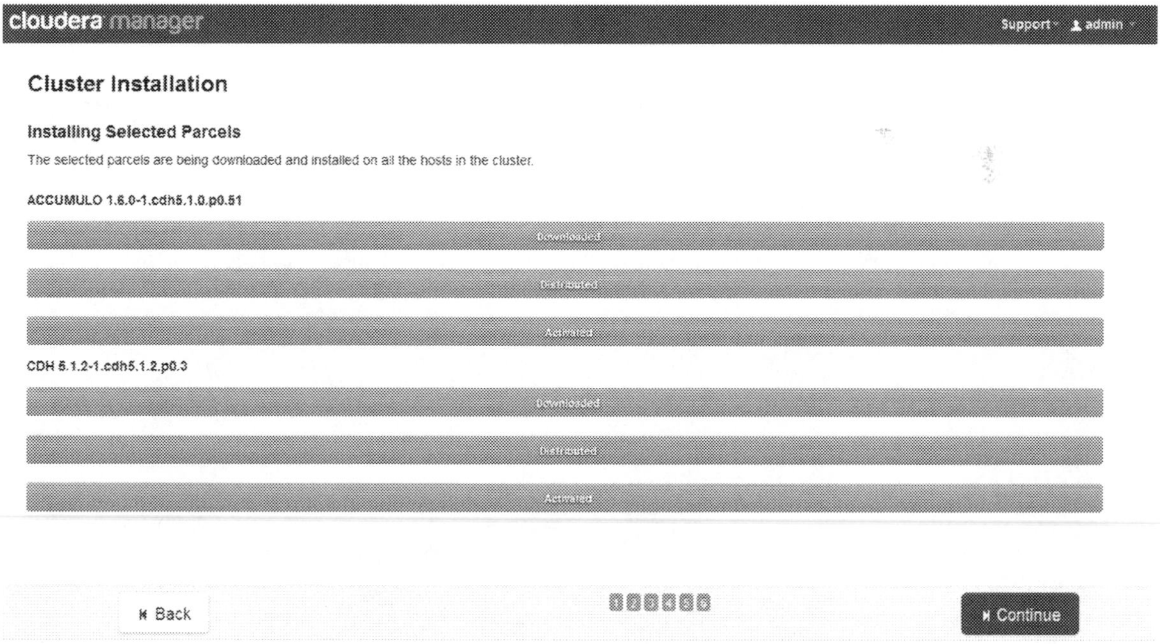

Figure 8-13. *Notice of successful parcel installation*

After I click Continue, the manager carries out a host inspection to determine whether there are any problems with the server. If there are none, I click Finish to advance to the first page of the cluster setup, as shown in Figure 8-14, where I then choose the CDH services to be installed. I choose to install Core Hadoop to reduce the load on my servers, then click Continue.

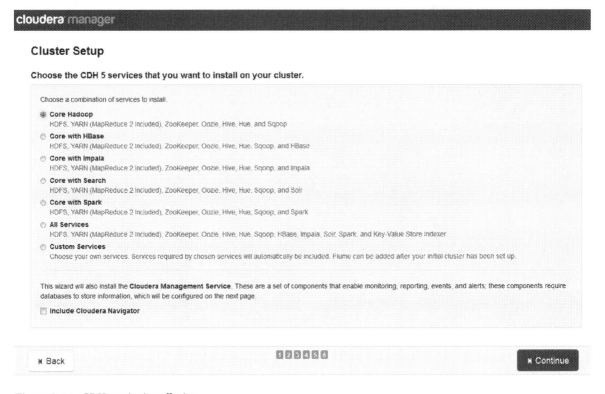

Figure 8-14. *CDH service installation*

Next, I assign the roles that determine where the services will run. For instance, I remember that ZooKeeper needs a quorum of instances and that it must be an odd number of instances so they can vote successfully. Figure 8-15 shows my assignments. I then click Continue.

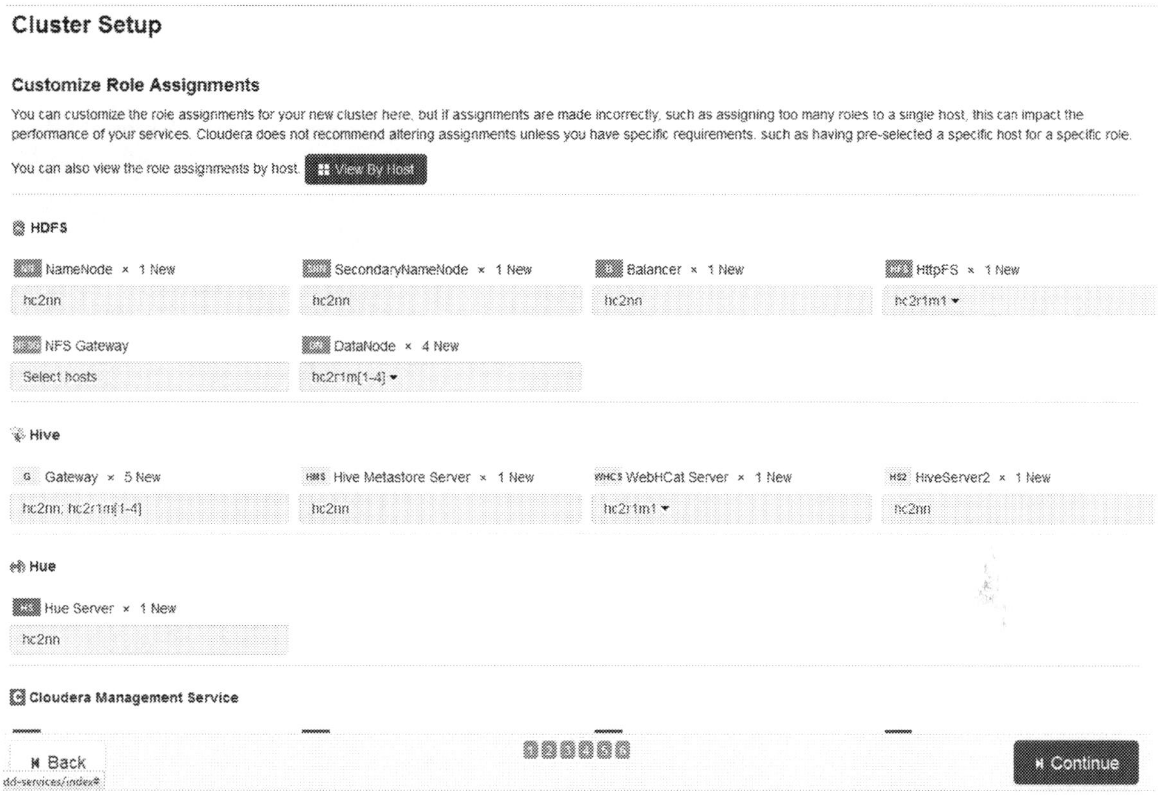

Cluster Setup

Customize Role Assignments

You can customize the role assignments for your new cluster here, but if assignments are made incorrectly, such as assigning too many roles to a single host, this can impact the performance of your services. Cloudera does not recommend altering assignments unless you have specific requirements, such as having pre-selected a specific host for a specific role.

You can also view the role assignments by host. ▦ View By Host

▨ **HDFS**

▦ NameNode × 1 New	▦ SecondaryNameNode × 1 New	▦ Balancer × 1 New	▦ HttpFS × 1 New
hc2nn	hc2nn	hc2nn	hc2r1m1 ▼

▦ NFS Gateway	▦ DataNode × 4 New		
Select hosts	hc2r1m[1–4] ▼		

▧ **Hive**

G Gateway × 5 New	HMS Hive Metastore Server × 1 New	WHCS WebHCat Server × 1 New	HS2 HiveServer2 × 1 New
hc2nn; hc2r1m[1–4]	hc2nn	hc2r1m1 ▼	hc2nn

◉ **Hue**

▦ Hue Server × 1 New
hc2nn

▣ **Cloudera Management Service**

◀ **Back**
dd-services/index#

▣▣▣▣▣▣

▶ **Continue**

Figure 8-15. *CDH role assignments*

On the subsequent Database Setup page (Figure 8-16), I choose whether to use the default embedded database (my choice) or to specify a custom database. After testing that the connections are successful, by clicking the Test Connections button, I click Continue.

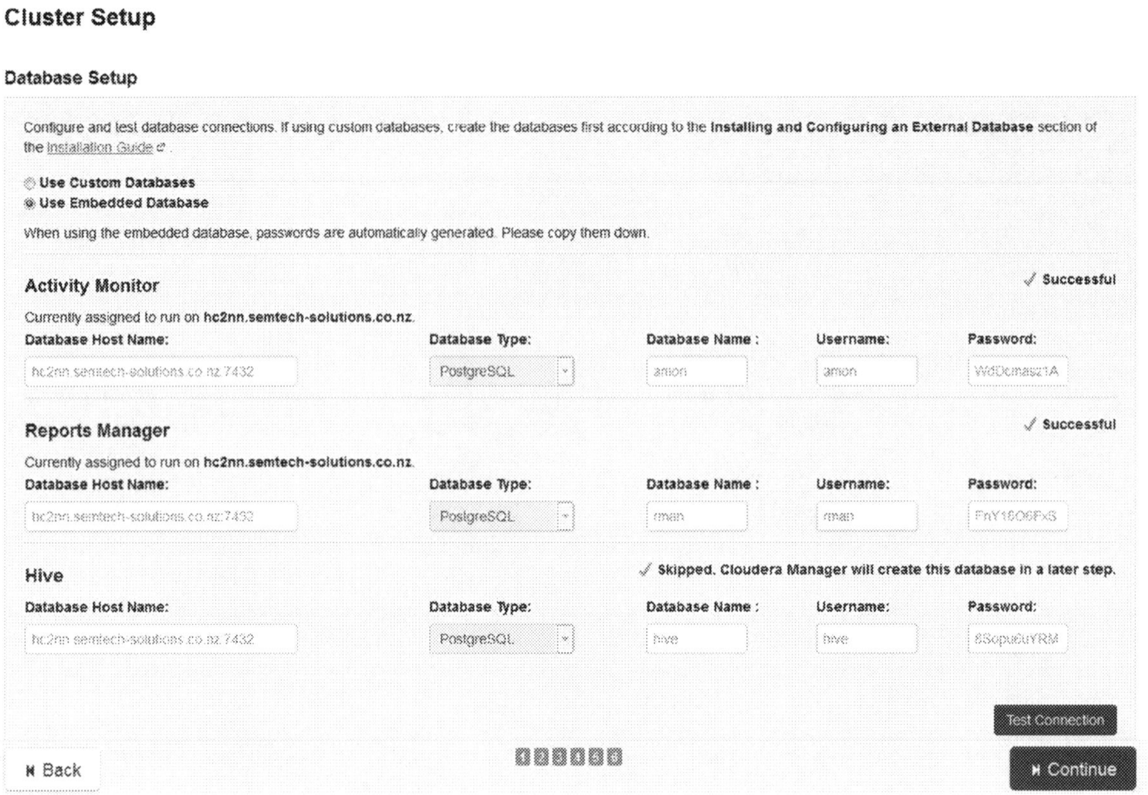

Cluster Setup

Database Setup

Configure and test database connections. If using custom databases, create the databases first according to the **Installing and Configuring an External Database** section of the Installation Guide ⌐ .

○ **Use Custom Databases**
◉ **Use Embedded Database**

When using the embedded database, passwords are automatically generated. Please copy them down.

Activity Monitor   ✓ **Successful**

Currently assigned to run on **hc2nn.semtech-solutions.co.nz**.

Database Host Name:	Database Type:	Database Name :	Username:	Password:
hc2nn.semtech-solutions.co.nz:7432	PostgreSQL	amon	amon	WdDcmasz1A

Reports Manager   ✓ **Successful**

Currently assigned to run on **hc2nn.semtech-solutions.co.nz**.

Database Host Name:	Database Type:	Database Name :	Username:	Password:
hc2nn.semtech-solutions.co.nz:7432	PostgreSQL	rman	rman	FnY1606FxS

Hive   ✓ **Skipped. Cloudera Manager will create this database in a later step.**

Database Host Name:	Database Type:	Database Name :	Username:	Password:
hc2nn.semtech-solutions.co.nz:7432	PostgreSQL	hive	hive	6SopubuYRM

Test Connection

◄ Back   □□□□□□   ► Continue

Figure 8-16. *Setting the CDH database connections*

At this point I'm given an opportunity to review the changes that will be made. I am happy with them, so I click Continue to install the cluster services. The Cluster Setup page lets me monitor the progress, as shown in Figure 8-17. When timeouts occur because there's a shortage of memory on the name node (as happens to me), I click Retry. Once the installations are complete, I click Continue.

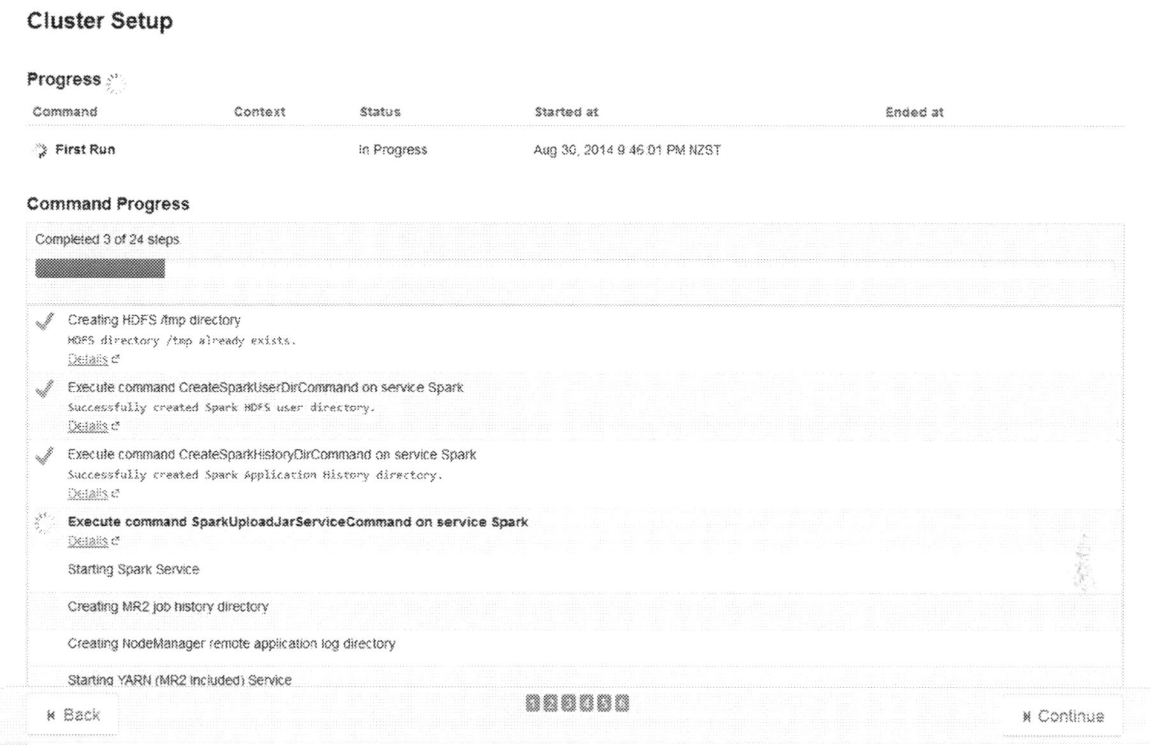

Cluster Setup

Progress

Command	Context	Status	Started at	Ended at
First Run		In Progress	Aug 30, 2014 9 46 01 PM NZST	

Command Progress

Completed 3 of 24 steps.

✓ Creating HDFS /tmp directory
HDFS directory /tmp already exists.
Details

✓ Execute command CreateSparkUserDirCommand on service Spark
Successfully created Spark HDFS user directory.
Details

✓ Execute command CreateSparkHistoryDirCommand on service Spark
Successfully created Spark Application History directory.
Details

Execute command SparkUploadJarServiceCommand on service Spark
Details

Starting Spark Service

Creating MR2 job history directory

Creating NodeManager remote application log directory

Starting YARN (MR2 Included) Service

◄ Back   ► Continue

Figure 8-17. *Monitoring the CDH cluster service installation*

Running Cloudera Cluster Manager

When the service installation is completed, I will see the Cluster Manager home screen (Figure 8-18). If I compare this to the Ambari home dashboard, I can see that the general approach is similar. Each cluster manager has a menu of services on the left of the display. Each supplier offers a slighty different list of products, though. For instance, the Ambari service list shows Tez, Falcon, Ganglia, and Nagios; the Cloudera list shows Hue. Both displays offer service graphs intended to show the state of resources—that is, memory, disk, and CPU.

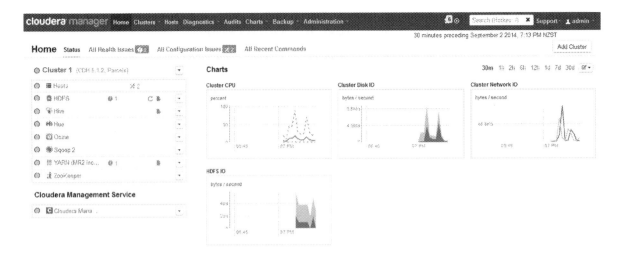

Figure 8-18. *Cloudera manager's home page*

I now progress through some of the Cloudera cluster manager screens to demonstrate its functionality.

Like the Hue application, the Cloudera cluster manager has an HDFS browser, shown in Figure 8-19. To access it, you click the HDFS service on the left of the home page and select the Browse File System option.

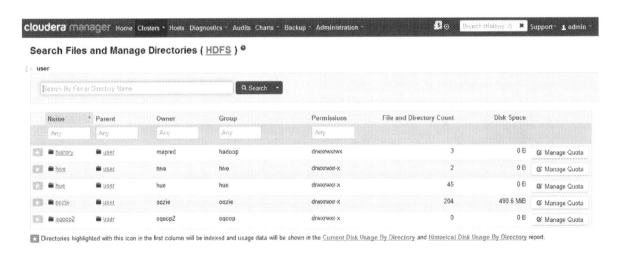

Figure 8-19. *Cloudera manager HDFS Browser*

Within the HDFS browser, you can navigate to file system elements by clicking them; the browser then displays a long list of file system details, including the object name, parent details, owner, group, permissions, an object count, and storage. Here, you can also specify quotas that allow you to set file or disk space limits for HDFS directories.

Selecting the Clusters drop-down menu from the top bar, then choosing Reports, provides a series of pre-defined reports. For example, Figure 8-20 reports the current disk usage by user.

Figure 8-20. *Reports of Cloudera cluster manager*

Beneath Cluster 1 on the left of Figure 8-18 is a list of options like Hosts, HDFS, and Hive. By selecting one of the service options from this menu, you can see a cluster manager screen that is dedicated to that service. On the screens for services like HDFS and Yarn, it is also possible to access a Hadoop web-based user interface for those services.

For instance, by selecting the cluster service YARN, followed by the Resource Manager Web user interface option, you will see the familiar Resource Manager web user interface, as shown in Figure 8-21.

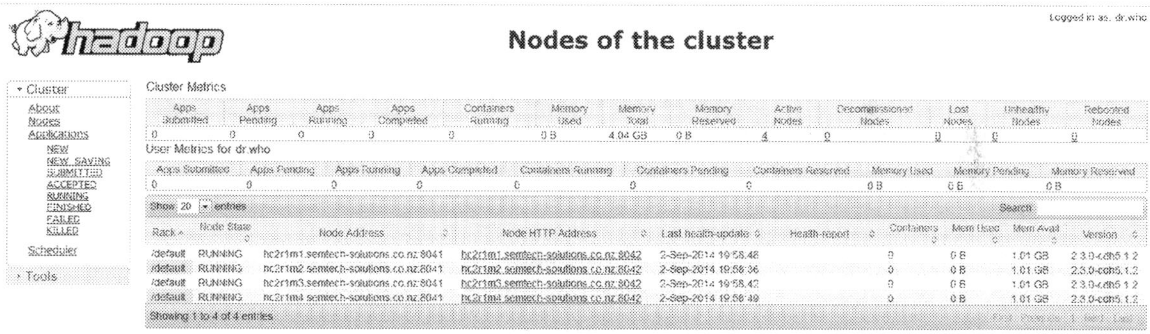

Figure 8-21. *User interface for Cloudera's Resource Manager*

Alternatively, by selecting a service option from the Cluster 1 menu, you can determine the state of that service. For instance, if you select HDFS, you'll see the resource charts, a status summary, list of health tests, and a health history, as shown in Figure 8-22.

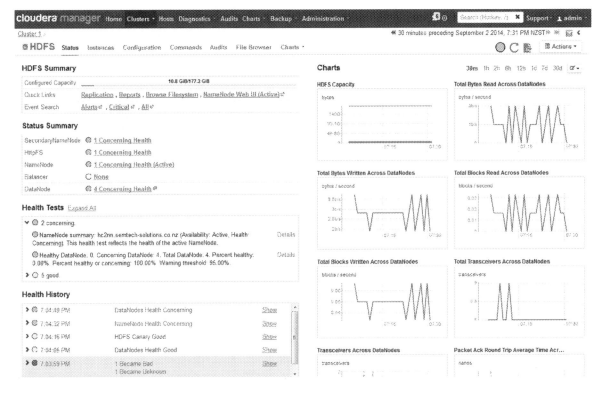

Figure 8-22. *Status of Cloudera services*

Next to Clusters in the top bar is the Hosts option. If you click that, you'll view a clickable list of cluster hosts, as shown in Figure 8-23. You can add hosts to the cluster on this page, as well as inspect the hosts, assign them to a rack, or decommission them. You can also start host roles here. Or, you can Click Disks Overview to obtain a detailed list of disk statuses, as shown in Figure 8-24.

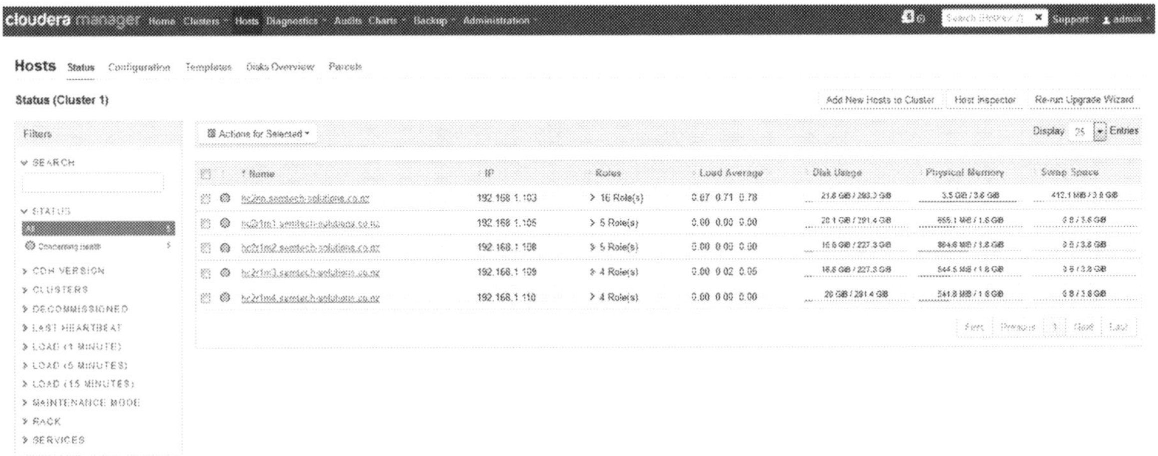

Figure 8-23. *List of Cloudera hosts*

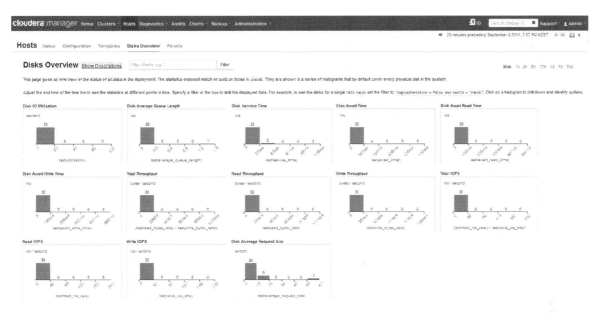

Figure 8-24. *Status of Cloudera hosts*

Click the Host page's Parcels option and you'll see the available parcels that can be installed. Parcels are provided by Cloudera from CDH4. They provide a mechanism for installing software updates without causing down time for the Hadoop cluster. They are actually a gzipped tar file bundle of software supplied by Cloudera in its own format, as are many of the downloaded and unpacked releases mentioned in this book. The difference is that Cloudera adds metadata to the parcel, which is extra information so that the Cloudera manager knows what to do with the parcel.

The Parcels display as shown in Figure 8-25 checks on the availability of new parcels via a menu option. The release of new software parcels to your cluster then just becomes a cycle of downloading those new parcels, distributing them to the cluster, and activating them—all accomplished from this single screen. If you remember the topics in Chapter 2, and all of the effort that went into installing and configuring CDH4, then you'll quickly see how this software release cycle makes the cluster manager worthwhile.

Figure 8-25. *Parcel installation menu*

In the top bar's Charts menu, you'll also find a Chart Builder option that uses an SQL-like language. With it, you can build your own reporting dashboard. Figure 8-26 shows a simple example of a dashboard that I've called "My Dash," which I built using the following SQL statement:

```
select jvm_heap_used_mb/1024, jvm_heap_committed_mb/1024 where category=ROLE and hostname="hc2nn.
semtech-solutions.co.nz"
```

Figure 8-26. *Chart builder for Cloudera cluster manager*

Given a few simple SQL-like statements, you can build a highly functional dashboard using the Cloudera cluster manager. Figure 8-26 shows the resource usage for the cluster server hc2nn, built using that simple SQL statement. This type of dashboard generation function certainly proves useful for your cluster monitoring and management.

The Cloudera cluster manager and CDH 5 stack offer a great deal more functionality than can be covered in this single chapter, but it will give you a sense of the services that the program can provide. Indeed, while developing this book, Cloudera has been my cluster manager for CDH5. For example, Cloudera enabled me to change the configuration of my cluster when tuning performance. The early-warning orange icons at the top of Figure 8-18 helped me investigate problems as they arose, and so avoid future entanglements. Also, the software parcels were especially useful when upgrading existing software and installing extra software. As I mentioned earlier, the costs in terms of licensing are handily offset by the savings in problem solving.

Apache Bigtop

Although not a cluster manager itself, Apache Bigtop aims to simplify installation and integration in its own way. Specifically, Bigtop is an attempt by the Apache Software Foundation to provide integration and smoke testing of the Apache Hadoop tool kit in order to provide an integrated Hadoop tool stack. Through Bigtop, Apache selects multiple Hadoop tools, each with its own release version, and uses a set of automated smoke tests to ensure that the set of applications works together as a stack. The result is a well-tested, high-quality, stack-based Hadoop product set.

Cloudera has recognized the value that the Apache Bigtop project is adding by basing its CDH releases on the Bigtop releases. Although Cloudera does its own testing for its CDH releases, that testing is based on a pre-tested Bigtop Hadoop stack product.

In this section I show how to source the Bigtop software and install it. Given that the smoke tests are part of the Apache Bigtop build, there isn't much to see. But I build Bigtop and extract the results of one smoke test as an example.

Installing Bigtop

To install Bigtop, I obtain the Linux yum repository file so that Bigtop software can be sourced, I run the following command on each machine in the cluster:

```
wget -O /etc/yum.repos.d/bigtop.repo http://archive.apache.org/dist/bigtop/bigtop-0.6.0/repos/
centos6/bigtop.repo

100%[============================================>] 172          --.-K/s    in 0s

2014-08-27 18:30:40 (16.7 MB/s) - "/etc/yum.repos.d/bigtop.repo" saved [172/172]
```

As for installations of the Ambari and Cloudera cluster managers, I make sure the server has CentOS 6 freshly installed. Next, I use the Linux yum command to install the components necessary for Bigtop on each machine:

```
yum install -y git cmake git-core git-svn subversion checkinstall build-essential dh-make debhelper
ant ant-optional autoconf automake liblzo2-dev libzip-dev sharutils libfuse-dev reprepro libtool
libssl-dev asciidoc xmlto ssh curl gcc gcc-c++ make fuse protobuf-compiler autoconf automake libtool
shareutils asciidoc xmlto lzo-devel zlib-devel fuse-devel openssl-devel python-devel libxml2-devel
libxslt-devel cyrus-sasl-devel sqlite-devel mysql-devel openldap-devel rpm-build create-repo redhat-
rpm-config wget
```

Now I can install Hadoop. For demonstration purposes, I do only a basic installation to get the name node and data nodes running. From that point, the installation process is the same as was shown in Chapter 2, but it is sourced from Bigtop. On the Name Node machine (hc2nn), I install the Name Node and Resource Manager components as the root user:

```
yum install hadoop-hdfs-namenode hadoop-yarn-resourcemanager
```

On the Data Node machines (hc2r1m1 to hc2r1m4), I install the data node and node manager components as the root user:

```
yum install hadoop-hdfs-datanode  hadoop-yarn-nodemanager
```

Now, I set up the configuration on each machine in the cluster under /etc/hadoop/conf. I present the necessary entries and changes briefly here, but you can find full details in Chapter 2.

I add the following to the core-site.xml file:

```
<property>
  <name>fs.default.name</name>
  <value>hdfs://hc2nn:8020</value>
</property>
```

I then modify the hdfs-site.xml file as follows:

```
<property>
  <name>dfs.replication</name>
  <value>2</value>
</property>

<property>
  <name>dfs.permissions.superusergroup</name>
  <value>hadoop</value>
</property>

<property>
  <name>dfs.replication</name>
  <value>2</value>
</property>
```

Next, I add the following changes to the yarn-site.xml file:

```
<property>
  <name>yarn.resourcemanager.address</name>
  <value>hc2nn:8032</value>
</property>

<property>
  <name>yarn.resourcemanager.scheduler.address</name>
  <value>hc2nn:8030</value>
</property>
```

```
<property>
  <name>yarn.resourcemanager.resource-tracker.address</name>
  <value>hc2nn:8031</value>
</property>

<property>
  <name>yarn.resourcemanager.admin.address</name>
  <value>hc2nn:8033</value>
</property>

<property>
  <name>yarn.resourcemanager.webapp.address</name>
  <value>hc2nn:8088</value>
</property>

<property>
  <name>yarn.resourcemanager.scheduler.class</name>
  <value>org.apache.hadoop.yarn.server.resourcemanager.scheduler.fair.FairScheduler</value>
</property>
```

I place the following entries into the slaves file to indicate where the slave data nodes reside:

```
hc2r1m1
hc2r1m2
hc2r1m3
hc2r1m4
```

To ensure that the Hadoop scheduler is correctly configured, I copy the fair-scheduler.xml file that I created in Chapter 5 and place it in the configuration directory /etc/hadoop/conf. Now I create the Linux and HDFS file system locations that are needed by the Hadoop server, and I set their ownership and permissions:

```
[root@hc1nn conf]# mkdir -p /var/lib/hadoop-hdfs/cache/hdfs/dfs/name
[root@hc1nn conf]# chown -R hdfs:hdfs /var/lib/hadoop-hdfs/cache/hdfs/dfs/name
[root@hc1nn conf]# chmod 700 /var/lib/hadoop-hdfs/cache/hdfs/dfs/name
```

I set the JAVA_HOME variable so that the Hadoop server knows where to find Java:

```
export JAVA_HOME=/usr/lib/jvm/java-1.6.0-openjdk.x86_64
```

As the hdfs user, I format the Hadoop file system without the Hadoop servers running:

```
[root@hc2nn conf]# su - hdfs
-bash-4.1$ hdfs namenode -format
```

Next, I create the directories used by YARN, setting the ownership and group membership to the YARN Linux user and group:

```
[root@hc1nn conf]# mkdir -p /var/log/hadoop-yarn/containers
[root@hc1nn conf]# mkdir -p /var/log/hadoop-yarn/apps
[root@hc1nn conf]# chown -R yarn:yarn /var/log/hadoop-yarn/containers
[root@hc1nn conf]# chown -R yarn:yarn /var/log/hadoop-yarn/apps
[root@hc1nn conf]# chmod 755 /var/log/hadoop-yarn/containers
[root@hc1nn conf]# chmod 755 /var/log/hadoop-yarn/apps
```

Then, I create the file system directories needed for staging. I use chown to set their ownership and group membership to YARN and the Linux chmod command to set the permissions:

```
[root@hc1nn conf]# mkdir -p /var/lib/hadoop-mapreduce/jobhistory/intermediate/donedir
[root@hc1nn conf]# mkdir -p /var/lib/hadoop-mapreduce/jobhistory/donedir
[root@hc1nn conf]# chown -R yarn:yarn  /var/lib/hadoop-mapreduce/jobhistory/intermediate/donedir
[root@hc1nn conf]# chown -R yarn:yarn /var/lib/hadoop-mapreduce/jobhistory/donedir
[root@hc1nn conf]# chmod 1777 /var/lib/hadoop-mapreduce/jobhistory/intermediate/donedir
[root@hc1nn conf]# chmod 750 /var/lib/hhostnameadoop-mapreduce/jobhistory/donedir
```

After carrying out these configuration file changes on all cluster nodes, I restart the servers using the root account. On the name node hc2nn, I enter:

```
service hadoop-hdfs-namenode start
service hadoop-yarn-resourcemanager start
```

On the data nodes, I enter:

```
service hadoop-hdfs-datanode start
service hadoop-yarn-nodemanager start
```

To confirm that Hadoop is up on the new cluster, I access the web interfaces for the name node and Resource Manager. I find the name node web interface at http://hc2nn:50070/, then I click Live Datanodes to show the list of active data nodes; Figure 8-27 shows the results.

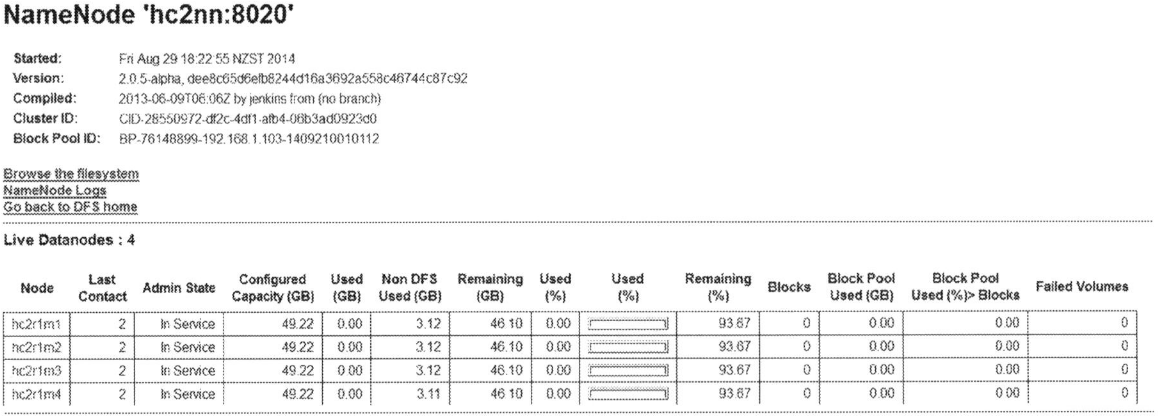

NameNode 'hc2nn:8020'

Started:	Fri Aug 29 18:22:55 NZST 2014
Version:	2.0.5-alpha, dee8c65d6efb8244d16a3692a558c46744c87c92
Compiled:	2013-06-09T06:06Z by jenkins from (no branch)
Cluster ID:	CID-28550972-df2c-4df1-afb4-06b3ad0923d0
Block Pool ID:	BP-76148899-192.168.1.103-1409210010112

Browse the filesystem
NameNode Logs
Go back to DFS home

Live Datanodes : 4

Node	Last Contact	Admin State	Configured Capacity (GB)	Used (GB)	Non DFS Used (GB)	Remaining (GB)	Used (%)	Used (%)	Remaining (%)	Blocks	Block Pool Used (GB)	Block Pool Used (%)> Blocks	Failed Volumes
hc2r1m1	2	In Service	49.22	0.00	3.12	46.10	0.00		93.67	0	0.00	0.00	0
hc2r1m2	2	In Service	49.22	0.00	3.12	46.10	0.00		93.67	0	0.00	0.00	0
hc2r1m3	2	In Service	49.22	0.00	3.12	46.10	0.00		93.67	0	0.00	0.00	0
hc2r1m4	2	In Service	49.22	0.00	3.11	46.10	0.00		93.67	0	0.00	0.00	0

Figure 8-27. *User interface for Bigtop name nodes*

To access the Resource Manager user interface, you go to http://hc2nn:8088/cluster. Click Scheduler in the left column to view the fair scheduler configuration on the new cluster, as shown in Figure 8-28.

Figure 8-28. *User interface for Bigtop Resource Manager*

Running Bigtop Smoke Tests

You can use Bigtop to quickly install a complete stack of Hadoop software if you don't want to use a cluster manager or a stack-based Hadoop release from one of the major suppliers. Even if you never plan to use Bigtop yourself, it is worth understanding how it works, because major Hadoop vendors like Cloudera use it.

To execute an Apache Bigtop smoke test as my example, I first install Apache Maven, which is used to build and run the tests.

░ **Note**   For full details on smoke tests, consult the Apache Bigtop website at `http://bigtop.apache.org`.

I then download the tar and zipped Maven package to /tmp on the server hc2nn by using a wget call to download:

```
[root@hc2nn tmp]# cd /tmp
[root@hc2nn tmp]# wget  http://supergsego.com/apache/maven/maven-3/3.2.3/binaries/apache-maven-
3.2.3-bin.tar.gz
```

I unzip the package as the root user and unpack it using the tar command:

```
[root@hc2nn tmp]# gunzip apache-maven-3.2.3-bin.tar.gz
[root@hc2nn tmp]# tar xvf apache-maven-3.2.3-bin.tar
```

Now, I move the unpacked Maven directory to /usr/local so that it resides in the correct place and create a generic symbolic link to point to the release, simplifying the configuration:

```
[root@hc2nn tmp]# mv apache-maven-3.2.3 /usr/local
[root@hc2nn tmp]# cd /usr/local
[root@hc2nn local]# ln -s apache-maven-3.2.3 apache-maven
[root@hc2nn local]# ls -ld apache-maven*
lrwxrwxrwx. 1 root root    18 Aug 30 08:23 apache-maven -> apache-maven-3.2.3
drwxr-xr-x. 6 root root 4096 Aug 30 08:21 apache-maven-3.2.3
```

I alter the Linux PATH variable so the Maven executable can be found from the command line; the type command shows that it is located under /usr/local/apache-maven/bin/mvn.

```
export PATH=/usr/local/apache-maven/bin:$PATH

[root@hc2nn tmp]# type mvn
mvn is /usr/local/apache-maven/bin/mvn
```

I check the Maven version to ensure that Maven is running without fault:

```
[root@hc2nn tmp]# mvn -version
Apache Maven 3.2.3 (33f8c3e1027c3ddde99d3cdebad2656a31e8fdf4; 2014-08-12T08:58:10+12:00)
```

Now, I set up some variables for the tests to define various Hadoop component paths:

```
export JAVA_HOME=/usr/lib/jvm/java-1.6.0-openjdk.x86_64
export HADOOP_HOME=/usr/lib/hadoop
export HADOOP_CONF_DIR=/etc/hadoop/conf
export ZOOKEEPER_HOME=/usr/lib/zookeeper
export PIG_HOME=/usr/lib/pig
export HADOOP_MAPRED_HOME=/usr/lib/hadoop-mapreduce
```

I have Hadoop and Pig installed for this example, but if you wanted to determine which other components were available for installation—say, Sqoop—you could use the yum list function, like this:

```
[root@hc2nn bigtop-tests]# yum list available | grep sqoop
hue-sqoop.x86_64                    2.5.1.5-1.el6                Bigtop
sqoop.noarch                        1.99.2.5-1.el6               Bigtop
sqoop-client.noarch                 1.99.2.5-1.el6               Bigtop
sqoop-server.noarch                 1.99.2.5-1.el6               Bigtop
```

To run the tests, I first install the test artifacts using the Maven (mvn) install command as root:

```
cd /home/hadoop/bigtop/bigtop/bigtop-tests/test-artifacts/

mvn -f pom.xml install
```

I look for the success banner to indicate that there are no errors:

```
[INFO] ------------------------------------------------------------------------
[INFO] BUILD SUCCESS
[INFO] ------------------------------------------------------------------------
[INFO] Total time: 08:26 min
[INFO] Finished at: 2014-08-30T08:44:38+12:00
[INFO] Final Memory: 20M/129M
[INFO] ------------------------------------------------------------------------
```

Now, I install the configuration components using the same Maven command and look for the same indication of success:

```
cd /home/hadoop/bigtop/bigtop/bigtop-tests/test-execution/conf

mvn -f pom.xml install

[INFO] ------------------------------------------------------------------------
[INFO] BUILD SUCCESS
[INFO] ------------------------------------------------------------------------
[INFO] Total time: 22.638 s
[INFO] Finished at: 2014-08-30T09:14:46+12:00
[INFO] Final Memory: 13M/112M
[INFO] ------------------------------------------------------------------------
```

Finally, I install the common component in the same way:

```
cd /home/hadoop/bigtop/bigtop/bigtop-tests/test-execution/common

mvn -f pom.xml install

[INFO] ------------------------------------------------------------------------
[INFO] BUILD SUCCESS
[INFO] ------------------------------------------------------------------------
[INFO] Total time: 2.301 s
[INFO] Finished at: 2014-08-30T09:15:30+12:00
[INFO] Final Memory: 9M/78M
[INFO] ------------------------------------------------------------------------
```

Now I can finally run the smoke test. I move to the desired location under Smokes and run the Maven verify command. For my test, I will run the Pig test case:

```
[root@hc2nn pig]# cd /home/hadoop/bigtop/bigtop/bigtop-tests/test-execution/smokes/pig
[root@hc2nn pig]# mvn verify
```

I won't reproduce the entire output here, but you can see that it is a series of Hadoop Pig scripts and test results, like this:

```
Running org.apache.pig.test.pigunit.pig.TestGruntParser
DUMP output;
A = LOAD 'input.txt' AS (query:CHARARRAY);
STORE output INTO '/path';
Tests run: 5, Failures: 0, Errors: 0, Skipped: 0, Time elapsed: 0.239 sec
```

This script shows that five test cases within the Pig test were run with no errors, and it demonstrates, briefly, that I can source a Hadoop stack from the Apache Bigtop project whose element programs have been integration-tested via a series of smoke tests.

You can download the Bigtop stack yourself and build it. You can also run the smoke tests on your own servers to prove that the installation will work. By running the smoke tests in your environment, you can prove that the whole Hadoop stack provided by the Bigtop project works as expected for you. Check the Apache Bigtop website (bigtop.apache.org) for further information.

Summary

Hadoop cluster managers will save you time and effort when you want to both install and upgrade your Hadoop environments. They automatically create the configuration on each cluster server. They also create a set of configuration files that are more complete and contain fewer errors than the hand-crafted alternative. Also, cluster managers offer the ability to globally modify your configuration attributes. The savings in time for that feature alone are huge if you think about manually changing the configuration files on each server in a cluster of even moderate size. Although I have examined only the Cloudera and Hortonworks cluster managers in this chapter, others are available from vendors such as MapR.

In short, the monitoring functionality of the Ambari and Cloudera cluster managers is impressive, and having a single place to check the state of the Hadoop cluster is very useful.

Although it's not a cluster manager, I included Apache Bigtop in this chapter for a number of reasons, but mainly because it is an attempt to integrate and smoke test a complete Hadoop stack. Cloudera uses Apache Bigtop to create its CDH stacks, for example. Because the work that the Bigtop project is doing is very worthwhile, I think that Hadoop users should be aware of it.

CHAPTER 9

Analytics with Hadoop

Analytics is the process of finding significance in data, meaning that it can support decision making. Decision makers turning to Hadoop's data for their answers will find numerous analytics options. For example, Hadoop-based databases like Apache Hive and Cloudera Impala offer SQL-like interfaces with HDFS-based data. For in-memory data processing, Apache Spark is available at a processing rate that is an order faster than Hadoop. For those who have had experience with relational databases, these SQL-like languages can be a simple path into analytics on Hadoop.

In this chapter, I will explain the building blocks of each Hadoop application's SQL-like language. The actions you need to take to transform your data will impact the methods you use to create your complex SQL. Just as in Chapter 4, when I introduced Pig user-defined functions (UDFs) to extend the functionality of Pig, in this chapter I will create and use a Hive UDF in an example using Hive QL. I begin with coverage of Cloudera Impala, move on to Apache Hive, and close the chapter with a discussion of Apache Spark.

Cloudera Impala

Released via an Apache license and provided as an open-source system, Cloudera Impala is a massively parallel processing (MPP) SQL query engine for Apache Hadoop. Forming part of Cloudera's data hub concept, Impala is the company's Hadoop-based database offering. Chapter 8 introduced the Cloudera cluster manager, which offers monitoring, security, and a well-defined upgrade process. Impala integrates with this architecture, and given that it uses HDFS, offers a low-cost, easily expandable, reliable, and robust storage system.

By way of example, I install Impala from the Cloudera CDH 4 stack, then demonstrate how to use it via the Hue application, as well as how to put Impala's shell tool and query language to work.

Installation of Impala

The chapter builds on work carried out in previous chapters on Cloudera's CDH 4 Hadoop stack. So, to follow this installation, you will need to have completed the installation of the CDH4 Hadoop stack in Chapter 2 and of the Hue application in Chapter 7. I install Impala manually on server hc1r1m1 and access it via the Hue tool that is already installed on the server hc1nn. I choose the server hc1r1m1 because I have a limited number of servers and I want to spread the processing load though my cluster.

So, the first step is to install some components required by Impala, as the Linux root user on the server hc1r1m1, by issuing the following Linux yum command. (These components may already be installed on your servers; if so, they will not be reinstalled. This step just ensures that they are available now):

```
[root@hc1r1m1 ~]# yum install python-devel openssl-devel python-pip
```

Next, I install a repository file under /etc/yum.repos.d on hc1r1m1 for Impala, so that the Linux yum command knows where to find the Cloudera Impala software. The repository file is downloaded from Cloudera's site by using the Linux wget command:

```
[root@hc1r1m1 ~]# cd /etc/yum.repos.d
[root@hc1r1m1 ~]# wget http://archive.cloudera.com/impala/redhat/6/x86_64/impala/cloudera-impala.repo
```

I can examine the contents of this downloaded repository file by using the Linux cat command:

```
[root@hc1r1m1 yum.repos.d]# cat cloudera-impala.repo

[cloudera-impala]
name=Impala
baseurl=http://archive.cloudera.com/impala/redhat/6/x86_64/impala/1/
gpgkey = http://archive.cloudera.com/impala/redhat/6/x86_64/impala/RPM-GPG-KEY-cloudera
gpgcheck = 1
```

Next, I install the Impala components and the Impala shell by using the yum command as the Linux root user:

```
[root@hc1r1m1 ~]# yum install impala impala-server impala-state-store impala-catalog impala-shell
```

These commands install the Impala Catalogue server, the Impala server, the Impala State Store server, and the Impala scripting shell. The Impala server runs on each node in an Impala cluster; it accepts queries and passes data to and from the files. The Impala scripting shell acts as a client to receive user commands and passes them to the server. Key to making an Impala cluster robust, the State Store server monitors the state of an Impala cluster and manages the workload when something goes wrong. The Catalog server manages metadata—that is, data about data—and passes details about metadata changes to the rest of the cluster.

As soon as the software is installed, it is time to configure it. I copy the Hive hive-site.xml, the HBase hbase-site.xml, and the Hadoop files core-site.xml and hdfs-site.xml to the Impala configuration area, which I find under/etc/impala/conf. The dot character (.) at the end of the cp (copy)command is just Linux shorthand for the current directory:

```
[root@hc1r1m1 conf]# cd /etc/impala/conf

[root@hc1r1m1 conf]# cp /etc/hive/conf/hive-site.xml .
[root@hc1r1m1 conf]# cp /etc/hadoop/conf/core-site.xml .
[root@hc1r1m1 conf]# cp /etc/hbase/conf/hbase-site.xml .
[root@hc1r1m1 conf]# cp /etc/hadoop/conf/hdfs-site.xml .
```

To specify the host and port number for the Hive metastore thrift API, as well as to specify a timeout value for access, I make the following changes to the hive-site.xml file in the Impala configuration area:

```
<!-- impala changes -->

<property>
  <name>hive.metastore.uris</name>
  <value>thrift://hc1r1m1:9083</value>
  <description>
    IP address (or fully-qualified domain name) and port of the metastore host
  </description>
</property>
```

```
<property>
  <name>hive.metastore.client.socket.timeout</name>
  <value>3600</value>
  <description>MetaStore Client socket timeout in seconds</description>
</property>
```

Then, I make the following changes to Impala's copy of the hdfs-site.xml file:

```
<!-- changes to impala -->

<property>
  <name>dfs.client.read.shortcircuit</name>
  <value>true</value>
</property>

<property>
  <name>dfs.domain.socket.path</name>
  <value>/var/run/hdfs-sockets/dn</value>
</property>

<property>
  <name>dfs.client.file-block-storage-locations.timeout</name>
  <value>4000</value>
</property>

<property>
  <name>dfs.datanode.hdfs-blocks-metadata.enabled</name>
  <value>true</value>
</property>
```

These changes boost performance with HDFS by bypassing (short-circuiting) the data node and accessing the files directly. They specify a domain socket path and a block storage location timeout. These changes, including the core-site.xml change that follows, provide a real performance boost to Impala's operation. So, I make the following changes to the Impala's copy of the core-site.xml file:

```
<!-- impala changes -->

<property>
  <name>dfs.client.read.shortcircuit</name>
  <value>true</value>
</property>
```

If the directory /var/run/hadoop-hdfs/ is group writeable, I make sure that the group is root. To check this, I use the Linux ls command to get a long listing of the directory:

```
[root@hc1r1m1 ~]# ls -ld  /var/run/hadoop-hdfs/
drwxr-xr-x. 2 hdfs hdfs 4096 Sep  7 09:21 /var/run/hadoop-hdfs/
```

The permissions string reads d rwx r-x r-x. The d means "directory," the first three characters are permissions for the owner, the next set is the group, and the final set is the world (or everybody else). So, the group membership string has r-x, or read and execute permissions; it is not writeable by the group because the w character is missing. Also, the string hdfs hdfs indicates that the directory is owned by the user and group hdfs. When I check, I make sure that the directory is either group writeable by the Linux root user or is not group writeable.

Now, I create the sockets directory and subdirectory under /var/run as the root user:

```
[root@hc1r1m1 impala]# mkdir -p /var/run/hdfs-sockets/dn
[root@hc1r1m1 impala]# ls -ld  /var/run/hdfs-sockets
drwxr-xr-x 3 root root 4096 Sep  7 09:58 /var/run/hdfs-sockets
```

I start the Impala services (the Impala State Store server, the Catalog server, and the Impala server) as root:

```
[root@hc1r1m1 ~]# service impala-state-store start
[root@hc1r1m1 ~]# service impala-catalog    start
[root@hc1r1m1 ~]# service impala-server      start
```

I check the Impala logs under /var/log/impala for errors. Errors can occur because of incorrect configuration after installation. For instance, if the sockets directory does not exist, the following error message would appear:

```
E0907 09:49:42.279753  4815 impala-server.cc:208] ERROR: short-circuit local reads is disabled
because
  - Impala cannot read or execute the parent directory of dfs.domain.socket.path
```

Or, if the configuration value dfs.client.file-block-storage-locations.timeout is either not specified or has too small a value, the following error message will be issued:

```
ERROR: block location tracking is not properly enabled because
  - dfs.client.file-block-storage-locations.timeout is too low. It should be at least 3000.
E0907 09:49:42.280009  4815 impala-server.cc:210] Aborting Impala Server startup due to improper
configuration
```

I plan to access Impala via Hue, so on the server hc1nn, I need to change the Hue configuration file hue.ini under /etc/hue/conf so that Hue knows where to find Impala. To do so, I set the server_host value in the [Impala] section of the file to the host where Impala is running, hc1r1m1:

```
server_host=hc1r1m1
```

After making the change, I restart the Hue server on the host hc1nn as the Linux root user:

```
[root@hc1nn ~]# service  hue restart
```

The interface is now be ready for use.

Impala User Interfaces

You can access the Impala server user interface via the URL hc1r1m1:25000. Figure 9-1 shows a basic example of this interface. The Impala State Store server user interface resides at port number 25010, and is reached using the URL hc1r1m1:25010. The Impala Catalogue server user interface is on port 25020. All three of these interfaces provide information such as configuration, log content, metrics, and session. Because I concentrate on Impala's analytics functionality using SQL, I only examine the Impala server interface and leave you to investigate the other options.

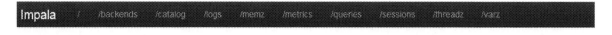

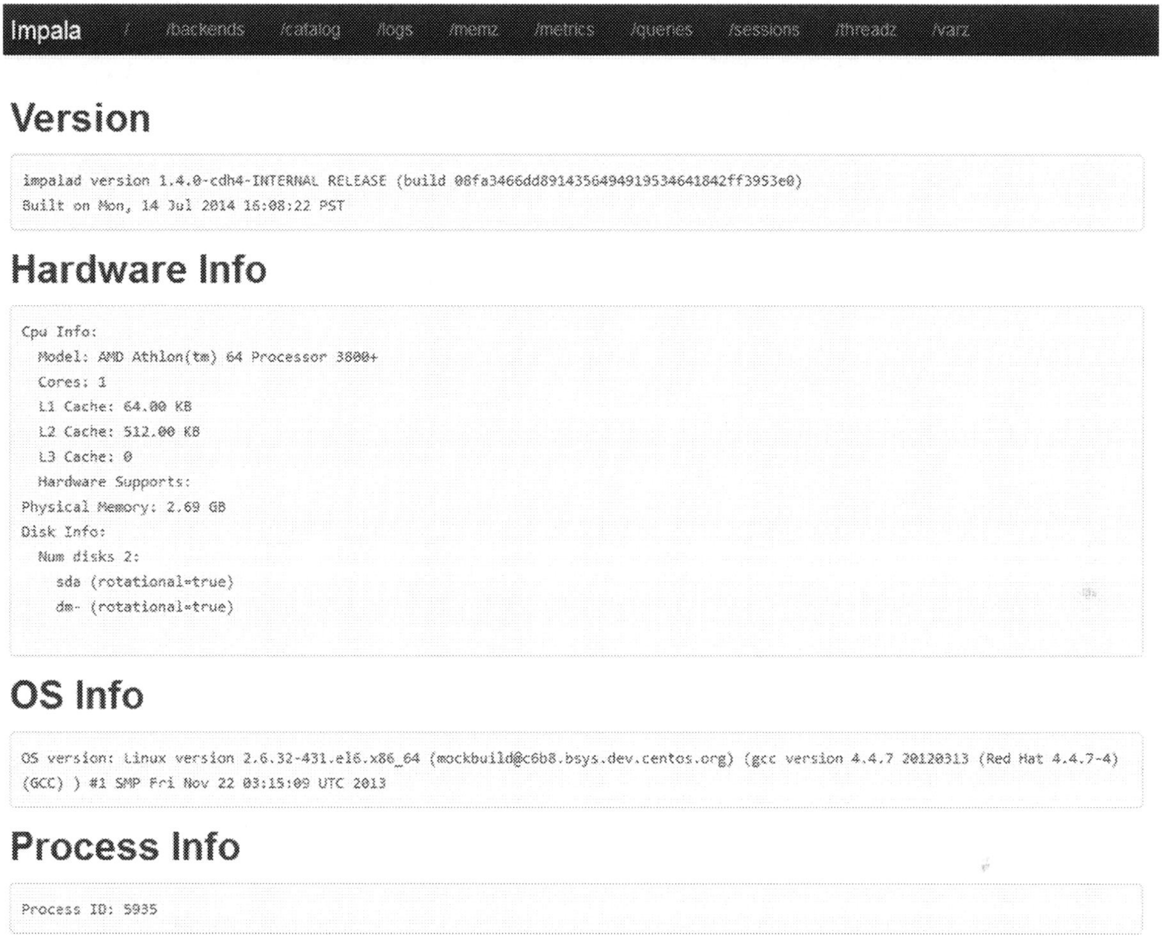

Impala
/
/backends
/catalog
/logs
/memz
/metrics
/queries
/sessions
/threadz
/varz
Version
impalad version 1.4.0-cdh4-INTERNAL RELEASE (build 08fa3466dd8914356494919534641842ff3953e0)
Built on Mon, 14 Jul 2014 16:08:22 PST
Hardware Info
Cpu Info:
Model: AMD Athlon(tm) 64 Processor 3800+
Cores: 1
L1 Cache: 64.00 KB
L2 Cache: 512.00 KB
L3 Cache: 0
Hardware Supports:
Physical Memory: 2.69 GB
Disk Info:
Num disks 2:
sda (rotational=true)
dm- (rotational=true)
OS Info
OS version: Linux version 2.6.32-431.el6.x86_64 (mockbuild@c6b8.bsys.dev.centos.org) (gcc version 4.4.7 20120313 (Red Hat 4.4.7-4) (GCC)) #1 SMP Fri Nov 22 03:15:09 UTC 2013
Process Info
Process ID: 5935

Figure 9-1. *Impala server user interface*

With the Hue server running and configured to access Impala, you can write Impala scripts from the Query Editor in the Hue interface, as shown in Figure 9-2. To open it, you use the URL hc1nn:8888; and from Hue, you click the stylized impala head (fourth icon from the left) in the top icon bar.

Figure 9-2. *Impala scripting via Hue user interface*

Uses of Impala

The real power of Impala lies not in its interfaces but in its query language and the options that it provides. For example, it can access HDFS-based data via external tables, and it offers standard SQL-based operations, such as filters, table joins, subqueries, inserts, and more. These terms are described next, in a step-by-step manner, and then their corresponding equivalents are examined for Hive. The database itself is highly scalable and robust, as it is built on top of HDFS.

At this point I need some data to process, so as to demonstrate Impala's SQL-based functionality. In Chapter 5, I uploaded a series of fuel consumption CSV data files to HDFS, under the HDFS directory /user/hue2/fuel_consumption/; the Hadoop file system ls command that follows shows that upload:

```
[hadoop@hc1r1m1 ~]$ hdfs dfs -ls /user/hue2/fuel_consumption
Found 16 items
-rw-r--r--   2 hadoop hue2     248956 2014-09-07 18:17 /user/hue2/fuel_consumption/MY1995-1999 Fuel
Consumption Ratings.csv
-rw-r--r--   2 hadoop hue2      45203 2014-09-07 18:17 /user/hue2/fuel_consumption/MY2000 Fuel
Consumption Ratings.csv
.....................
Consumption Ratings.csv
-rw-r--r--   2 hadoop hue2      77452 2014-09-07 18:17 /user/hue2/fuel_consumption/MY2013 Fuel
Consumption Ratings.csv
-rw-r--r--   2 hadoop hue2      77186 2014-09-07 18:17 /user/hue2/fuel_consumption/MY2014 Fuel
Consumption Ratings.csv
```

Database Creation

To create a database in Impala, I use the CREATE DATABASE command. I enter the following SQL command in Hue's Impala Query Editor to create the fuel database:

```
CREATE DATABASE fuel ;
```

After clicking the Execute button to form this text, I find that the database drop-down menu on the Hue's Impala user interface has a new option: fuel. To use this database in a SQL script, I can now specify the USE option:

```
USE fuel ;
```

Alternatively, I could use the database name before the table name, as shown in this SELECT command:

```
SELECT * FROM fuel.customer ;
```

External Table Creation

A Hive *external table* is a table where you specify the location for data storage rather than using the default value. For example, by using CREATE EXTERNAL TABLE, I can create an external table against an HDFS directory that contains comma-separated files (CSV). When the table is dropped the data is not deleted. The following code creates an external table called "consumption" in the fuel database and that table can then be used to investigate trends in vehicle fuel consumption:

```
CREATE EXTERNAL TABLE fuel.consumption
(
  myear STRING,
  manufacturer STRING,
  model STRING,
  fclass STRING,
  enginesz STRING,
  cylinders STRING,
  transmission STRING,
  fuel STRING,
  consumption1 STRING,
  consumption2 STRING,
  consumption3 STRING,
  consumption4 STRING,
  avefuel STRING,
  co2 STRING
)
ROW FORMAT DELIMITED FIELDS TERMINATED BY ','
LOCATION '/user/hue2/fuel_consumption/';
```

Because a CSV file uses commas as the column separators, the row following the end parenthesis [)] uses a DELIMITED option to process the data:

```
ROW FORMAT DELIMITED FIELDS TERMINATED BY ','
```

This indicates that the columns are delimited by commas, and so the columns in the table will match the columns in the CSV file. The final line of the command specifies the location of the data, and the table name is specified in the form database.table (here, fuel.consumption).

When I want to remove a table, I use the DROP TABLE command. For example, I can remove the "consumption" table from the fuel database as follows:

```
DROP TABLE fuel.consumption ;
```

External tables are useful for data feeds based on files. Rather than load the files into a database table, I could can create an external table against the files. Then the data can be loaded from the external table as an initial step in an ETL chain before the data is processed further.

Table Creation

The same information can be used to create an internal table with just a few tweaks of the SQL. I simply remove the EXTERNAL keyword and leave out the DELIMITED and LOCATION options, as follows:

```
CREATE TABLE fuel.consumption2
(
  myear STRING,
  manufacturer STRING,
  model STRING,
  fclass STRING,
  enginesz STRING,
  cylinders STRING,
  transmission STRING,
  fuel STRING,
  consumption1 STRING,
  consumption2 STRING,
  consumption3 STRING,
  consumption4 STRING,
  avefuel STRING,
  co2 STRING
);
```

For instance, to show which tables exist for the fuel database, I can use the SQL SHOW TABLES command:

```
SHOW TABLES ;
```

The result is a single row called "consumption." To examine the structure of a table, I use the DESCRIBE keyword:

```
DESCRIBE fuel.consumption;
```

The result now is a series of output data rows that represent the columns in the table:

```
0    myear          string
1    manufacturer   string
2    model          string
3    fclass         string
4    enginesz       string
5    cylinders      string
```

6	transmission	string
7	fuel	string
8	consumption1	string
9	consumption2	string
10	consumption3	string
11	consumption4	string
12	avefuel	string
13	co2	string

The SELECT Statement

The SELECT keyword plucks data from a database table. For example, I can select data from the "consumption" table of the fuel database. The asterisk (*) indicates that all column data within the table should be selected. (As this example contains a lot of data, I provide only a couple of rows of the output to give an idea of the results):

```
SELECT * from fuel.consumption ;
```

1	1995	ACURA	INTEGRA		SUBCOMPACT	1.8	4	A4	X	10.2
	7	28		1760	202					
2	1995	ACURA	INTEGRA		SUBCOMPACT	1.8	4	M5	X	9.6
	7	29	40	1680	193					

If I need only a portion of the data, I can replace the asterisk (*) with specific columns names, such as "myear" and "manufacturer":

```
SELECT  myear, manufacturer  from fuel.consumption ;
1       1995    ACURA
2       1995    ACURA
```

The WHERE Clause

The WHERE clause is used with SELECT, INSERT, and DELETE statements to filter the results of a request. Here are some examples of SELECT statements with their WHERE clauses serving as filters for obtaining data by manufacturer or year:

```
SELECT * from fuel.consumption WHERE manufacturer = 'ACURA' ;

SELECT * from fuel.consumption WHERE myear = '1995' AND manufacturer = 'AUDI'   ;

SELECT * from fuel.consumption WHERE myear = '1995' OR manufacturer = 'AUDI'   ;
```

The first statement limits the selected rows to those where "manufacturer" is equal to ACURA. The second limits the selected rows to where "myear" is 1995 *and* the "manufacturer" is AUDI. The final example limits the selected rows to where "myear" is 1995 *or* the "manufacturer" is AUDI.

The Subquery

SELECT statements can be nested as subqueries. Here are two examples of how that is done. The first is to use a SELECT statement as a subquery in the FROM clause of an outer SELECT statement, as follows:

```
SELECT rd.* FROM
(
    SELECT
      myear,manufacturer,model,enginesz,cylinders
    FROM
     fuel.consumption

)  rd ;
```

The outer SELECT statement uses a derived table called "rd" and selects all columns from rd using the column list rd.*. The contents of the parentheses form the derived table named rd, which uses a subquery that selects the following five columns from the external table fuel.consumption:

```
myear,manufacturer,model,enginesz,cylinders
```

This first example uses a subquery as a way to transform table data to form a derived table. For example, you might want to filter table data or join a group of tables together to form a larger data set.

The second example uses a subquery in the WHERE clause of an SQL statement. In this manner, an outer SELECT statement selects all column data (fl.*) from the table fuel.consumption, with a subquery in its WHERE clause to filter on the engine size. The result is a table showing data for which the engine size is greater than average:

```
SELECT
  fl.*
FROM
  fuel.consumption  fl
WHERE
  fl.enginesz > ( SELECT AVG(st.enginesz) FROM  fuel.consumption  st  )
```

In this second example, the subquery in parentheses selects the average (AVG) engine size value from the instance of the fuel.consumption table with an alias of "st." The outer query filters the contents of the fuel.consumption table with an alias of "fl," choosing only those rows where the fl.enginesz column is greater than the average value.

Table Joins

Table joins allow you to join the data in one table to the data in a second table if there are comparable columns in each table. For instance, the first table could contain a list of people, with each person having a unique identity number. A second table might contain a list of addresses plus associated personal identity numbers. By joining the two tables on the identity numbers, you are able to determine a person's name and address.

The following SQL example selects data from the earlier fuel.consumption table where the columns "myear," "manufacturer," and "model" match another table called fuel.consumption3. Aliases have been used here for the two tables—rd1 and rd2; this means there's less typing when you are specifying the table columns and so there's no confusion about which table a column belongs to. The SELECT statement will only output a data row if a matching row is found on the "myear," "manufacturer," and "model" columns of both tables:

```
SELECT
  rd1.*
FROM
  fuel.consumption   rd1,
  fuel.consumption3  rd2
WHERE
  rd1.myear =  rd2.myear AND
  rd1. manufacturer =  rd2. manufacturer AND
  rd1. model =  rd2. model
```

The INSERT Statement

The INSERT statement allows you to insert a single row into an Impala table, or you can combine INSERT with a SELECT statement to insert multiple rows. These commands are designed to move large data volumes.

In a first example of this command, the INSERT statement inserts a single row into the fuel database table called consumption2. The number and type of column values specified in parentheses must match the table definition. Because the data values are strings, they are shown within single quotes:

```
INSERT INTO fuel.consumption2 VALUES  ('1995','ACURA','INTEGRA','SUBCOMPACT','1.8','4','A4',
'X','10.2','7','28','40','1760','202')
```

Although this single INSERT statement moves a single row of data into the table fuel.consumption2, a second example offers better performance by using a SELECT statement to populate the table fuel.consumption2 via a bulk insert:

```
INSERT INTO TABLE fuel.consumption2 SELECT * FROM  fuel.consumption3
```

This second version is a simple example; no extra filters or table joins have been used. But the size of the subtable could be large; with a simple statement, you could copy a large volume of data.

Note: When executing an INSERT statement, you may receive an error message similar to the following:

```
AnalysisException: Unable to INSERT into target table (fuel.consumption2) because Impala does not
have WRITE access to at least one HDFS path: hdfs://hc1nn/user/hive/warehouse/fuel.db/consumption2
```

This error occurs if the referenced database and/or table name on HDFS is not owned by the HDFS impala user. To fix this, you change the ownership with a statement like the following:

```
[hadoop@hc1r1m1 ~]$ hdfs dfs -chown -R impala:supergroup /user/hive/warehouse/fuel.db/
```

When run as the hadoop user, this statement recursively changes ownership of the HDFS fuel.db directory to the impala user and changes the group to a supergroup. Then, you drop the table and re-create it to update the Impala metadata for the table. After these steps, your INSERT statement should work.

This has been a very short introduction to Cloudera Impala SQL. For full details, check the Impala web site at http://www.cloudera.com/content/cloudera/en/documentation/cloudera-impala. Additionally, try to create your own SQL statements by joining these simple building blocks. For instance, you can create SELECT statements with WHERE clauses. And you can add table joins to your SQL statements and consider adding subqueries.

Apache Hive

Like Cloudera Impala, Apache Hive offers an SQL-type language called Hive QL that can be used to manipulate Hive-based tables. The functionality of Hive QL can be extended by creating user-defined functions (UDF), as you'll see in an example shortly.

In this section, I use Hive version 0.10, which was installed in Chapter 7 along with the Hue application. As you remember, Hue was installed on the server hc1nn and has the URL http://hc1nn:8888/. Though I don't mention Hue again in this chapter, I use the Beeswax Hive user interface to enter the scripts at the Hue URL. Here, I walk you, step by step, through table creation, SELECT statements, joins, and WHERE clauses. To make the examples a bit more interesting, I have sourced some real UK trade CSV files to use as data.

▨ **Note**   For more information and in-depth documentation on Hive and Hue, see the Apache Hive website at hive.apache.org.

Database Creation

To begin the example, I create a database to contain this Hive work, using the CREATE DATABASE command. I name the database and specify to create it only if it does not already exist:

```
CREATE DATABASE  IF NOT EXISTS  trade;
```

I set the current database with the USE command; in this case, I set it to trade:

```
USE trade;
```

External Table Creation

External tables are a useful step in an ETL chain because they offer the opportunity to move raw data files from an HDFS directory into a Hive staging table. From the staging table you can transform the data for its journey to its final state. Before I demonstrate table creation, though, I need to move the data files that I downloaded from the UK government data site (data.gov.uk) from the Linux file system on hc1nn to HDFS. (If you want to obtain the same data set to run these examples, you can source it from http://data.gov.uk/dataset/financial-transactions-admin-spend-ukti.)

To start the move, I create the /data directory on HDFS as the Linux hadoop user:

```
[hadoop@hc1nn data]$ hdfs dfs -mkdir /data
```

I then move the data files under the Linux directory /home/hadoop/data/uk_trade to this HDFS directory via a copyFromLocal HDFS command:

```
[hadoop@hc1nn uk_trade]$ pwd
/home/hadoop/data/

[hadoop@hc1nn uk_trade]$ ls  uk_trade
ukti-admin-spend-apr-2011.csv  ukti-admin-spend-jun-2012.csv
ukti-admin-spend-apr-2012.csv  ukti-admin-spend-mar-2011.csv
.......
[hadoop@hc1nn data]$ hdfs dfs -copyFromLocal uk_trade /data
```

```
[hadoop@hc1nn data]$ hdfs dfs -ls /data/uk_trade

Found 22 items
-rw-r--r--   2 hadoop hadoop    355466 2014-09-16 18:09 /data/uk_trade/UKTI_FEBRUARY_2013.csv
-rw-r--r--   2 hadoop hadoop    231177 2014-09-16 18:09 /data/uk_trade/ukti-admin-spend-apr-2011.
csv
.......
```

The necessary CSV-based data set now resides in HDFS under the directory /data/uk_trade, so I can start using the data in Hive Query Language (Hive QL) statements. For example, the following CREATE TABLE statement creates the rawtrans (raw transaction) table in the trade database. (Again, the IF NOT EXISTS clause ensures it is created only if it does not exist.) The table is linked to the HDFS directory /data/uk_trade/, which contains the UK trade expense information in CSV format via a LOCATION clause. As the data is in CSV format, the external table specifies that the fields are delimited by commas, as specified by the ROW FORMAT DELIMITED FIELDS TERMINATED BY clause:

▓ **Note**   As a rule, reserved words appear in uppercase to make the examples clearer, although Hive QL is not case-sensitive.

```
CREATE TABLE IF NOT EXISTS
trade.rawtrans
(
dept STRING,
entity STRING,
paydate STRING,
exptype STRING,
exparea STRING,
supplier STRING,
trans STRING,
amount DOUBLE
)
ROW FORMAT DELIMITED FIELDS TERMINATED BY ','
LOCATION '/data/uk_trade';
```

The columns in the table are separated by commas and are bound by parentheses. The column name is in lowercase (for instance, dept), followed by the data type in uppercase (such as STRING). The name of the table to be created is "rawtrans," and it resides in the trade database. So the name trade.rawtrans refers to the rawtrans table in the trade database.

The number of columns in the CSV files must match the columns in the SELECT statement. Also, it is good practice to use meaningful names to represent each column—it avoids confusion later.

I can now access the data in the external table, as a simple COUNT(*) shows. This returns the result that there are 18,976 rows in this external table:

```
SELECT COUNT(*) FROM trade.rawtrans
```

Hive UDFs

Hive makes possible the creation of *user-defined functions* (UDFs), with which you can extend and customize the functionality of Hive. To demonstrate the process, I create a simple date-conversion function.

Suppose that the date columns in the CSV data that was collected have the wrong format for Hive; specifically, the dates follow the format dd/MM/yyyy, when they need to have a format of yyyy-MM-dd. Using Java date methods, I can create a simple Java-based UDF to change the date format. When compiled and loaded into a library, this function can then be embedded in the Hive QL statements.

To create a Hive UDF, I must install the Scala sbt interactive build tool on the server to compile the Java UDF package. I create the example UDF on the server hc1nn, so I need to install the sbt program on that server by using the Linux root account. I download an rpm package for sbt from the scala-sbt.org website to the /tmp directory on hc1nn, and I install it from there. The following command moves to the /tmp directory and downloads the sbt.rpm package by using wget:

```
[root@hc1nn ~]# cd /tmp
[root@hc1nn ~]#wget http://repo.scala-sbt.org/scalasbt/sbt-native-packages/org/scala-sbt/sbt/0.13.1/
sbt.rpm
[root@hc1nn ~]# rpm -ivh sbt.rpm
```

The final command, rpm, installs the sbt.rpm package with options I for install and v for verify. I also install the Java OpenJDK 1.6 development package to support this compilation as the root Linux user (because I want access to tools like jar and jps). I use the openJDK because I can install it via the yum command, and I don't have to go through a registration process to get it.

```
[root@hc2nn ~]# yum install  java-1.6.0-openjdk-devel
```

I compile the new Hive UDF function as the Linux hadoop user, so I use su to change to that account:

```
[root@hc2nn ~]# su - hadoop
```

Next, I need to set up a directory structure that will hold the UDF code, so initially I create the directories hive/udf in the hadoop account home directory to hold my Apache Hive UDF code. Next, I move to that new udf directory:

```
[hadoop@hc2nn ~]$ mkdir -p hive/udf
[hadoop@hc2nn ~]$ cd  hive/udf
```

At this level, I have created a file called build.sbt that the sbt tool will use to aid in the compilation of the UDF. It describes details like the function name, the version, the organization that it belongs to, and the version of Scala installed. Here's the contents of the file displayed by using the Linux cat command; I have added line numbers to aid understanding:

```
[hadoop@hc2nn udf]$ cat build.sbt

01  name := "DateConv"
02
03  version := "0.1"
04
05  organization := "nz.co.semtechsolutions"
06
07  scalaVersion := "2.10.4"
08
```

```
09   resolvers += "CDH4" at "https://repository.cloudera.com/artifactory/cloudera-repos/"
10
11   libraryDependencies += "org.apache.hadoop" %  "hadoop-core"        % "0.20.2"      % "provided"
12
13   libraryDependencies += "org.apache.hive"   %  "hive-exec"          % "0.10.0"      % "provided"
```

Each line of this file is separated from the next with a blank line. Notice that the organization name is the reverse of my company's domain name and the Java package name to which the UDF will belong also uses the same naming standard. The version of Scala used is defined, as are library dependencies for Hadoop and Hive.

Having created the build sbt file, I now need a directory structure to contain the UDF code. It must match the structure of the UDF package name, starting with the directories src/main/java. The first line of the Java UDF file contains a package name:

```
package nz.co.semtechsolutions.hive.udf;
```

And so a directory structure must be created to match this, using the mkdir command. Using the -p option causes all subdirectories in the path to be created at the same time. I then move down to the lowest point in the directory structure that I have created, the udf directory:

```
[hadoop@hc2nn udf]$ mkdir -p src/main/java/nz/co/semtechsolutions/hive/udf
[hadoop@hc2nn udf]$ cd src/main/java/nz/co/semtechsolutions/hive/udf
```

I have created a UDF Java file called DateConv.java that contains the Java code for the UDF function. The following Linux cat command shows the contents of the Java:

```
[hadoop@hc2nn udf]$ cat DateConv.java

1       package nz.co.semtechsolutions.hive.udf;
2
3       import org.apache.hadoop.hive.ql.exec.UDF;
4       import org.apache.hadoop.io.Text;
5       import java.text.SimpleDateFormat;
6       import java.util.Date;
7
8       class DateConv extends UDF
9       {
10
11          public Text evaluate(Text s)
12          {
13
14                  Text to_value = new Text("");
15
16                  if (s != null)
17                  {
18                      try
19                      {
20
21                          SimpleDateFormat incommingDateFormat = new SimpleDateFormat
                            ("dd/MM/yyyy");
22                          SimpleDateFormat convertedDateFormat = new SimpleDateFormat
                            ("yyyy-MM-dd");
23
```

```
24                              Date parsedate = incommingDateFormat.parse( s.toString() );
25
26                              to_value.set( convertedDateFormat.format(parsedate) );
27
28                          }
29                          catch (Exception e)
30                          {
31                              to_value = new Text(s);
32                          }
33                      }
34                      return to_value;
35              }
36      }
```

The package name is defined at line 1, while import statements to import Hive, Hadoop, and Java functionality exist between lines 3 and 6.

```
1       package nz.co.semtech-solutions.hive.udf;

3       import org.apache.hadoop.hive.ql.exec.UDF;
4       import org.apache.hadoop.io.Text;
5       import java.text.SimpleDateFormat;
6       import java.util.Date;
```

The class DateConv that is the UDF function name is defined at line 8; it extends an existing class UDF.

```
8       class DateConv extends UDF
```

At line 11, the public class evaulate is defined, which takes a Text parameter and returns a Text value:

```
11          public Text evaluate(Text s)
```

Finally, the main functionality of the UDF occurs between lines 21 and 26 in the try/catch section of the code. The input date string is converted from the format dd-MM-yyyy to the format yyyy-MM-dd. (This is a somewhat contrived example that takes only a single date format, but it gives an idea of what can be achieved with Hive UDFs.)

```
21                              SimpleDateFormat incommingDateFormat = new
                                SimpleDateFormat("dd/MM/yyyy");
22                              SimpleDateFormat convertedDateFormat = new
                                SimpleDateFormat("yyyy-MM-dd");
23
24                              Date parsedate = incommingDateFormat.parse( s.toString() );
25
26                              to_value.set( convertedDateFormat.format(parsedate) );
```

Having created the Java file that will form the new UDF function, I move back to the top of the directory structure by using the Linux cd command and invoke the sbt command to compile the code:

```
[hadoop@hc2nn udf]$ cd /home/hadoop/hive/udf/

[hadoop@hc2nn udf]$ sbt
[info] Set current project to DateConv (in build file:/home/hadoop/hive/udf/)
```

I enter the sbt command `compile` at the sbt> prompt to compile the code, followed by the `package` command to package the code into a jar file. Finally, the sbt `exit` command causes the sbt build session to finish:

```
> compile
[success] Total time: 3 s, completed Sep 16, 2014 7:50:39 PM

 > package
[success] Total time: 1 s, completed Sep 16, 2014 7:50:59 PM

> exit
```

The jar library containing the new Hive UDF code is contained in the target/scala-{version} directory. It is called dateconv_2.10-0.1.jar, as the Linux long listing shows:

```
[hadoop@hc2nn udf]$ ls -l target/scala-2.10/dateconv_2.10-0.1.jar
-rw-rw-r-- 1 hadoop hadoop 1579 Sep 16 19:36 target/scala-2.10/dateconv_2.10-0.1.jar
```

I can use the Java `jar` command to show the contents of the library. For example, the following command passes the options vtf to the `jar` command and takes the library as a parameter. The v option means verbose, the t option means show the table of contents, and the f option allows the jar file name to be specified. The output shows the structure of the jar file and shows that it contains the compiled class file DateConv.class:

```
[hadoop@hc1nn udf]$ jar vtf target/scala-2.10/dateconv_2.10-0.1.jar
   288 Tue Sep 16 19:36:32 NZST 2014 META-INF/MANIFEST.MF
     0 Tue Sep 16 19:36:32 NZST 2014 nz/
     0 Tue Sep 16 19:36:32 NZST 2014 nz/co/
     0 Tue Sep 16 19:36:32 NZST 2014 nz/co/semtechsolutions/
     0 Tue Sep 16 19:36:32 NZST 2014 nz/co/semtechsolutions/hive/
     0 Tue Sep 16 19:36:32 NZST 2014 nz/co/semtechsolutions/hive/udf/
   899 Tue Sep 16 19:36:24 NZST 2014 nz/co/semtechsolutions/hive/udf/DateConv.class
```

Now that the Hive UDF jar file has been created, I can add it to a Hive shell session so that I can use the new Hive UDF function in Hive Query language (Hive QL). The following add jar command in my Apache Hive session registers the jar file with the session:

```
[hadoop@hc1nn udf]$ hive

hive> add jar /home/hadoop/hive/udf/target/scala-2.10/dateconv_2.10-0.1.jar;

Added /home/hadoop/hive/udf/target/scala-2.10/dateconv_2.10-0.1.jar to class path
Added resource: /home/hadoop/hive/udf/target/scala-2.10/dateconv_2.10-0.1.jar
```

The full name of the DateConv UDF function is co.nz.semtechsolutions.hive.udf.DateConv. This is a long name based on the package name. It would be much more convenient and quicker to just refer to the function as DateConv. That is what the next command does: it registers the temporary function name DateConv based on the long name:

```
hive> create temporary function DateConv as 'nz.co.semtechsolutions.hive.udf.DateConv';
OK
Time taken: 0.02 seconds
```

Now, the DateConv Hive UDF function is ready to be used in a Hive QL query by using just its short name, DateConv. It will take a date with a structure of dd-MM-yyyy and return a date formatted to yyyy-MM-dd for use in a Hive table. Next, I show how to invoke this function.

Table Creation

Suppose, as the next step in your ETL chain, you want to extract monthly purchase totals by supplier. You could, as shown earlier, create a table with CREATE TABLE, using the IF NOT EXISTS clause to prevent duplication; for example, I create the following table:

```
CREATE TABLE IF NOT EXISTS
trade.suppliertot
(
payyear INT,
paymonth INT,
supplier STRING,
totamount DOUBLE
);
```

Called "suppliertot" and residing in the trade database, this table would have four columns: one for the year of a transaction, one for the month, one for the the supplier name, and one for the total amount of the purchases from that supplier (indicated by the comma-separated list within parentheses). The table would not, however, contain any data yet, as it has no LOCATION clause to link it to an HDFS directory.

Alternatively, I could create the table via a SELECT statement, which would automatically populate it with data. To do that, I drop the table first and then run the CREATE statement, as follows:

```
DROP TABLE trade.suppliertot;


CREATE TABLE IF NOT EXISTS
trade.suppliertot
AS
SELECT
  year(DateConv (paydate) ) as payyear,
  month(DateConv (paydate) ) as paymonth,
  supplier,
  SUM(amount) as totamount
FROM
  trade.rawtrans
GROUP BY
  year(DateConv (paydate) ) ,
  month(DateConv (paydate) ) ,
  supplier ;
```

This statement uses the DateConv UDF that was created previously to convert date strings with the format dd/MM/yyyy to the format yyyy-MM-dd; here, it is used against the "paydate" column. I use the same CREATE TABLE IF NOT EXISTS option to create the table trade.suppliertot. However, I replace the list of columns in parentheses with an AS SELECT statement that has four columns that takes data from the table trade.rawtrans via a FROM clause.

I can confirm that this second table now contains the data by issuing the statement SELECT COUNT(*) FROM trade.suppliertot. This returns the result 391, indicating that the trade.suppliertot table contains 391 rows. The COUNT(*) is a special aggregation function that returns the number of rows in a table.

The SELECT Statement

The SQL SELECT statement starts with the keyword SELECT, which is followed by a comma-separated list of columns. The source of the data is identified in the FROM clause, which is followed by the name of the table from which it is selecting data.

In continuing the second example begun in the previous section, I use the the table with the name trade. rawtrans. The SELECT statement selects three columns: "paydate," "supplier," and "amount." The "paydate" column, however, is transformed via the DateConv UDF. The payyear and paymonth values are derived from the "paydate" column via the year and month functions in Hive. The "amount" column is sum totaled with the aggregating function SUM.

```
SELECT
  year( DateConv (paydate) ) as payyear,
  month(DateConv (paydate) ) as paymonth,
  supplier,
  SUM(amount) as totamount
FROM
  trade.rawtrans
GROUP BY
  year( DateConv (paydate) ) ,
  month(DateConv (paydate) ) ,
  supplier ;
```

Because the aggregating function SUM is used, a GROUP BY clause is needed to specify which columns are to be sum totaled. The general rule is that all columns that are not aggregated (i.e. payyear, paymonth, and supplier) must be in the GROUP BY clause. This example groups data by year, month, and supplier.

The WHERE Clause

When you are running a SELECT statement, you might want to filter the data returned, either because there is too much data or because you would like to filter on one of the columns. By combining the WHERE clause, column names, simple operators, and logical expressions like AND and OR, you can build complex filters.

Consider the following example, which adds a filter via the WHERE keyword to the SELECT statement from the earlier example:

```
SELECT
  year( DateConv (paydate) ) as payyear,
  month(DateConv (paydate) ) as paymonth,
  supplier,
  SUM(amount) as totamount
FROM
  trade.rawtrans
WHERE
  supplier NOT LIKE 'UK Trade%'  AND
  supplier NOT LIKE 'Corporate%'
GROUP BY
  year( DateConv (paydate) ) ,
  month(DateConv (paydate) ) ,
  supplier ;
```

The value of the column supplier is filtered so that it is "not like" UK Trade% and "not like" Corporate%. This means that the only columns that are selected are those in which the supplier name does not start with the strings UK Trade and Corporate.

The percent (%) character is a wild card that matches any data in the column; when placed at the end of the string, it matches everything from that point to the end of the string. Strings are expressed in single quotes and the string is case sensitive.

I could also use the WHERE clause as follows:

```
WHERE
  supplier =  'ADETIQ LTD'  OR
  supplier =  'ADS GROUP LTD'
```

This clause then explicity filters the supplier column on the values that are equal to ADETIQ LTD or ADS GROUP LTD, and returns the two data rows, as shown:

```
2013    2       ADETIQ LTD          783.84
2013    2       ADS GROUP LTD       15549.0
```

The next clause filters the supplier column to equal ADETIQ LTD and not equal (<>) the value ADS GROUP LTD, as follows:

```
WHERE
  supplier =  'ADETIQ LTD'  AND
  supplier <> 'ADS GROUP LTD'
```

And so it returns a single row:

```
2013    2       ADETIQ LTD          783.84
```

The Subquery

The SELECT statements can also be used in subqueries. Subqueries are handy when you want to reduce or filter data from a table before using it. They are coupled for Hive QL in both the FROM clause and the WHERE clause. That is, in the FROM section of the parent SELECT statement, you simply enclose your subquery in parentheses.

For example, I could select all columns from the trade.rawtrans table, filtering the supplier column to values containing the string INDIA. This then becomes a derived table with an alias of b, as follows:

```
SELECT
  DateConv (b.paydate) as paydate,
  b.supplier,
  b.amount
FROM
  (
     SELECT  a.*  FROM  trade.rawtrans a WHERE a.supplier LIKE '%INDIA%'
  ) b ;
```

And this returns the column data:

```
2013-02-01      THE INDIA SHOP (IMPORTS) LTD    1000.0
2013-02-08      LIVING MEDIA INDIA LTD          4109.3
2013-02-22      UK INDIA BUSINESS COUNCI        4125.0
```

Notice that the example uses aliases (a, b) in both SELECT statements. The subquery uses an alias (a) for the table trade.rawtrans, and so its columns are referenced using this alias (a.supplier). The derived table b contains the table built by the subquery in parentheses. The columns from the derived table are also referenced using an alias (b.supplier). (This aliasing avoids confusion as to which table, real or derived, a column belongs to.)

You can also use subqueries in the WHERE clause of a SELECT statement (from Hive 0.13), as my example shows here:

```
SELECT
  DateConv (b.paydate) as paydate,
  b.supplier,
  b.amount
FROM
  trade.rawtrans b
WHERE
  b.supplier IN ( SELECT supplier FROM trade.uksupplier );
```

The data from the trade.rawtrans table with alias b is being filtered against the supplier name. The subquery in the WHERE clause is checking that the supplier name from trade.rawtrans exists in the UK supplier list table trade.uksupplier by using the SQL IN clause.

Table Joins

In the real world, you're seldom pulling data from just one table. What happens if you have data in two tables and you wish to build a SELECT statement using data from both tables? You can use *table joins* to merge the data from multiple tables to form a compound data set. Of course, you will need to know which columns exist in each table and that the same data exists in each table so that the rows can be joined.

For instance, suppose two derived tables (a and b) each contain a column named "supplier"; that means I can join them on that column. The SELECT statement that follows uses those two derived tables: the first (aliased a) selects the "department" and "supplier" columns from the trade.rawtrans table. The second derived table (aliased b) selects the "supplier" and "amount" columns from the same table. Even though they are taking data from the same table, they are treated as two different derived tables.

The DISTINCT keyword is used in both subqueries to remove duplicates from the data. The derived tables are then joined using the JOIN keyword. They are joined on the "supplier" key using the ON keyword. Note: only "equal" joins are accepted. For instance, you could not say ON (a.amount > b.amount).

```
SELECT
 a.dept,
 a.supplier,
 b.amount
FROM
(
    SELECT DISTINCT
      c.dept,c.supplier
    FROM
      trade.rawtrans c
) a
```

```
JOIN
(
     SELECT DISTINCT
       d.supplier,d.amount
     FROM
       trade.rawtrans d
) b
ON ( a.supplier = b.supplier ) ;
```

The INSERT Statement

You can use the INSERT statement to add rows to your table. However, it's a good idea to check the structure of your table before you consider inserting data. To check the trade.suppliertot table, for example, I would use the DESCRIBE command:

```
hive> DESCRIBE trade.suppliertot ;
OK
payyear           int
paymonth        int
supplier           string
totamount        double
Time taken: 0.317 seconds
```

This shows that my table has two integer columns named "payyear" and "paymonth," followed by a string column named "supplier" and a double (real) column named "totamount." So, the table has four columns. I would then check to see what university suppliers exist in the data by using the following SELECT statement:

```
SELECT * FROM trade.suppliertot WHERE supplier LIKE 'UNIVERSITY%' ;

2013    2       UNIVERSITY OF EAST LONDON      550.0
2013    2       UNIVERSITY OF THE ARTS LONDON  550.0
```

So, I have used a SELECT statement to select all (*) columns from the table called "suppliertot" in the database named "trade." I added a WHERE clause that searches for suppliers whose name starts with the word "UNIVERSITY". This shows me that there are two such rows in the table. Therefore, there are two ways to change the data in this table: the first is to load data from the HDFS file system, and the second is to insert rows from a SELECT statement.

If I should decide to use the second approach, the following statement inserts rows into the trade.suppliertot table from the SELECT statement on rows 2 and 3 below. Notice that the WHERE clause is the same as that above, so it is the two UNIVERSITY rows that will be affected. Notice also that a combination of table rows and hard-coded values have been selected. The payyear and paymonth values have been selected from the table, while the hard-coded values 'UNIVERSITY OF SEMTECH' and 700.0 have been set in column positions 3 and 4 of the SELECT statement:

```
INSERT INTO TABLE trade.suppliertot
  SELECT payyear,paymonth,'UNIVERSITY OF SEMTECH',700.0 FROM
    trade.suppliertot WHERE supplier LIKE 'UNIVERSITY%' ;
```

Running the same SELECT statement to check the university supplier rows in the table now yields four rows. The year and month were selected from the existing rows, while the supplier and total values were hard-coded:

```
SELECT * FROM trade.suppliertot WHERE supplier LIKE 'UNIVERSITY%' ;
```

```
2013    2        UNIVERSITY OF EAST LONDON         550.0
2013    2        UNIVERSITY OF THE ARTS LONDON     550.0
2013    2        UNIVERSITY OF SEMTECH             700.0
2013    2        UNIVERSITY OF SEMTECH             700.0
```

I could also use an OVERWRITE clause in an INSERT statement that will cause existing rows to be overwritten. Running the INSERT statement again, using the clause INSERT OVERWRITE, causes all four university rows to be changed:

```
INSERT OVERWRITE TABLE trade.suppliertot
  SELECT payyear,paymonth,'UNIVERSITY OF SEMTECH',950.0 FROM
    trade.suppliertot WHERE supplier LIKE 'UNIVERSITY%' ;
```

```
SELECT * FROM trade.suppliertot WHERE supplier LIKE 'UNIVERSITY%' ;
```

```
2013    2        UNIVERSITY OF SEMTECH    950.0
2013    2        UNIVERSITY OF SEMTECH    950.0
2013    2        UNIVERSITY OF SEMTECH    950.0
2013    2        UNIVERSITY OF SEMTECH    950.0
```

As this resulting data now shows, all of the rows have the same values. They have all been overwritten by the INSERT statement.

Organization of Table Data

Simply retrieving data is but one aspect of analytics; organizing the data is another. The following SELECT statement will produce a list of suppliers and the number of transactions associated with them in the table trade.rawtrans:

```
SELECT supplier, COUNT(*) FROM trade.rawtrans GROUP BY supplier ;
```

Although the count of transactions is grouped (GROUP BY) by each supplier so the name and total count for each supplier is displayed, the statement does not provide any control over the order in which the data is presented. This is where the ORDER BY clause can be useful. With it, you can order your data in several ways, such as by supplier name, or present the count values in ascending or descending order. For example, this command will display the supplier transaction count list in reverse alphabetical order:

```
SELECT supplier, COUNT(*) FROM trade.rawtrans GROUP BY supplier ORDER BY supplier  DESC ;
```

The DESC clause means "descending"; that is, the supplier names starting with Z will be at the top while those starting with A will be at the bottom of the list.

That's fine, but what if you want to find suppliers that meet specific criteria, such as those with more than 1,000 transactions? For this, you must use the HAVING clause:

```
SELECT
  supplier, COUNT(*)
FROM
  trade.rawtrans
GROUP BY
  supplier
HAVING COUNT(*) > 1000
ORDER BY
  supplier DESC ;
```

Notice that the HAVING clause operates on the COUNT(*) column and uses a greater than (>) operator.

So, with just nine lines of SQL, it is possible to generate the transaction volumes for suppliers by using COUNT(*) and GROUP BY. The list is sorted in reverse order, with an ORDER BY clause. Finally, the HAVING clause is used to find the highest volume suppliers. So, by combining these terms in one statement, it is possible to extract some very useful information from the raw data. The rows with more than 1,000 transactions are shown as follows:

```
UK Trade & Investment - Trade Development                 1158
UK Trade & Investment - Sectors Group                     1503
UK Trade & Investment - Regional Directorate              2134
UK Trade & Investment - International Group                1038
UK Trade & Investment - Defence and Security Organisation 2970
UK Trade & Investment - Business Group                    1229
```

When combined with Cloudera Impala and Apache Hive, simple SQL statements become even more powerful. As you have seen, you can extend the functionality of Apache Hive by using UDFs, while Impala integrates with Cloudera's enterprise data hub. For further information on Apache Hive QL, see the Apache language reference at https://cwiki.apache.org/confluence/display/Hive/LanguageManual.

The power of these SQL-like languages comes partly from their functionality but also from the fact that they are familiar to people who have had exposure to relational databases. Apache Hive and Impala Cloudera are both Hadoop HDFS-based databases. In the next section, I briefly demonstrate Apache Spark data processing and use an SQL statement in Spark to show that SQL can also be used to manipulate Spark-based data. (In Chapter 10, I present Talend and Pentaho, which are also used to form ETL and manipulate data using a visual object-based approach.)

Apache Spark

Apache Spark is a cluster computing system that offers very fast in-memory distributed processing. You can develop applications in Java, Python, and Scala or use the built-in scripting shell for ad hoc script development. With the capability to scale to a very large degree (2000 nodes), Spark is also able to cache data for memory-based analytics.

Spark can run in local mode or use cluster managers, such as Mesos, YARN, or Spark. The cluster manager manages the executor processes on the worker nodes. The executors run applications on worker nodes and process application data. Spark uses a resilient distributed data set RDD) data model for data processing.

So while a Hadoop cluster provides a distributed batch processing system for handling very large volumes of data, Spark works in real time and is much faster. Like Hadoop, it offers a robust distributed processing model; however, Hadoop uses HDFS, while Spark is a memory-based system. Spark can also integrate with Hadoop, pulling data from and saving data to HDFS.

Installation of Spark

By way of example, I install Spark onto a 64-bit cluster using the CDH5 name node machine hc2nn and the data nodes hc2r1m1 to hc2r1m4. Spark works on a master-slave model, so I use the name node machine hc2nn as the master and the date node machines as the slaves. Unless stated otherwise, I carry out the installation as the Linux root user.

My first step is to set up a suitable repository file under the directory /etc/yum.repos.d on each machine so that the Linux yum command knows where and how to source the installation packages:

```
[root@hc2r1m1 ~]# cd /etc/yum.repos.d
[root@hc2r1m1 yum.repos.d]# cat  cloudera-cdh5.repo

[cloudera-cdh5]
# Packages for Cloudera's Distribution for Hadoop, Version 5, on RedHat or CentOS 6 x86_64
name=Cloudera's Distribution for Hadoop, Version 5
baseurl=http://archive.cloudera.com/cdh5/redhat/6/x86_64/cdh/5/
gpgkey = http://archive.cloudera.com/cdh5/redhat/6/x86_64/cdh/RPM-GPG-KEY-cloudera
gpgcheck = 1
```

The repository file (cloudera-cdh5.repo) tells yum to look at the repository URL http://archive.cloudera.com/cdh5/redhat/6/x86_64/cdh/5/ when installing the software. After setting up the repository file on each machine, I'm ready to install the Spark services on all machines. This command installs the Spark Master server, History server, and worker servers, as well as core and Python modules:

```
[root@hc2r1m1 ~]# yum install spark-core spark-master spark-worker spark-history-server spark-python
```

I install these components on each node, then set up the configuration under /etc/spark/conf/. I remember to make these changes on all servers unless instructed otherwise. Initially, I set up the slave files so that Spark knows where the slaves will run:

```
[root@hc2r1m1 ~]# cd /etc/spark/conf/

[root@hc2r1m4 conf]# cat slaves
# A Spark Worker will be started on each of the machines listed below.
hc2r1m1
hc2r1m2
hc2r1m3
hc2r1m4
```

Next, I edit the file spark-env.sh and set the value of the STANDALONE_SPARK_MASTER_HOST variable to be the full name of the master host:

```
export STANDALONE_SPARK_MASTER_HOST=hc2nn.semtech-solutions.co.nz
```

> **Note** If you set this value incorrectly—for instance, using a host short name—you may encounter this error:
>
> ```
> 14/09/09 18:20:52 ERROR remote.EndpointWriter: dropping message [class akka.actor.SelectChildName]
>   for non-local recipient [Actor[akka.tcp://sparkMaster@hc2nn:7077/]]
>   arriving at [akka.tcp://sparkMaster@hc2nn:7077] inbound
>   addresses are [akka.tcp://sparkMaster@hc2nn.semtech-solutions.co.nz:7077]
> ```

With the correct value for the variable set, I start the Spark Master and History servers on the master node:

```
[root@hc2nn ~]# service spark-master   restart
[root@hc2nn ~]# service spark-history-server restart
```

Finally, I start the Spark workers on all of the data nodes:

```
[root@hc2r1m1 ~]# service spark-worker restart
```

That's it; I have just started a basic Spark cluster! I now have the choice of user interfaces to monitor the Spark cluster. In the configuration file spark-env.sh, the following default variables define the master and worker user interface ports:

```
export SPARK_MASTER_WEBUI_PORT=18080
export SPARK_WORKER_WEBUI_PORT=18081
```

The Spark Master user interface can be found at hc2nn:18080; Figure 9-3 shows its appearance before any applications are run. Notice the Spark Master URL at the top of the page; that's needed to run applications later.

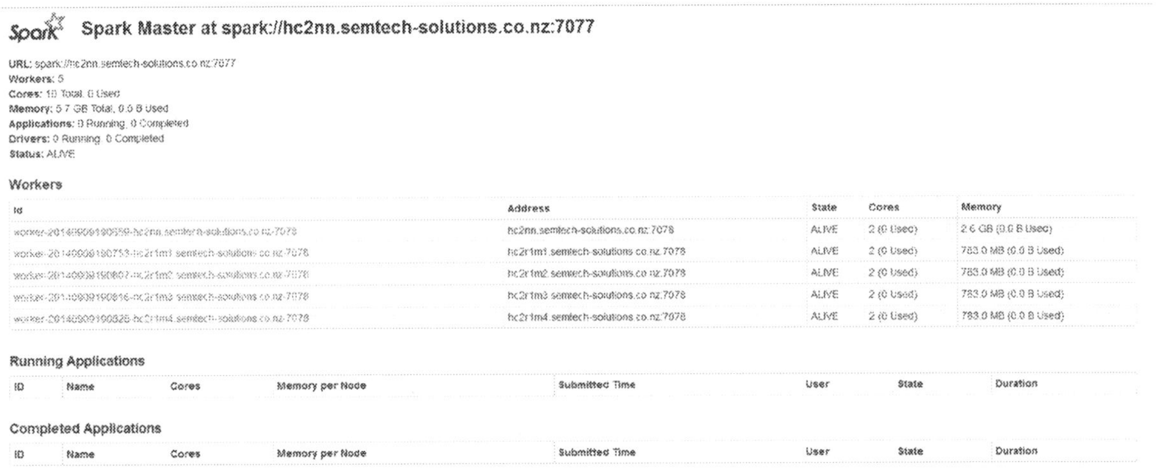

Spark **Spark Master at spark://hc2nn.semtech-solutions.co.nz:7077**

URL: spark://hc2nn.semtech-solutions.co.nz:7077
Workers: 5
Cores: 10 Total, 0 Used
Memory: 5.7 GB Total, 0.0 B Used
Applications: 0 Running, 0 Completed
Drivers: 0 Running, 0 Completed
Status: ALIVE

Workers

Id	Address	State	Cores	Memory
worker-20140909190559-hc2nn.semtech-solutions.co.nz-7078	hc2nn.semtech-solutions.co.nz:7078	ALIVE	2 (0 Used)	2.6 GB (0.0 B Used)
worker-20140909180753-hc2r1m1.semtech-solutions.co.nz-7078	hc2r1m1.semtech-solutions.co.nz:7078	ALIVE	2 (0 Used)	783.0 MB (0.0 B Used)
worker-20140909190807-hc2r1m2.semtech-solutions.co.nz-7078	hc2r1m2.semtech-solutions.co.nz:7078	ALIVE	2 (0 Used)	783.0 MB (0.0 B Used)
worker-20140909190816-hc2r1m3.semtech-solutions.co.nz-7078	hc2r1m3.semtech-solutions.co.nz:7078	ALIVE	2 (0 Used)	783.0 MB (0.0 B Used)
worker-20140909190826-hc2r1m4.semtech-solutions.co.nz-7078	hc2r1m4.semtech-solutions.co.nz:7078	ALIVE	2 (0 Used)	783.0 MB (0.0 B Used)

Running Applications

ID	Name	Cores	Memory per Node	Submitted Time	User	State	Duration

Completed Applications

ID	Name	Cores	Memory per Node	Submitted Time	User	State	Duration

Figure 9-3. *Spark Master server's user interface*

The interface also lists the Spark workers and the machines that they are running on, as well as some information about the state, cores, and memory available to each worker. (In Figure 9-3, you'll notice that I have also run a worker on the name node, just to increase the processing capacity in this example.) The area at the bottom of the screen in Figure 9-3 provides details on running applications, as well as completed applications. In Figure 9-3, none are running, so the area is blank.

Uses of Spark

In this section, I use an example to demonstrate the Spark shell, an interactive Spark scripting session, and a Spark application (provided with the installation) to show you how jobs can be submitted, as well as how they appear in the Spark user interface. The Spark shell can be used interactively to run ad hoc scripts against Spark cluster-based data. Running a Spark application, as you will see, allows you to run a job on a Spark cluster by using in-memory processing in batch mode.

My SQL-based example shows that Spark-based information can be accessed using SQL. Although Spark processing is not included in Chapter 10's discussion of Talend and Pentaho, both tools integrate with Spark to offer visual object-based data manipulation.

The first step in running a Spark script interactively in the Spark shell is to set the master URL displayed on the Spark Master user interface to be the Spark URL:

```
[root@hc2nn ~]# spark-shell --master spark://hc2nn.semtech-solutions.co.nz:7077
```

As shown in Figure 9-4, the Spark shell application is now visible on the Spark Master user interface. The Running Applications section in Figure 9-4 lists the appplication's ID and name, as well as the number of cores and the memory available to the application. I can also check the application's submission time, state, user, and duration of its run.

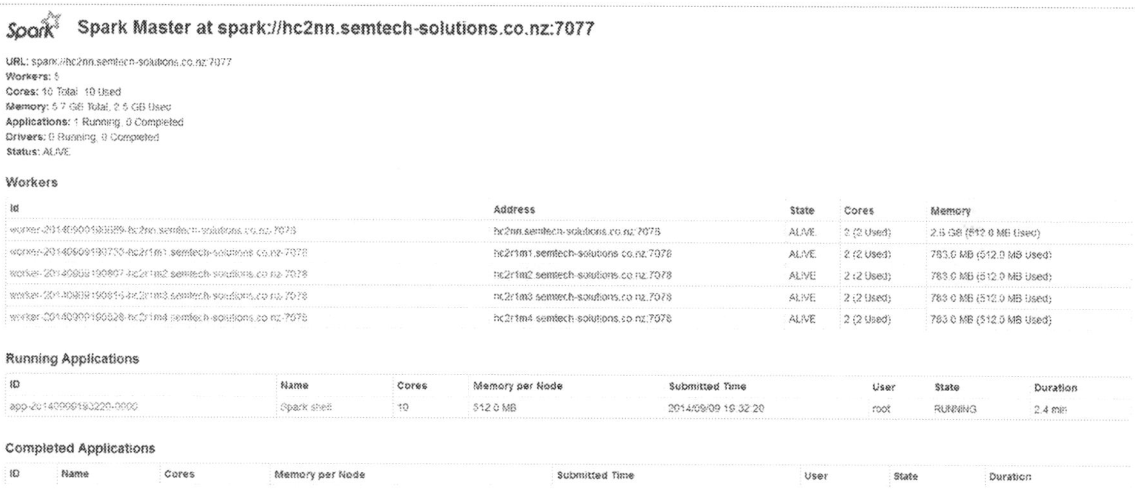

Figure 9-4. *Spark Master interface with applications listed*

If I click one of the worker node IDs listed in the Workers section of the Spark Master interface, I can drill down for more information, as shown in Figure 9-5. This detailed view shows the cores and memory available on the worker node plus the executor for the running Spark shell on that node. I click the Back to Master link to return to the Spark Master interface.

Spark Worker at hc2nn.semtech-solutions.co.nz:7078

ID: worker-20140909190659-hc2nn.semtech-solutions.co.nz-7078
Master URL: spark://hc2nn.semtech-solutions.co.nz:7077
Cores: 2 (2 Used)
Memory: 2.6 GB (512.0 MB Used)

Back to Master

Running Executors 1

ExecutorID	Cores	Memory	Job Details	Logs
3	2	512.0 MB	**ID:** app-20140909193220-0000 **Name:** Spark shell **User:** root	stdout stderr

Figure 9-5. *Spark Worker user interface*

The work of a given application is spread across the cluster over a series of worker executors on each node. To reach the application user interface, I simply click an application listed on the Spark Master interface that is in the Running Applications section, as shown in Figure 9-4. The list of executors for this application is is then shown in Figure 9-6.

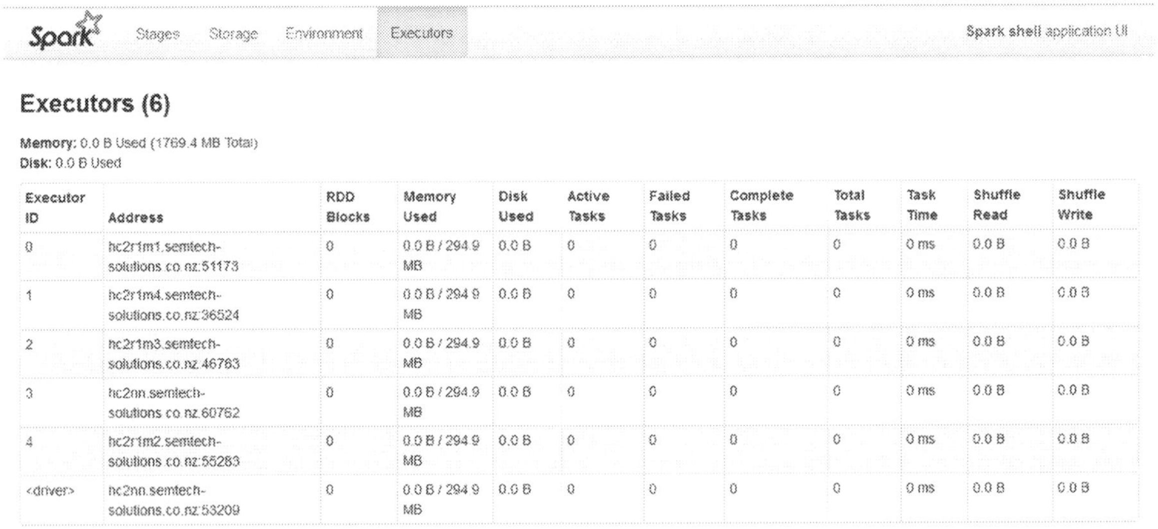

Spark   Stages   Storage   Environment   Executors   Spark shell application UI

Executors (6)

Memory: 0.0 B Used (1769.4 MB Total)
Disk: 0.0 B Used

Executor ID	Address	RDD Blocks	Memory Used	Disk Used	Active Tasks	Failed Tasks	Complete Tasks	Total Tasks	Task Time	Shuffle Read	Shuffle Write
0	hc2r1m1.semtech-solutions.co.nz:51173	0	0.0 B / 294.9 MB	0.0 B	0	0	0	0	0 ms	0.0 B	0.0 B
1	hc2r1m4.semtech-solutions.co.nz:36524	0	0.0 B / 294.9 MB	0.0 B	0	0	0	0	0 ms	0.0 B	0.0 B
2	hc2r1m3.semtech-solutions.co.nz:46763	0	0.0 B / 294.9 MB	0.0 B	0	0	0	0	0 ms	0.0 B	0.0 B
3	hc2nn.semtech-solutions.co.nz:60762	0	0.0 B / 294.9 MB	0.0 B	0	0	0	0	0 ms	0.0 B	0.0 B
4	hc2r1m2.semtech-solutions.co.nz:55283	0	0.0 B / 294.9 MB	0.0 B	0	0	0	0	0 ms	0.0 B	0.0 B
<driver>	hc2nn.semtech-solutions.co.nz:53209	0	0.0 B / 294.9 MB	0.0 B	0	0	0	0	0 ms	0.0 B	0.0 B

Figure 9-6. *Spark application interface*

Executors are the tasks based on the cluster worker nodes that process and store an application's data on the Spark cluster. Figure 9-6 shows that each executor has a unique ID and address. It also shows the state of the executor in terms of memory and disk, plus the task time.

So, when the interactive Spark shell is running, what can you do with it? To demonstrate a simple script, I read a Linux-based CSV file from HDFS, run a line count on it in memory, and then do a string search on it. I also confirm the results by checking the output against Linux commands.

The top few lines of the data from my CSV file using the Linux head command are as follows:

```
[root@hc2nn fuel_consumption]# head scala.csv

MODEL,MANUFACTURER,MODEL,VEHICLE CLASS,ENGINE SIZE,CYLINDERS,TRANSMISSION,FUEL,FUEL
CONSUMPTION,,,,FUEL,CO2 EMISSIONS
YEAR,,,,(L),,,TYPE,CITY (L/100 km),HWY (L/100 km),CITY (mpg),HWY (mpg),(L/year),(g/km)
2014,ACURA,ILX,COMPACT,2,4,AS5,Z,8.6,5.6,33,50,1440,166
2014,ACURA,ILX,COMPACT,2.4,4,M6,Z,9.8,6.5,29,43,1660,191
2014,ACURA,ILX HYBRID,COMPACT,1.5,4,AV7,Z,5,4.8,56,59,980,113
2014,ACURA,MDX 4WD,SUV - SMALL,3.5,6,AS6,Z,11.2,7.7,25,37,1920,221
```

I first copy this file to the /tmp directory on HDFS, so that Scala can access it, by using the HDFS file system copyFromLocal command:

```
[root@hc2nn fuel_consumption]# hdfs dfs -copyFromLocal scala.csv /tmp/scala.csv
```

Note that when the Spark shell is used, the special variable sc is created. Called a *Spark Context*, the variable describes the connection to the Spark cluster. So, at my Scala shell script prompt (scala >), I use the sc variable in the following command to read the scala.csv file into memory:

```
scala> val myFile = sc.textFile("/tmp/scala.csv")

14/09/09 19:55:21 INFO storage.MemoryStore: ensureFreeSpace(74240) called with curMem=155704,
maxMem=309225062
14/09/09 19:55:21 INFO storage.MemoryStore: Block broadcast_1 stored as values to memory (estimated
size 72.5 KB, free 294.7 MB)
myFile: org.apache.spark.rdd.RDD[String] = MappedRDD[3] at textFile at <console>:12
```

The next command produces a line count on the file, now represented by the variable myFile, in memory:

```
scala> myFile.count()

14/09/09 19:55:41 INFO spark.SparkContext: Job finished: count at <console>:15, took 3.174464234 s
res1: Long = 1069
```

The result indicates that there are 1,069 lines in the file. The Spark-based line count can be checked against the original file on the Linux file system. To do so, I use the Linux wc (word count) command with a -l switch to confirm the count of 1,069 lines:

```
[root@hc2nn fuel_consumption]# wc -l scala.csv
1069 scala.csv
```

The following Spark shell Scala command counts the number of instances of the string "ACURA" in the in-memory file:

```
scala> myFile.filter(line => line.contains("ACURA")).count()
14/09/09 19:58:10 INFO spark.SparkContext: Job finished: count at <console>:15, took 2.815524655 s
res0: Long = 12
```

The result is 12 lines; checking that total by using the Linux grep command piped to the same wc command gives the same result:

```
[root@hc2nn fuel_consumption]# grep ACURA scala.csv | wc -l
12
```

Interactively typing the shell commands is useful for short, simple tasks. For larger scripts, you can use the spark-submit command to submit applications to the Spark cluster. For instance, I use one of the examples in the spark-examples application library available under /usr/lib/spark/examples/lib/ in the CDH5 release (which is supplied with the Spark install):

```
[root@hc2nn ~]# cd /usr/lib/spark/examples/lib/
[root@hc2nn lib]# ls -l spark-examples_2.10-1.0.0-cdh5.1.2.jar
-rw-r--r-- 1 root root 734539 Aug 26 15:07 spark-examples_2.10-1.0.0-cdh5.1.2.jar
```

I can execute example applications from this library on the cluster and monitor them using the master user interface. For instance, the following code uses the spark-submit command to run the SparkPi example program from the examples library. It sets the memory to be used on each worker at 700 MB and the total cores to be used at 10. It uses the same master Spark URL to connect to the cluster and specifies 10,000 tasks/iterations:

```
[root@hc2nn ~]# spark-submit \
  --class org.apache.spark.examples.SparkPi \
  --master spark://hc2nn.semtech-solutions.co.nz:7077  \
  --executor-memory 700M \
  --total-executor-cores 10 \
  /usr/lib/spark/examples/lib/spark-examples_2.10-1.0.0-cdh5.1.2.jar \
  10000
```

Checking the Spark Application interface for this application in Figure 9-7 shows the application details and the progress of the job, such as the duration of the application run. I can also see that the default scheduling mode of FIFO (first in, first out) is being used. As with YARN, I can set up a fair scheduler for Spark. In the Active Stages section, the blue bar shows the progress of the application run.

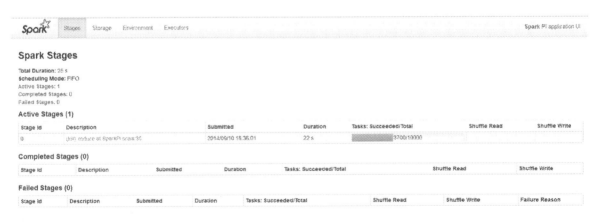

Figure 9-7. *Spark Application interface shows details of the job*

By clicking the Executors tab at the top of the Spark Application interface, I can examine details of the application task, as shown in Figure 9-8. I can see the spread of tasks across the cluster nodes and the task times by node, as well as a list of tasks, the nodes they run on, their execution times, and their statuses. The Summary Metrics section also contains minimum, maximum, and percentile information for details like serialization, duration, and delay.

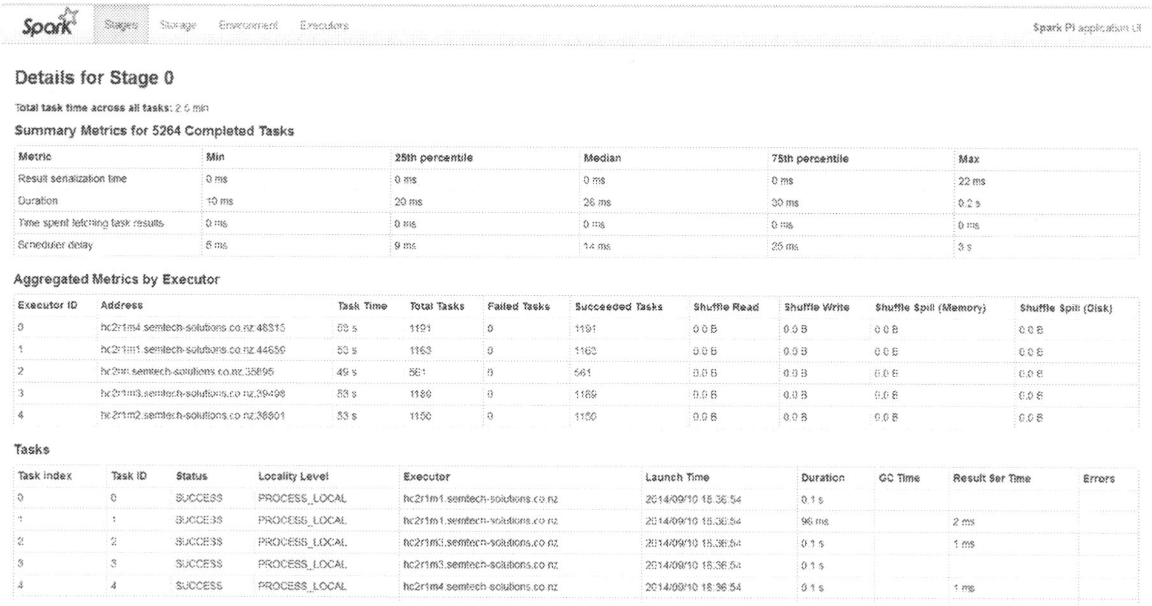

Figure 9-8. *Spark Application interface shows job executors*

As you can see from these simple examples, the Spark cluster is easy to install, set up, and use. To investigate Spark in greater detail, visit the Apache Software Foundation website for Apache Spark at spark.apache.org.

Spark SQL

Rather than using the default Spark context object sc, you can create an SQL context from the default Spark context and process CSV data using SQL. Spark SQL is an incubator project, however; it is not a mature offering at this time. For instance, its APIs may yet change, and the version in CDH5 no longer reflects Spark SQL's latest functions. Despite that, it does offer some interesting features for memory-based SQL cluster processing. I use a simple example of CSV file processing using Spark SQL, based on a schema-based RDD example at https://spark.apache.org/docs/1.0.0/sql-programming-guide.html.

■ **Note**   For the latest details on Spark SQL, see the Spark website at spark.apache.org.

This example uses the same CSV file as was used in a previous example. The basic steps are to create a SQL Context object in the Spark shell session to make Spark SQL functionality available, then import the data into a table and run some SQL against it. Here is the CSV file data in the file /tmp/scala.csv. I have removed its header rows so that it contains only raw data:

```
[root@hc2nn fuel_consumption]# head scala.csv

2014,ACURA,ILX,COMPACT,2,4,AS5,Z,8.6,5.6,33,50,1440,166
2014,ACURA,ILX,COMPACT,2.4,4,M6,Z,9.8,6.5,29,43,1660,191
2014,ACURA,ILX HYBRID,COMPACT,1.5,4,AV7,Z,5,4.8,56,59,980,113
2014,ACURA,MDX 4WD,SUV - SMALL,3.5,6,AS6,Z,11.2,7.7,25,37,1920,221
```

To use SQL in Spark, I enter the following command into my Spark shell:

```
scala> val sqlContext = new org.apache.spark.sql.SQLContext(sc)
```

This creates a SqlContext from the default Spark context sc, using the SQLContext class. Next, I import the sqlContext library functionality so that it is available for the rest of the script:

```
scala> import sqlContext._
```

I define the schema using a case class to represent all of the comma-separated fields in the data file line. This defines the number of fields, their name, order, and type:

```
scala> case class Vehicle(year: Int,manufacturer: String, model: String, vclass: String, engine:
Double, cylinders: Int, fuel: String, consumption: String, clkm: Double, hlkm: Double, cmpg: Int,
hmpg: Int, co2lyr: Int, co2gkm: Int)
```

░ **Note**   The column types are case-sensitive; for instance, "int" will cause an error while "Int" will not.

The data types used here are Int, String, and Double to represent the data vaues in the CSV file. The order is important, as it should match the data in the CSV file row. Also, reserved words need to be avoided, so I have used the column name vclass to describe my vehicle class.

Now, I create an RDD from the vehicle record, import the CSV file /tmp/scala.csv, split the file by the comma character, and convert the data columns by type to match the columns in the schema above:

```
scala> val vehicle = sc.textFile("/tmp/scala.csv").map(_.split(",")).map(p => Vehicle(
p(0).trim.toInt,
p(1),
p(2),
p(3),
p(4).trim.toDouble,
p(5).trim.toInt,
p(6),
p(7),
```

```
p(8).trim.toDouble,
p(9).trim.toDouble,
p(10).trim.toInt,
p(11).trim.toInt,
p(12).trim.toInt,
p(13).trim.toInt
))
```

That's quite a complicated statement, but if I break it down, it will seem simpler. The `textFile` option loads the CSV file as a text file. The first map option splits the columns in the text file by comma. The next `map` option maps the data columns from the previous step into the columns of the vehicle class that was just defined. So, the vehicle RDD contains the comma-separated data from the file.

I register the vehicle RDD as a table called "vehicle" so that SQL can be executed against the table:

```
scala> vehicle.registerAsTable("vehicle")
```

At this point, the data has been imported into the RDD and the RDD has been registered as a table, so I am ready to execute some SQL against the table. I want to select details of Aston Martin cars from the data. The following statement creates a schema RDD called "aston" that contains the data from the SELECT statement:

```
scala> val aston = sql( "SELECT year, manufacturer, model, vclass, engine FROM  vehicle WHERE
manufacturer = 'ASTON MARTIN' ")
```

The SELECT statement takes the year, manufacturer, model, class, and engine size columns from the vehicle table. It filters the data, selecting only those where the manufacturer's name is Aston Martin.

When printed, the resulting aston schema RDD appears as a string. One line is printed for each row that is matched from the table. The five columns from the table that match the columns in the SQL are embedded in the results string:

```
scala> aston.map(  t => "year: " + t(0) + " manufacturer " + t(1) + " model " + t(2) + " class " +
t(3)  + " engine " + t(4)  ).collect().foreach(println)
```

That string prints the following data:

```
year: 2014 manufacturer ASTON MARTIN model DB9              class MINICOMPACT engine 5.9
year: 2014 manufacturer ASTON MARTIN model RAPIDE           class SUBCOMPACT  engine 5.9
year: 2014 manufacturer ASTON MARTIN model V8 VANTAGE       class TWO-SEATER  engine 4.7
year: 2014 manufacturer ASTON MARTIN model V8 VANTAGE       class TWO-SEATER  engine 4.7
year: 2014 manufacturer ASTON MARTIN model V8 VANTAGE S     class TWO-SEATER  engine 4.7
year: 2014 manufacturer ASTON MARTIN model V8 VANTAGE S     class TWO-SEATER  engine 4.7
year: 2014 manufacturer ASTON MARTIN model VANQUISH         class MINICOMPACT engine 5.9
```

Thus, the results show seven matching Aston Martin data rows with their model and class details.

From an analytics point of view, Spark SQL gives analysts SQL-based access to Spark data in memory. In processing terms, it is much faster than traditional Map Reduce processing. Also, people with a background in relational databases will be comfortable using SQL to interrogate their data.

Summary

Analysts, whether they be managers, testers, or researchers, need to find meaning in their data. They need to be able to move, load, extract, and transform all or parts of their data to create meaning. Although the simple examples for Impala, Hive, and Spark in this chapter may not yield any revelations in your company's data, they do demonstrate the available building blocks of analytics and provide an overview of the capabilities of these tools.

In this chapter I have shown that you can represent data on HDFS as a database table and use Hive QL or Impala SQL to query and transform that data. If you combine the steps in this chapter with tools like Sqoop and Flume (covered in Chapter 6), you can start to build ETL chains to source, move, and modify your data, step by step.

If you find that you need real-time processing rather than batch processing, you might consider using Apache Spark. Following the example installation in this chapter, you can start using Spark on your cluster. The Spark SQL example also shows how to process your Spark cluster based in memory data using SQL.

The next chapter covers the ETL tools Pentaho and Talend, which can be used to visually manipulate Hadoop- and Spark-based data. They integrate with Map Reduce and Hadoop base tools like Pig, Sqoop, and Hive, and can be used to create and schedule ETL-based chains using a combination of the Hadoop tools that have been introduced so far.

CHAPTER 10

■ ■ ■

ETL with Hadoop

Given that Hadoop-based Map Reduce programming is a relatively new skill, there is likely to be a shortage of highly skilled staff for some time, and those skills will come at a premium price. ETL (extract, transform, and load) tools, like Pentaho and Talend, offer a visual, component-based method to create Map Reduce jobs, allowing ETL chains to be created and manipulated as visual objects. Such tools are a simpler and quicker way for staff to approach Map Reduce programming. I'm not suggesting that they are a replacement for Java or Pig-based code, but as an entry point they offer a great deal of pre-defined functionality that can be merged so that complex ETL chains can be created and scheduled. This chapter will examine these two tools from installation to use, and along the way, I will offer some resolutions for common problems and errors you might encounter.

Pentaho Data Integrator

In this first half of the chapter I explain how to source and install the Pentaho Data Integration (PDI) application. Offering tools to analyze, visualize, explore, report, and predict in the same platform, PDI can work as a stand-alone tool or can be downloaded into Pentaho Business Analytics. Pentaho offers enhanced functionality, features, and professional support for PDI. The open-source version is called Kettle. PDI is downloaded as a generic zipped package that can be installed on either Windows or Linux. Here's how to install PDI and use it with Hadoop.

■ **Note**   For complete details on PDI, see the company's website at www.pentaho.com/product/data-integration.

Installing Pentaho

You can download the installation package for the Pentaho Data Integrator (PDI) from the following URL:

```
http://sourceforge.net/projects/pentaho/files/Data%20Integration/5.1/pdi-ce-5.1.0.0-752.zip/
download.
```

With an installation package that you can use for either Lunix or Windows, the zipped file is 580 MB, so it takes quite a while to download. By way of example, after I downloaded and extracted the package, I installed the Windows package on my C: drive, as shown in Figure 10-1. As you can see in Figure 10-1, the software installs into a directory called "data-integration." Note the directory structure, the start-up scripts, and the plug-ins directory that have been marked with red boxes. On Windows, you would start the application using the Spoon.bat script; on Linux, you would use the Spoon.sh script.

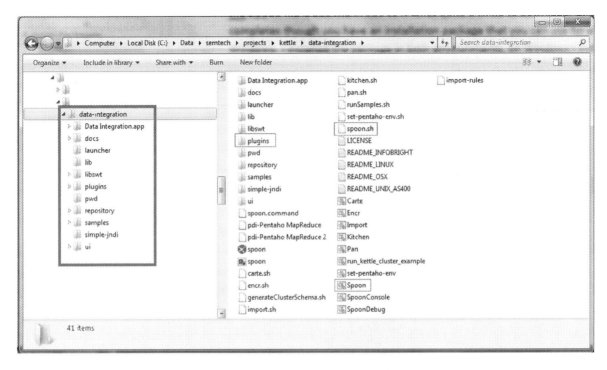

Figure 10-1. *Pentaho Data Integrator's installation structure*

Before you start the application, the big data plug-in needs to know what Hadoop configuration you are using. Figure 10-2 shows the structure of the PDI plugins directory, and specifically, the big data plug-in. It shows that the pentaho-big-data-plugin resides within the plugins directory. This directory contains a file called plugin.properties (Figure 10-2), which needs to be altered so that PDI knows which Hadoop version it is connecting to. To determine which PDI plugin version to use, you check the Pentaho URL http://wiki.pentaho.com/display/BAD/Configuring+Pentaho+for+your+Hadoop+Distro+and+Version, which provides a mapping of PDI plugin to Hadoop version.

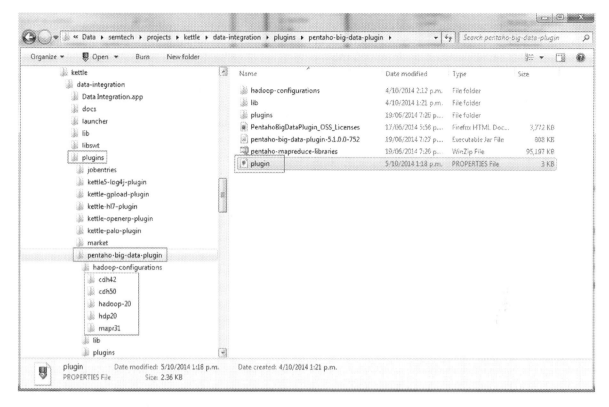

Figure 10-2. *Pentaho Data Integrator, showing big data plug-in structure*

The subdirectories shown in the hadoop-configurations directory indicate which Hadoop configuration values are supported by the pentaho-big-data-plugin. By changing the following line in the file plugin.properties, you set the configuration:

```
active.hadoop.configuration=cdh50
```

For example, the setting shown in Figure 10-2 shows that I have set the pentaho-big-data-plugin for PDI to use Cloudera's CDH5 (cdh50). Because I have limited memory available on my CDH5 cluster, I decide to run PDI on a Windows machine and access Hadoop on Linux remotely.

You also need to copy the Hadoop configuration files to the PDI plug-in hadoop directory. From the Cloudera CDH5 manager home page, you select the YARN (MR2 Included) option, then select the Actions drop-down menu, followed by Download Client configuration. The zipped file that is downloaded contains the files core-site.xml, hadoop-env.sh, hdfs-site.xml, hive-site.xml, mapred-site.xml, and yarn-site.xml. Because I am using the CDH5 (cdh50) configuration, I copy these files to the following PDI directory: data-integration\plugins\pentaho-big-data-plugin\hadoop-configurations\cdh50.

Pentaho requires Sun/Oracle Java 1.7, which is available at https://java.com/en/download/index.jsp. Be sure to download and install this on Windows; the cmd.exe Window session output shows my Java installation as follows:

```
C:\Users\mikejf12>java -version
java version "1.7.0_67"
Java(TM) SE Runtime Environment (build 1.7.0_67-b01)
Java HotSpot(TM) 64-Bit Server VM (build 24.65-b04, mixed mode)
```

■ **Note**   If you plan to access MySQL with PDI, you also need to install a MySQL jar file from `http://dev.mysql.com/downloads/connector/j/` called `mysql-connector-java-5.1.32-bin.jar` into the PDI directory data-integration\lib.

I am using CDH5, so I also need to add the following to the copy of yarn-site.xml in the PDI plug-in directory that was copied from Hadoop. This helps the Pentaho libraries locate the YARN server:

```
<!-- added for pentaho pdi -->

  <property>
    <name>yarn.resourcemanager.hostname</name>
    <value>hc2nn</value>
  </property>
```

Running the Data Integrator

You start PDI on Windows by using the Spoon.bat script, which starts the Spoon client application, as shown in Figure 10-3. This is where you create your Map Reduce jobs.

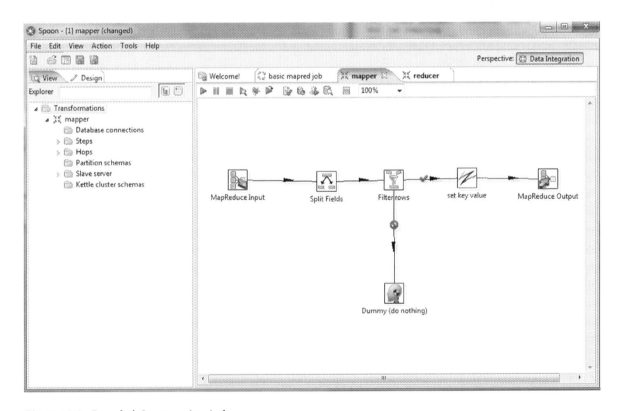

Figure 10-3. *Pentaho's Spoon main window*

To be able to show the functionality of PDI, I have created some tasks, including a Map Reduce job called "basic mapred job" and the associated Map and Reduce tranformations called "mapper" and "reducer," respectively. As you can see, these are already open and displayed in Figure 10-3. Also, note that the Explorer pane functionality on the left of the figure changes depending on the current job or transformation that is displayed in the right pane.

The Explorer pane on the left of the window has View and Design tabs so that you can either see the content of the task that you are working on or select functions to add to your job or transformation. The Working pane on the right currently contains that Map Reduce job I just mentioned, called "basic mapred job" and the two transformations called "mapper" and "reducer," which it calls. If you drag the functionality icons from the Explorer pane to the Working pane on the right, you can then configure and connect the functions to the flow of the task. (There will be more about those functions later in the chapter.)

The Explorer view of the current mapper transformation, displayed in Figure 10-4, shows that it contains Map Reduce inputs and outputs, a filter, a field splitter, a set key, and a dummy task. (I'll explain all of these items in more detail later.) The view also shows the task workflow hops that will be executed. As Figure 10-4 illustrates, you can configure a slave server, meaning one instance of PDI uses another as a slave and gets that instance to run a task. (To learn more about slave servers and how they work, see the Pentaho website at www.pentaho.com.)

Figure 10-4. *Pentaho's Explorer view*

Figure 10-5 provides a taste of the functionality that's available in the Design view. The two columns on the left list the functions available for transformations (note the expanded big data section), while the two columns on the right show the functions for jobs, again with an expanded big data view. You can construct transactions in a logical, step-by-step manner by using these building blocks, then include the transactions in jobs that you schedule within PDI.

Figure 10-5. *Pentaho Explorer's Design view*

Creating ETL

Now that you have a sense of the PDI interface, it's time to examine an example of a Map Reduce task to see how PDI functions. I create an ETL example by starting with mapper and reducer transformations, and follow with the Map Reduce job itself. By following my steps you'll learn how each module is configured, as well as gain some tips on how to avoid pitfalls.

To create my PDI Map Reduce example, I first need some data. The HDFS file (rawdata.txt) should look familiar—parts of it were used in earlier chapters. Here, I use fuel consumption details for various vehicle models over a number of years. The data file is CSV-based and resides under HDFS at /data/pentaho/rdbms/. I use the Hadoop file system cat command to dump the file contents and the Linux head command to limit the data output:

```
[hadoop@hc2nn ~]$ hdfs dfs -cat /data/pentaho/rdbms/rawdata.txt | head -5
```

```
1995,ACURA,INTEGRA,SUBCOMPACT,1.8,4,A4,X,10.2,7,28,40,1760,202
1995,ACURA,INTEGRA,SUBCOMPACT,1.8,4,M5,X,9.6,7,29,40,1680,193
1995,ACURA,INTEGRA GS-R,SUBCOMPACT,1.8,4,M5,Z,9.4,7,30,40,1660,191
1995,ACURA,LEGEND,COMPACT,3.2,6,A4,Z,12.6,8.9,22,32,2180,251
1995,ACURA,LEGEND COUPE,COMPACT,3.2,6,A4,Z,13,9.3,22,30,2260,260
```

Using Map Reduce, I want to create a count of manufacturer and model groupings from the data. The first step is to create two transformations: a mapper and a reducer. The *mapper* receives the file lines and strips those lines into fields. It filters out any empty lines and creates a key from the fields I am interested in. Finally, it outputs a single record per line of the compound key and a value of 1. The *reducer* receives the output from the mapper, sorts data by the key, then sums the values by key. It then outputs a sorted list of summed values for each key.

Figure 10-6 shows the structure of the mapper transformation.

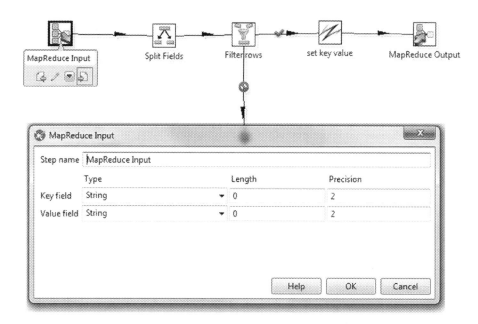

Figure 10-6. *Input step of mapper transformation*

Each Map Reduce transformation must start with a Map Reduce Input and end with a Map Reduce Output. To set up the sequence, I simply click the Design tab in the Explorer pane of the main PDI interface, then drag the components from the Design view to the Working pane on the right. To connect the components into a workflow, I click a component to open a drop-down menu below it, click the rightmost green arrow icon (bordered in red in Figure 10-6), and then drag a workflow to the next component to connect them. The workflow arrow indicates the direction of flow and shows whether the action is unconditional or if it occurs only when the result is True or False. By double-clicking a component, such as Map Reduce Input, I can open its configuration as shown in Figure 10-6. Here, I can see that the input component has inputs called "key" and "value," with fields described as "string" from the HDFS file data.

Double-clicking the Split Fields icon opens its configuration, as shown in Figure 10-7. This component receives a file line containing a comma-separated set of file fields that need to be split into separate values in order to be manipulated. That is what this step does: it splits the string-based value field by using a comma as a separator and it creates 14 new fields, Field 1 to Field 14.

Field splitter												

Step name Split Fields

Field to split value

Delimiter ,

Enclosure

Fields

#	New field	ID	Remove ID?	Type	Length	Precision	Format	Group	Decimal	Currency	Nullif	Default	Trim type
1	Field1		N	String									none
2	Field2		N	String									none
3	Field3		N	String									none
4	Field4		N	String									none
5	Field5		N	String									none
6	Field6		N	String									none
7	Field7		N	String									none
8	Field8		N	String									none
9	Field9		N	String									none
10	Field10		N	String									none
11	Field11		N	String									none
12	Field12		N	String									none
13	Field13		N	String									none
14	Field14		N	String									none

Help   OK   Cancel

***Figure 10-7.** Split Fields step of mapper transformation*

The next step in the flow is Filter Rows, a decision point where null rows are discarded based on the value of Field 1. As you can see in Figure 10-8, the empty rows go to a dummy step whose flow has a False state. Valid rows move on to the Set Key Value step whose flow has a True state, as indicated by the green tick.

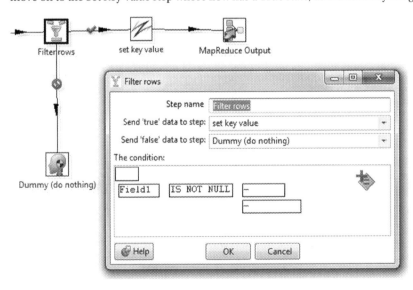

***Figure 10-8.** Filter Rows step of mapper transformation*

After filtering any bad rows out of the data, my next move is to build a compound key from the extracted data fields via the Set Key Values step. This step creates a combined comb_key key value from Fields 2 and 3, separating the string values with a dash. (Actually, this is created as a user-defined Java expression. It also creates a comb_value value field with a value of 1, as shown in Figure 10-9).

Figure 10-9. *Java expression for Set Key Values step of mapper transformation*

These values are then passed to the MapReduce Output step, which assigns the comb_key and comb_value variables to the mapper output variables key and value, as shown in Figure 10-10.

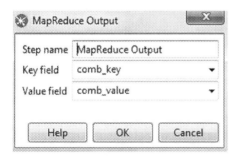

Figure 10-10. *Output step of mapper transformation*

That completes the definition of the *mapper* transformation that creates a key and value from the incoming data.

As described earlier, the *reducer* transformation accepts the incoming key / value pair and sorts it, then groups by key values, sums the value, and finally outputs the results. Figure 10-11 shows the structure of the reducer transformation. The Input step is the same as for the mapper transformation, but Figure 10-12 provides greater detail on the Sort Rows step. Here, I define a single field: the sort key field (at the bottom). I also reduce the value in the sort size field because I didn't have much data and I wanted to save memory.

Figure 10-11. *Steps in reducer transformation*

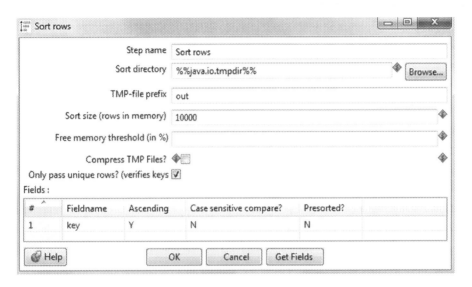

Figure 10-12. *Sort Rows step of reducer transformation*

Figure 10-13 shows the Group By step, which groups the data by the key value and sums it by the numeric value to create a new summed value variable called summed_val. The aggregate value is defined as the sum function working on the value variable.

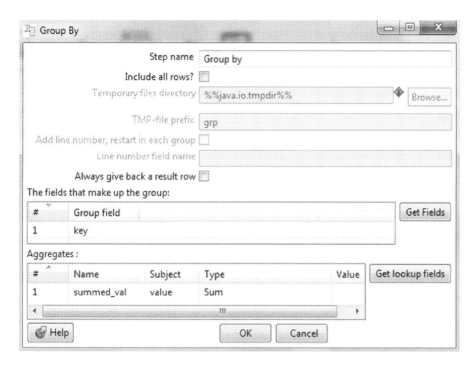

Figure 10-13. Group By step of reducer transformation

Finally, the reducer Output step produces the key/value data pair as the compound key value that was created by the mapper and the summed value variable summed_val, as shown in Figure 10-14.

Figure 10-14. Output step of reducer transformation

With the mapper and reducer transformations now defined, I can create a Map Reduce job, as shown in Figure 10-15. A job always starts with a Start step, which I have pulled over from the General folder. I also add a Map Reduce step, which uses the new mapper and reducer transformations.

START   Pentaho MapReduce

Figure 10-15. *Map Reduce job setup*

Double-clicking the Start step gives me the option to schedule the job. As you can see in Figure 10-16, this Map Reduce job is scheduled to run daily at 13:20.

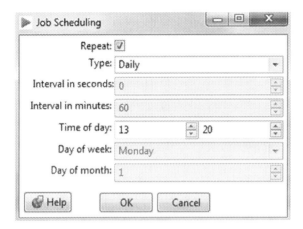

Figure 10-16. *Job scheduling for Map Reduce job*

Figure 10-17 shows the Mapper configuration for the Map Reduce job called "pmr1." The mapper value is defined as the newly created mapper transformation, while the input and output values are defined as using the mapper transformation MapReduce Input and Output steps.

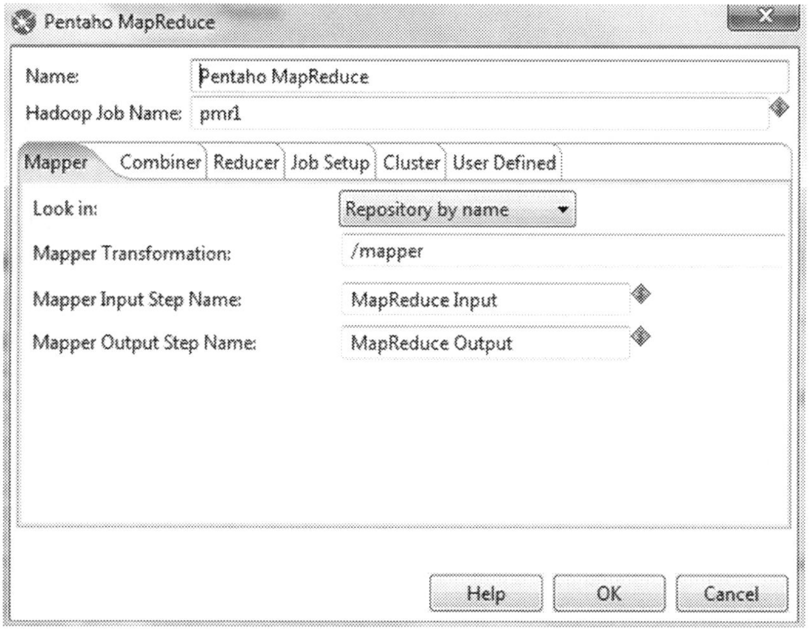

Figure 10-17. *Mapper configuration for job*

Clicking the Reducer tab displays the job reducer configuration, as shown in Figure 10-18. The reducer step is defined as the reducer transformation.

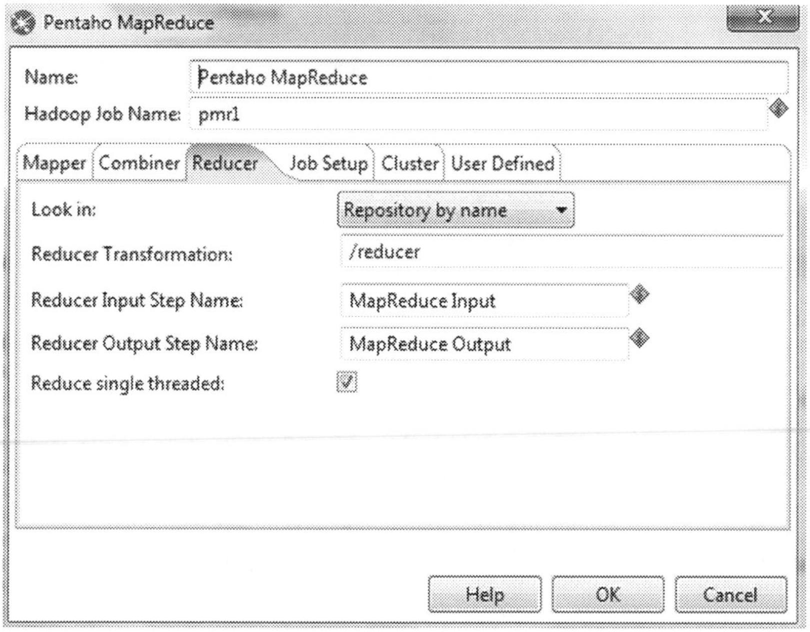

Figure 10-18. *Reducer configuration for job*

Next, I click the Job Setup tab to specify the input and output paths for the job data, a shown in Figure 10-19. The input data file is stored on HDFS at /data/pentaho/rdbms, as explained earlier.

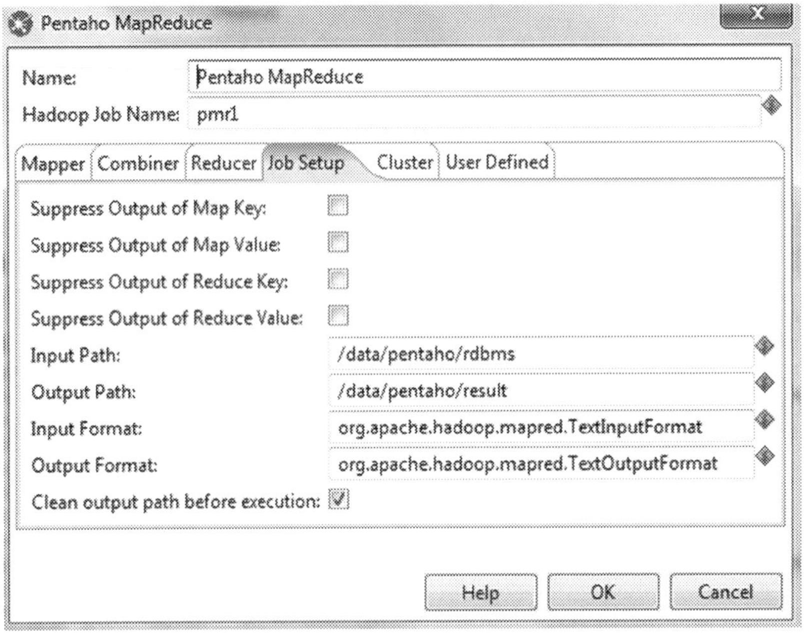

Figure 10-19. *Job Setup tab for job pmr1*

The input and output data formats for this job are defined as Hadoop Map Reduce based Java classes, such as org.apache.hadoop.mapred.TextOutputFormat. The Clean option is selected so that the job can be rerun. That is, each time the job runs, it will clean out the results directory.

Lastly, I define the connection to the Hadoop cluster using the Cluster tab. As you can see in Figure 10-20, the only fields that I have changed in this tab are the hostnames and ports, so that Pentaho knows which hosts to connect to (hc2nn) for HDFS and Map Reduce. I have also specified the ports, 8020 for HDFS and 8032 for the Resource Manager (which is actually labeled as the Job Tracker, but this is a CDH5 cluster using YARN).

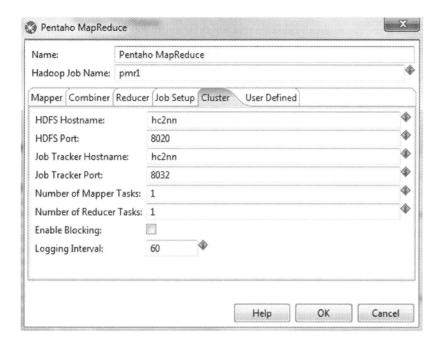

Figure 10-20. *Cluster tab for job pmr1*

Now that the job is fully specified, I can run the Map Reduce job I've called "pmr1" against YARN and I can monitor it via Pentaho PDI and the Resource Manager user interface. When it finishes, I can check the data on HDFS. Clicking the green run arrow, highlighted with a red box in Figure 10-21, causes the Execute a Job job configuration window to pop up.

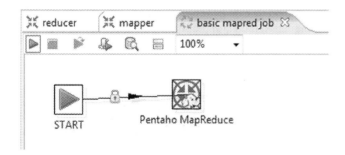

Figure 10-21. *Main window for Map Reduce job*

Figure 10-22 shows the Execute a Job window, in which I specify the logging levels and metrics so the Pentaho Map Reduce job can be run. Note that the Logging Level drop-down menu is highlighted and set to Basic Logging in Figure 10-22. If an error occurs, I can rerun the job and increase the logging level. (There's more about logging levels in the main PDI interface as well.)

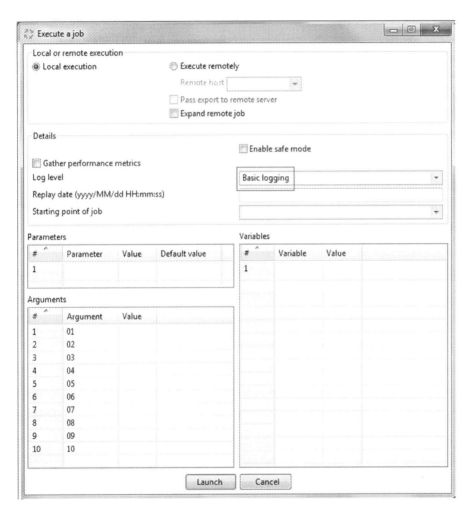

Figure 10-22. *Execute a Job window*

Clicking the Launch button executes the job and produces the basic-level log output that is shown in Figure 10-23.

```
2014/10/08 18:02:44 - Spoon - Asking for repository
2014/10/08 18:02:45 - RepositoriesMeta - Reading repositories XML file: C:\Users\mikejf12\.kettle\repositories.xml
2014/10/08 18:02:45 - Version checker - OK
2014/10/08 18:02:49 - Spoon - Connected to metastore : 1, added to delegating metastore
2014/10/08 19:49:20 - Spoon - Starting job...
2014/10/08 19:49:23 - basic mapred job - Start of job execution
2014/10/08 19:49:23 - basic mapred job - Starting entry [Pentaho MapReduce]
2014/10/08 19:49:25 - Pentaho MapReduce - Cleaning output path: hdfs://hc2nn:8020/data/pentaho/result
2014/10/08 19:49:25 - Pentaho MapReduce - Configuring Pentaho MapReduce job to use Kettle installation from /opt/pentaho/mapreduce/5.1.0.0-5.1.0.0-752-cdh50
2014/10/08 19:49:42 - basic mapred job - Finished job entry [Pentaho MapReduce] (result=[true])
2014/10/08 19:49:42 - basic mapred job - Job execution finished
2014/10/08 19:49:42 - Spoon - Job has ended.
2014/10/09 18:23:44 - Spoon - Spoon
2014/10/09 18:26:59 - Spoon - Spoon
```

Figure 10-23. *Results of job run*

I have also monitored this job via my Hadoop Resource Manager interface on the URL http://hc2nn.semtech-solutions.co.nz:8088/cluster/apps. This URL allows me to watch the job's progress until it is finished and monitor log files, if necessary. As I know that the job has finished, there must be an existing part file under the results directory that contains the results data. To see that output, I run this command from the Linux hadoop account:

```
[hadoop@hc2nn ~]$ hdfs dfs -cat /data/pentaho/result/part-00000 | head -10
```

```
ACURA-1.6 EL      2
ACURA-1.6EL       6
ACURA-1.7EL       12
ACURA-2.2CL       2
ACURA-2.3 CL      2
ACURA-2.3CL       2
ACURA-2.5TL       3
ACURA-3.0 CL      1
ACURA-3.0CL       2
ACURA-3.2 TL      1
```

I use the Hadoop file system cat command to dump the contents of the HDFS-based results part file, and then the Linux head command to limit the output to the first 10 rows. What I see, then, is a summed list of vehicle makes and models.

PDI's visual interface makes it possible for even inexperienced Hadoop users to create and schedule Map Reduce jobs. You don't need to know Map Reduce programming and can work on client development machines. Simply by selecting graphical functional icons, plugging them together, and configuring them, you can create complex ETL chains.

Potential Errors

Nothing in life goes perfectly, so let's addresses some errors you may encounter during a job creation.

For instance, while working on the example just given, I discovered that a MySQL connector jar file had not been installed into the PDI library directory when I tried to connect PDI to MySQL. I received the following error message:

```
Driver class 'org.gjt.mm.mysql.Driver' could not be found, make sure the 'MySQL' driver (jar file)
is installed.
```

Remember, if you plan to access MySQL with PDI, you also need to install a MySQL jar file called `mysql-connector-java-5.1.32-bin.jar` from http://dev.mysql.com/downloads/connector/j/ into the PDI directory data-integration\lib.

When PDI uses the big data plug-in, it copies libraries and configuration files to a directory called /opt/pentaho on HDFS. Therefore, you need to make sure the user account you're using for PDI has the correct permissions. For my example, I was running PDI from a client Windows machine which employed the user ID from the Windows session to access HDFS. I received the following error message:

```
2014/09/30 18:26:20 - Pentaho MapReduce 2 - Installing Kettle to /opt/pentaho/mapreduce/5.1.0.0-
5.1.0.0-752-cdh42
2014/09/30 18:26:28 - Pentaho MapReduce 2 - ERROR (version 5.1.0.0, build 1 from 2014-06-19_19-02-57
by buildguy) : Kettle installation failed
2014/09/30 18:26:28 - Pentaho MapReduce 2 - ERROR (version 5.1.0.0, build 1 from 2014-06-19_19-02-57
by buildguy) : org.apache.hadoop.security.AccessControlException:
Permission denied: user=mikejf12, access=WRITE, inode="/":hdfs:hadoop:drwxr-xr-x
```

The error was caused because the Windows account (mikejf12) did not have directory access on HDFS. You can resolve this type of problem by using the HDFS chown and chmod commands to grant access on HDFS as the commands below show:

```
[hadoop@hc2nn ~]$ hdfs dfs -chown mikejf12 /opt/pentaho
[hadoop@hc2nn ~]$ hdfs dfs -chmod 777 /opt/pentaho
[hadoop@hc2nn ~]$ hdfs dfs -ls /opt
Found 1 items
drwxrwxrwx   - mikejf12 hadoop          0 2014-10-25 16:02 /opt/pentaho
```

Unfortunately, deadlines prevented me from resolving an error that occurred on my Linux CDH 4.6 cluster when I tried to run a PDI Map Reduce job. I knew that it was not the fault of PDI, but in fact was a configuration problem with the cluster, probably YARN. Here's the error message I received:

```
2014/10/01 18:08:56 - Pentaho MapReduce 2 - ERROR (version 5.1.0.0, build 1 from 2014-06-19_19-02-57
by buildguy) : Unknown rpc kind RPC_WRITABLE
2014/10/01 18:08:56 - Pentaho MapReduce 2 - ERROR (version 5.1.0.0, build 1 from 2014-06-19_19-
02-57 by buildguy) : org.apache.hadoop.ipc.RemoteException(java.io.IOException): Unknown rpc kind
RPC_WRITABLE
```

This is a running cluster, but it is not quite configured in the way that PDI needs. If a cluster is configured with CDH5 manager, then it seems to work, so the difference between the two configurations must hold the clue to the solution.

The following error occurred when I tried to run the example PDI application on Centos Linux:

```
# A fatal error has been detected by the Java Runtime Environment:
#
#  SIGSEGV (0xb) at pc=0x80a3812b, pid=4480, tid=3078466416
```

I resolved it by stopping the application from showing the welcome page at startup. To do so, I simply added the following line to the file $HOME/.kettle/.spoonrc of the user running PDI:

```
ShowWelcomePageOnStartup=N
```

If the wrong type is specified for a key field, there will be an error message generated similar to the following:

```
{"type":"TASK_FAILED","event":{"org.apache.hadoop.mapreduce.jobhistory.TaskFailed":{"taskid":"task_1
412385899407_0008_m_000000","taskType":"MAP","finishTime":1412403861583,"error":",
Error: java.io.IOException: org.pentaho.hadoop.mapreduce.converter.TypeConversionException: \n

Error converting to Long: 1995,ACURA,INTEGRA,SUBCOMPACT,1.8,4,A4,X,10.2,7,28,40,1760,202\n

For input string: \"1995,ACURA,INTEGRA,SUBCOMPACT,1.8,4,A4,X,10.2,7,28,40,1760,202\"\n\n
```

In this case, a string key was incorrectly being treated as a value.

An error in the configuration of the PDI Map Reduce job can cause the following error message:

```
commons.vfs.FileNotFoundException: Could not read from
"file:///yarn/nm/usercache/mikejf12/appcache/application_1412471201309_0001/
container_1412471201309_0001_01_000013/job.jar"

/yarn/nm/usercache/mikejf12/appcache/application_1412471201309_0001/
container_1412471201309_0001_01_000001
because it is a not a file.
```

Although it looks like some kind of Hadoop configuration error, it is not. It was again caused by setting the wrong data type on Map Reduce variable values. Just follow the example installation and configuration in this section and you will be fine.

Finally, a lack of available memory on the Hadoop Resource Manager host Linux machine produces an error like the following:

```
2014-10-07 18:08:57,674 INFO [RMCommunicator Allocator] org.apache.hadoop.mapreduce.v2.app.
rm.RMContainerAllocator:
Reduce slow start threshold not met. completedMapsForReduceSlowstart 1
```

To resolve a problem like this, try reducing the Resource Manager memory usage in the CDH Manager so that it does not exceed that available.

Now that you understand how to develop a Map Reduce job using Pentaho, let's see how to create a similar job using Talend Open Studio. The illustrative example uses the same Hadoop CDH5 cluster as a data source and for processing.

Talend Open Studio

Talend offers a popular big data visual ETL tool called Open Studio. Like Pentaho, Talend gives you the ability to create Map Reduce jobs against existing Hadoop clusters in a logical, step-by-step manner by pulling pre-defined modules from a palette and linking them in an ETL chain to create a Map Reduce based job. I describe how to source, install, and use Open Studio, as well as to create a Pig-based Map Reduce job. Along the way, I point out a few common errors and their solutions.

Installing Open Studio for Big Data

You can find Open Studio, as well as a number of other big data offerings, on Talend's website at www.talend.com, including a big data sand box and big data Studio and Enterprise editions. For the chapter's example, I use the free, 30-day trial version downloaded from the Talend website rather than the sandbox version. This is because I plan to connect Talend to my existing Hadoop cluster and I will tackle any problems as they arise. However, you may find the sand box version useful because it contains sample code and a fully working Hadoop cluster. Also, I create a Pig-based Map Reduce job because the full Java-based Map Reduce functionality is available only in the Enterprise product.

When I attempt to install these big data ETL tools, I always try to install them on Windows machines first, as I hope to use them as clients connecting to my Linux-based Hadoop clusters. A shell-based error prevented me from doing this at this time, so instead I install the Talend software on the Centos 6 Linux host hc1nn and I configure it to connect to the CDH5 Hadoop cluster on nc2nn. (See the "Potential Errors" section for details on this error, which calls for a fix to be added to future Cloudera releases.)

For this installation, I download the Talend Open Studio for Big Data 5.5 from the URL www.talend.com/download. I select the Big Data tab and download the Open Studio software, as shown in Figure 10-24. (I added red indicator boxes to the options that I need.) The download took an hour for me; the length of download time depends on your bandwidth.

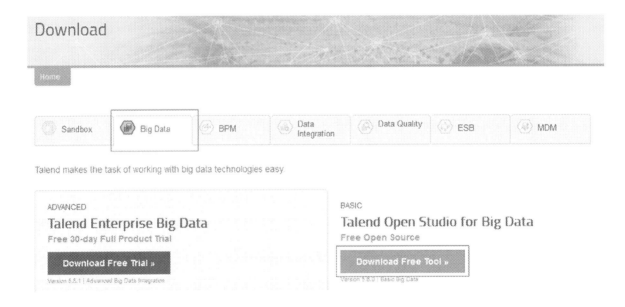

Figure 10-24. *Software download for Talend*

I place the software in a directory called talend in the Linux hadoop user account's home directory, using the Linux pwd command:

```
[hadoop@hc1nn talend]$ pwd
/home/hadoop/talend
```

The downloaded zip file is 1010 MB and needs to be unzipped before use. The ls -lh command gives a long file listing with sizes in a more readable form:

```
[hadoop@hc1nn talend]$ ls -lh TOS_BD-r118616-V5.5.1.zip
-rw-r--r-- 1 hadoop hadoop 1010M Oct 13 18:22 TOS_BD-r118616-V5.5.1.zip
```

When unpacked with the unzip command, the software resides in a directory called TOS_BD-r118616-V5.5.1:

```
[hadoop@hc1nn talend]$ unzip  TOS_BD-r118616-V5.5.1.zip

[hadoop@hc1nn talend]$ ls
TOS_BD-r118616-V5.5.1  TOS_BD-r118616-V5.5.1.zip

[hadoop@hc1nn talend]$ cd  TOS_BD-r118616-V5.5.1
```

There are a lot of files in this directory, but for this example, all I need to do is run the shell file (.sh). It determines the architecture of the machine that it resides on and runs the correct binary. For instance, I am running Talend on a 64-bit Centos Linux host.

Running Open Studio for Big Data

To start Talend Open Studio, I issue the following command:

```
[hadoop@hc1nn TOS_BD-r118616-V5.5.1]$ ./TOS_BD-linux-gtk-x86.sh &
```

The "and" character (&) means that the job is running in the background so I can enter further commands in the Linux session, if necessary. The command brings up the Project Chooser window, where I can either create or select a Talend project, as shown in Figure 10-25. For this example, I select my project called "bd1," and click Open.

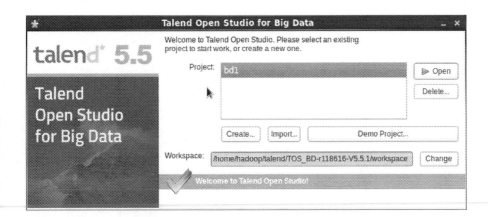

Figure 10-25. *Project chooser window for Talend*

To be able to demonstrate the Designer perspective, which is displayed by default, previously I had created the ETL job called "tmr1." This allows me to explain the Open Studio interface features. Open Studio for Big Data opens to the main interface, shown in Figure 10-26. The searchable palette on the right provides a list of dragable modules. For instance, searching for "thdfs" and "tpig" provides a list of HDFS and Pig modules. The center of the interface contains the current open jobs as tabs; in this case, tmr1, the Pig-based Map Reduce job I had created previously, is shown. The arrows indicate data flows or conditions, while the icons represent functional modules, such as tHDFSConnection_1. Because I have already successfully run tmr1, it shows some extra information about how the fast data rows were processed (i.e., one row in 166.12 seconds).

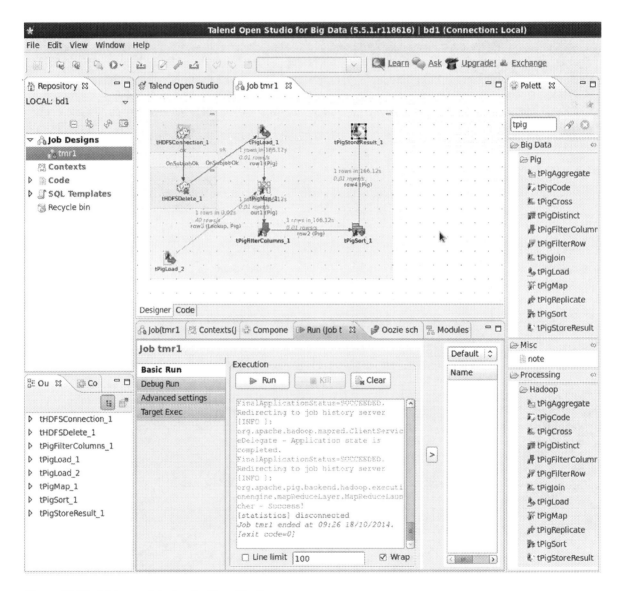

Figure 10-26. *Main user interface for Open Studio*

The top-left side of the Figure 10-26 interface shows the local repository for the project bd1; from here, I can double-click the tmr1 job to open it. At the bottom of the interface is a designer and code section. The Code tab enables me to examine the Java code that Talend generates from the job file; the Designer tab allows me to both configure each step of the job by selecting it and to run the job once the configuration is completed.

Before I proceed to use the Open Studio interface, I take a moment to consider the test data that this example job will use. For instance, I have stored two CSV-based data files in the HDFS directory /data/talend/rdbms/, as the following Hadoop file system `ls` command shows:

```
[hadoop@hc2nn ~]$ hdfs dfs -ls /data/talend/rdbms
Found 2 items
-rw-r--r--   3 hadoop supergroup    1381638 2014-10-10 16:36 /data/talend/rdbms/rawdata.txt
-rw-r--r--   3 hadoop supergroup       4389 2014-10-18 08:17 /data/talend/rdbms/rawprices.txt
```

The first file, called rawdata.txt, contains the vehicle model fuel consumption data that has been used in previous chapter examples, while the second file, called rawprices.txt, contains the matching model prices. The combined Hadoop file system `cat` command and the Linux `head` commands list the first 10 rows of each file, as follows:

```
[hadoop@hc2nn ~]$ hdfs dfs -cat /data/talend/rdbms/rawdata.txt | head -10

1995,ACURA,INTEGRA,SUBCOMPACT,1.8,4,A4,X,10.2,7,28,40,1760,202
1995,ACURA,INTEGRA,SUBCOMPACT,1.8,4,M5,X,9.6,7,29,40,1680,193
1995,ACURA,INTEGRA GS-R,SUBCOMPACT,1.8,4,M5,Z,9.4,7,30,40,1660,191
1995,ACURA,LEGEND,COMPACT,3.2,6,A4,Z,12.6,8.9,22,32,2180,251
1995,ACURA,LEGEND COUPE,COMPACT,3.2,6,A4,Z,13,9.3,22,30,2260,260
1995,ACURA,LEGEND COUPE,COMPACT,3.2,6,M6,Z,13.4,8.4,21,34,2240,258
1995,ACURA,NSX,TWO-SEATER,3,6,A4,Z,13.5,9.2,21,31,2320,267
1995,ACURA,NSX,TWO-SEATER,3,6,M5,Z,12.9,9,22,31,2220,255
1995,ALFA ROMEO,164 LS,COMPACT,3,6,A4,Z,15.7,10,18,28,2620,301
1995,ALFA ROMEO,164 LS,COMPACT,3,6,M5,Z,13.8,9,20,31,2320,267

[hadoop@hc2nn ~]$ hdfs dfs -cat /data/talend/rdbms/rawprices.txt | head -10

ACURA,INTEGRA,44284
ACURA,INTEGRA,44284
ACURA,INTEGRA GS-R,44284
ACURA,LEGEND,44284
ACURA,LEGEND COUPE,44284
ACURA,LEGEND COUPE,44284
ACURA,NSX,32835
ACURA,NSX,32835
ACURA,2.5TL,44284
ACURA,3.2TL,44284
```

For my example, I plan to use only columns 2 and 3 from the first file, which contain the manufacturer and model details, and the price information from the second file. (Note that these prices are test data, not real prices.)

Creating the ETL

Examining a completed job is a good way to understand Open Studio's modules and workflow. Figure 10-27 shows job tmr1, the Pig-based Map Reduce job I created for this example. It starts with a connection to HDFS called tHDFSConnection_1, on the condition that the connection works control is passed to an HDFS delete step called tHDFSDelete_1, which clears the results directory for the job. If that is okay, then control is passed to a tPigLoad step called tPigLoad_1, which loads the rawdata.txt file from HDFS. At the same time, another load step called tPigLoad_2 loads the rawprices.txt file from HDFS. The data from these files is then passed to a module called tPigMap_1, which will combine the data.

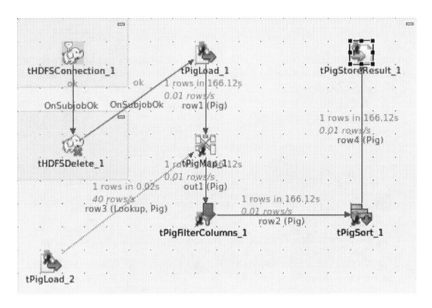

Figure 10-27. *Work flow for Pig native Map Reduce job*

Figure 10-27 shows that the tPigMap_1 step combines the data from the two files, while the tPigFilterColumns_1 step removes the column data that is not of interest. The tPigSort_1 step sorts the data, then the tPigStoreResult_1 step saves the sorted data to HDFS.

To create your own jobs, you select modules in the palette and drag them to the center jobs pane. Then, you right-click the icons to connect the modules via conditional arrows or arrows that represent data flows (as was done with Pentaho earlier). Figure 10-28 shows the creation of a conditional flow between steps of a job. If the tHDFSConnection_1 step works, then control will pass to the deletion step. Figure 10-29 illustrates the creation of a specific data flow between the tPigLoad_1 and tPigMap_1 steps.

Figure 10-28. *Conditional job flow*

Figure 10-29. *Data flow for Pig Map Reduce job*

Now that you have a sense of the job as a whole, take a closer look at how it's put together. From this point, I walk slowly through the configuration of each step of the tmr1 example job, as well as point out some common troublespots. At the end, I run the job and display the results from the HDFS results directory.

As long as the Component tab in the Designer window is clicked, I can select each step in a job and view its configuration. For the HDFS connection, for example, the Hadoop version is defined as Cloudera and CDH5. The URI for the Name Node connection is defined via the cluster Name Node host hc2nn, using the port number 8020. The user name for the connection is defined as the Linux hadoop account; these choices are shown in Figure 10-30.

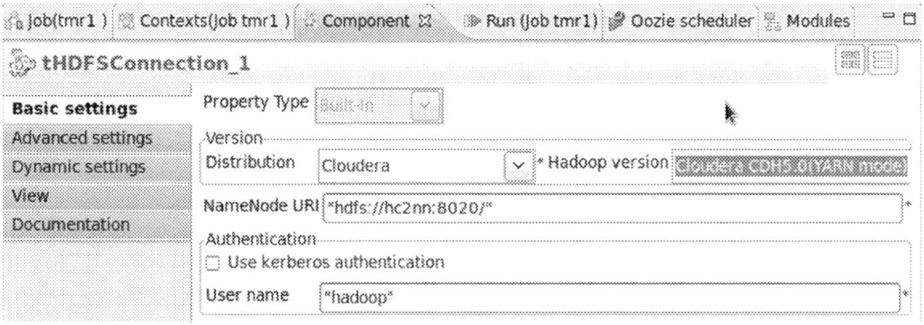

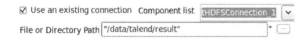

Figure 10-30. *The trm1 job, with HDFS connection*

The HDFS delete step shown in Figure 10-31 just deletes the contents of the /data/talend/result/ HDFS directory so that this Talend job can be rerun. If this succeeds, then control passes to the next step. The ellipses (. . .) button to the right of the Path field allows me to connect to HDFS and browse for the delete location.

Figure 10-31. *HDFS delete step for trm1 job*

My next step is to load the rawdata.txt file (tPigLoad_1). Remember that even though the Hadoop cluster may be fully configured via XML-based site configuration files, the Talend job carries configuration information, as shown in Figure 10-32. Note also that the Map/Reduce icon has been selected here, telling Talend that this will be a Map Reduce job. The same CDH5 cluster information has been specified. However, this time the host- and port-based addresses have been set for the Resource Manager, the Job History server, and the Resource Manager scheduler. The port values have been suggested by Talend as default values, and they match the default values chosen by the CDH5 Cluster Manager installer. Again, I use the Linux hadoop account for the connection.

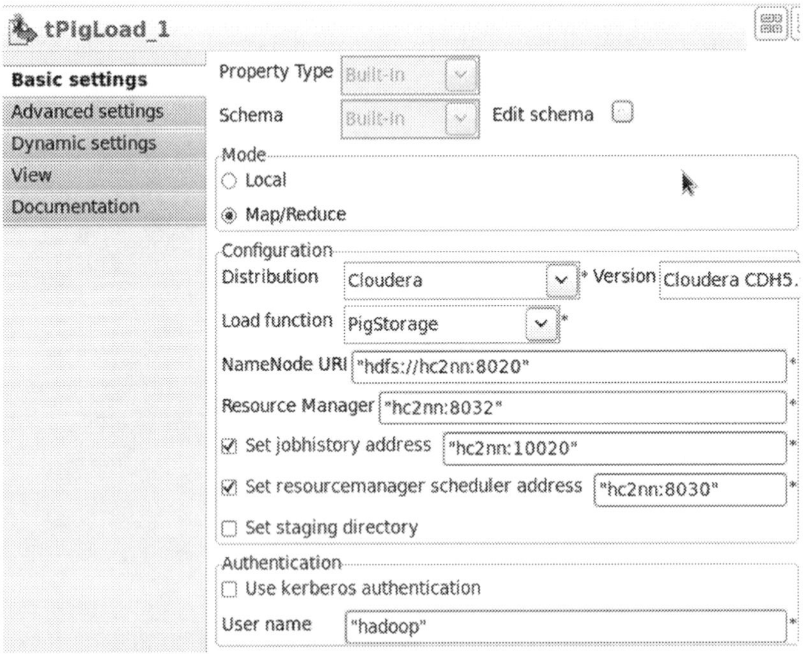

Figure 10-32. *Loading the HDFS raw data file*

At this step I encountered an error message. Initially failing to set the Resource Manager scheduler address caused the Resource Manager-based job to time out and fail (see the "Potential Errors" section for more detail).

When loading a data file, you must also specify the schema to indicate what columns are in the incoming data, what they should be called, and what data types they have. For this example, I click the Edit Schema button (shown at the top of Figure 10-32) to open the Schema window, shown in Figure 10-33.

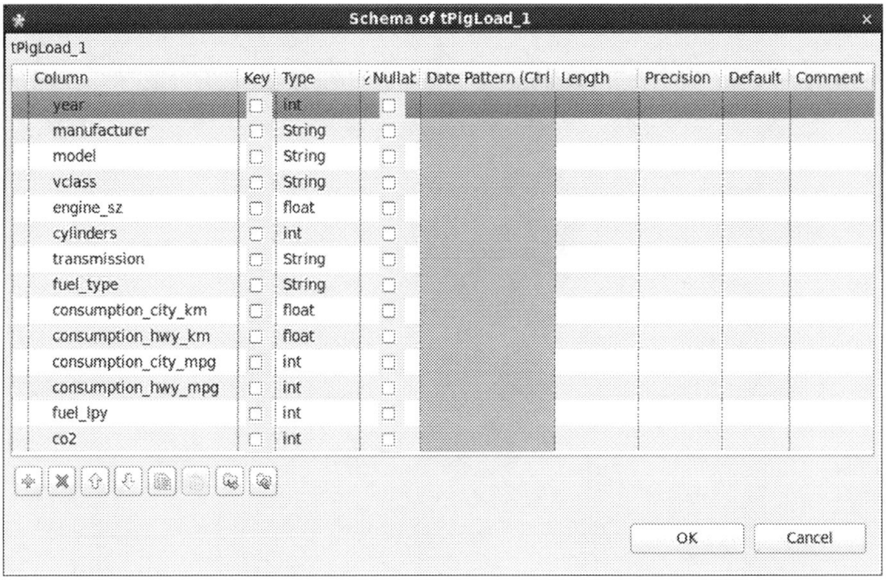

Figure 10-33. *Schema window for trm1 job*

I use the green plus icon at the bottom left of the window to manually specify the column names and data types. I try to make the names meaningful so that they accurately represent the data they contain. I do not add any keys, and my data does not contain null values, so I leave those fields blank. The schema for the raw prices file from the tPigLoad_2 step shows just three columns, the last of which is the vehicle price (see Figure 10-34).

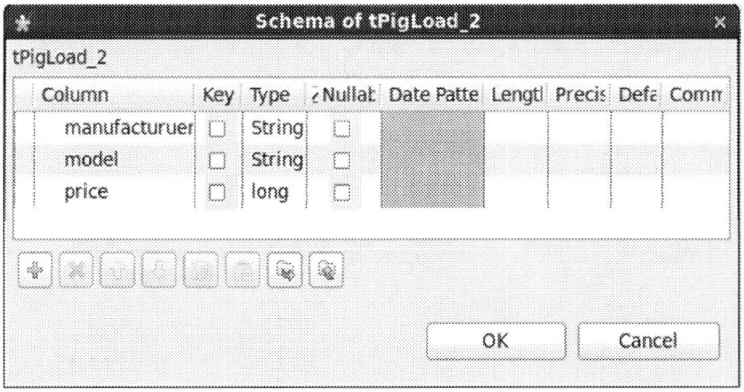

Figure 10-34. *Three-column setup for trm1 job*

The tPigMap_1 step takes the data from the loaded rawdata.txt and rawprices.txt files and combines them on the manufacturer and model names. It then outputs the combined schema as a data flow called "out1," shown in Figure 10-35. The arrows in Figure 10-35 show the flow of columns between the incoming and outgoing data sources. The schemas at the bottom of the window show the incoming and outgoing data. Note: the example does not map all the incoming columns, as in a later step I will filter out the columns that are not needed from the resulting data set.

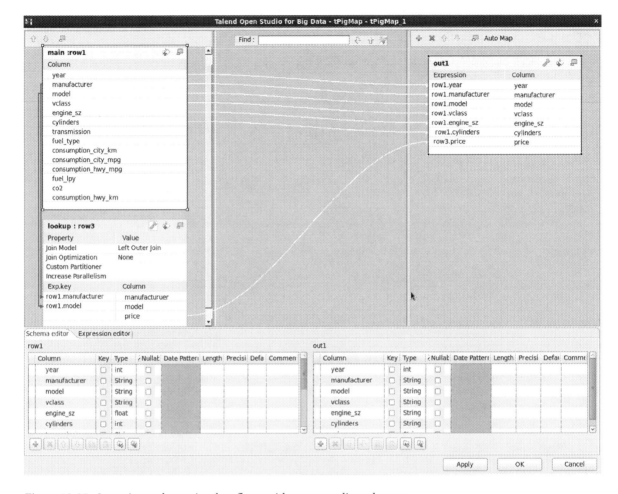

Figure 10-35. *Incoming and outgoing data flows, with corresponding schemas*

The column filtering step separates out the columns in the data source to just those required for the resulting data set. In this case, the columns have been reduced to the vehicle manufacturer, model, and price information, as shown Figure 10-36.

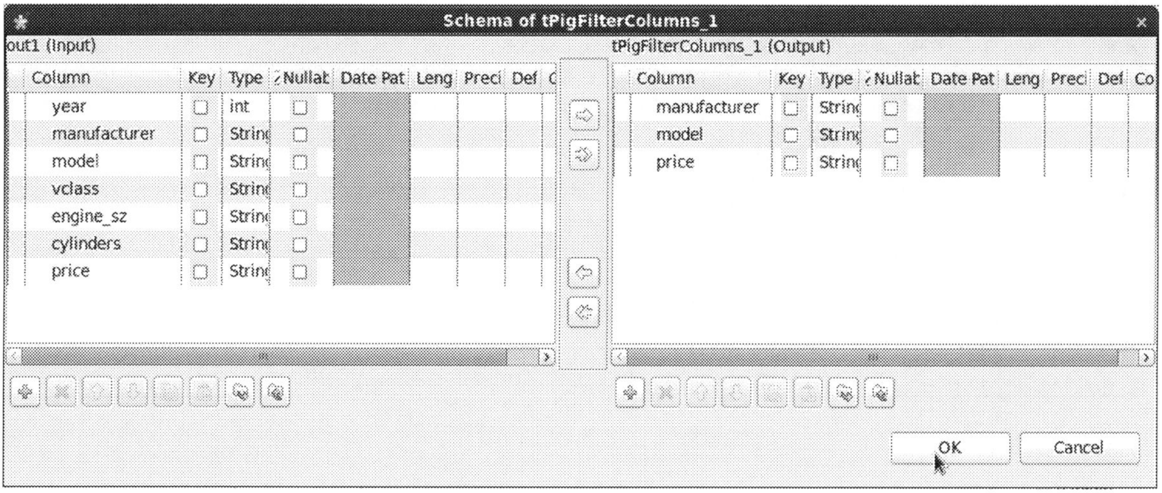

Figure 10-36. *Filter Columns step to eliminate unneeded columns*

The next step is to sort the data and specify the order in which the columns will be sorted; Figure 10-37 shows a compounded sorting key of manufacturer, model, and price.

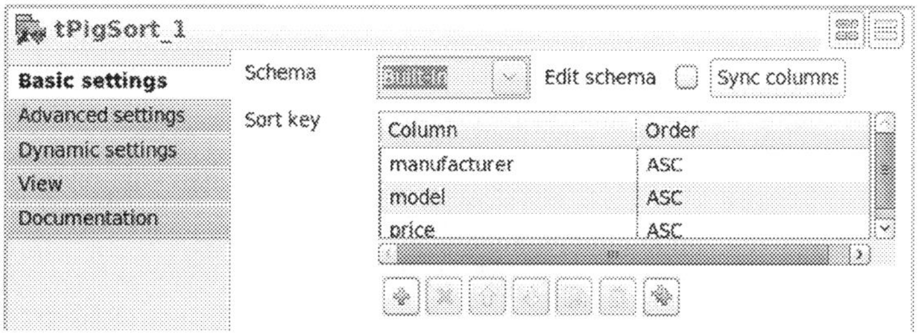

Figure 10-37. *Sort step to put columns in desired order*

Finally, the data is stored via a results step, which specifies where the Map Reduce job will store its data and what the field separator will be, as shown in Figure 10-38. I use the default storage method of PigStorage and instruct the data to be stored in the directory /data/talend/result/. Because this is a Map Reduce job, a successful outcome will create a part file.

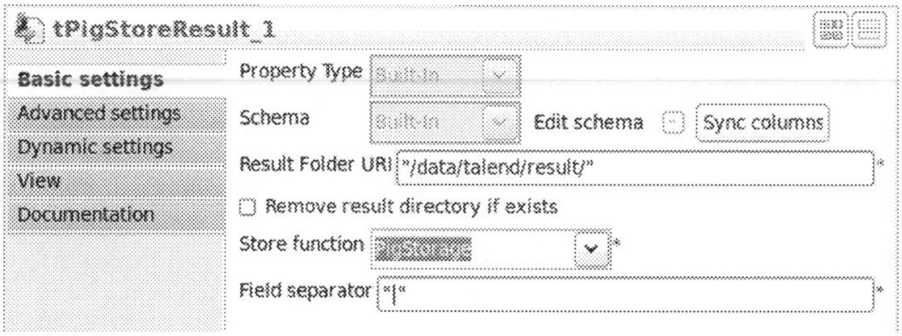

Figure 10-38. *Results step provides for storage of the data*

Having successfully created the job, I can now run it from the Designer window's Run tab. I click the Run button with the green run arrow to start the job. The Basic Run window shows minimal job output, while the Debug Run option displays a little more output in case of error. The output is color coded; green lines are good and red lines show an error. See Figure 10-39 for the results of my job example.

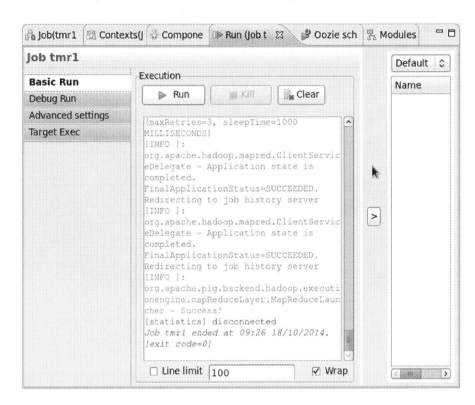

Figure 10-39. *Job is run and results are shown*

Remember that it is possible to track these jobs from the Resource Manager interface, as shown in Figure 10-40. For my cluster, the Name Node hostname is hc2nn, and the Resource Manager http port number is 8088, so the URL is hc2nn:8088.

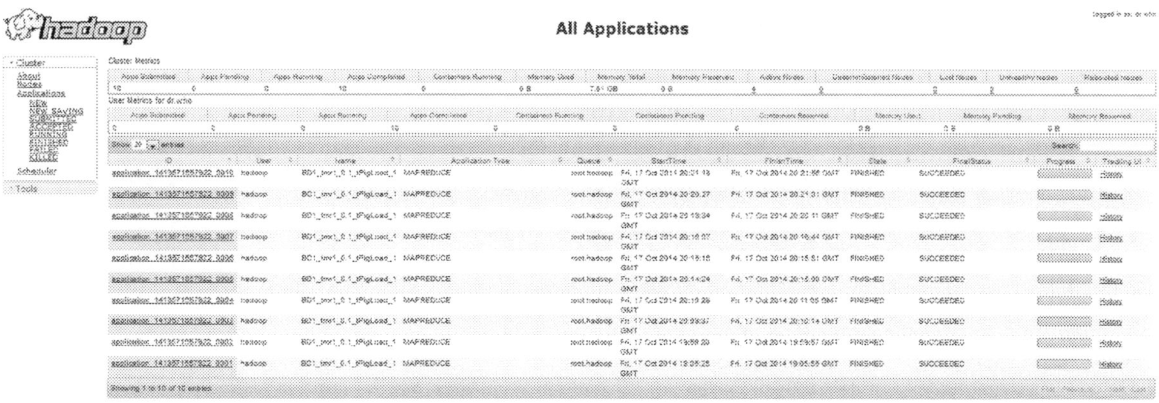

Figure 10-40. Resource Manager interface for monitoring the job

___

■ **Note**   If a Talend Map Reduce job hangs, the only place you can investigate the details of the job log is by using the Resource Manager user interface.

___

Figures 10-39 and 10-40 both show that running this Talend job was a success, meaning I can find its output data on HDFS in the target directory—/data/talend/result/, in this case. I use the HDFS file system command ls to create a file listing of that directory showing the resulting data files:

```
[hadoop@hc2nn ~]$ hdfs dfs -ls /data/talend/result
Found 2 items
-rw-r--r--   3 hadoop supergroup          0 2014-10-18 09:21 /data/talend/result/_SUCCESS
-rw-r--r--   3 hadoop supergroup     441159 2014-10-18 09:21 /data/talend/result/part-r-00000
```

As would be expected, there is a _SUCCESS file and a part file that contains the data. Dumping the contents of the part file by using the HDFS file system cat command shows me that the vehicle manufacturer, model, and price data has been sorted and placed in the part file:

```
[hadoop@hc2nn ~]$ hdfs dfs -cat /data/talend/result/part-r-00000 | head -10

ACURA|1.6 EL|44284
ACURA|1.6 EL|44284
ACURA|1.6 EL|44284
ACURA|1.6 EL|44284
ACURA|1.6EL|44284
ACURA|1.6EL|44284
ACURA|1.6EL|44284
ACURA|1.6EL|44284
ACURA|1.6EL|44284
ACURA|1.6EL|44284
```

All of the pricing information is the same because this is a small sample of the pricing test data. The output data has been formatted with separator vertical ("|") characters.

This simple example reflects only a small portion of the available Talend functionality, but it shows the potential for building Map Reduce jobs using a drag-and-drop approach. Remember also that there is still more functionality in the Talend Enterprise application. With the Enterprise application, you can build traditional Map Reduce jobs by using a Map Reduce job type and specifying the functionality of the mapper and reducer components of the job.

Potential Errors

Here are some of the errors I encountered when developing this example, as well as their solutions. Check the Talend forum at www.talendforge.org/forum for past issues encountered and to ask questions if you cannot find a solution to your problem.

When I installed Talend on a Windows 7 machine, it crashed on startup, with the following error message:

```
Java was started but returned exit code = 1
C:\Windows\system32\javaw.exe
-Xms154m
-Xmx2536m
-XX:MaxPermSize=3256m
```

This problem was caused by a Windows patch called kb2977629. My solution was to remove the patch by following these steps:

1. Select the Start button.

2. Click Control Panel.

3. Click Programs.

4. Under Programs and Features, Click View Installed Updates.

5. Search for kb2977629 and uninstall.

6. Restart Windows.

There was a permissions problem with HDFS when I was running Talend from Linux, which caused the following error message:

```
cause:org.apache.hadoop.security.AccessControlException: Permission denied:
user=hadoop, access=WRITE, inode="/tmp/hadoop-yarn":yarn:supergroup:drwxr-xr-x
```

To fix the problem, I use the Linux su command to change the user to the YARN Linux user, and then used the HDFS file system chmod command to change the permissions of the directory, as follows:

```
[root@hc2nn ~]# su - yarn
[yarn@hc2nn ~]$  hdfs dfs -chmod -R 777 /tmp/hadoop-yarn
```

The following memory-based error occurred because the maximum memory specified for Application Master component on YARN was less than the level that was needed.

```
PriviledgedActionException as:hadoop (auth:SIMPLE) cause:java.io.IOException:
org.apache.hadoop.yarn.exceptions.InvalidResourceRequestException:
Invalid resource request, requested memory < 0, or requested memory > max configured,
requestedMemory=1536, maxMemory=1035
```

I solved this problem on YARN by changing the value of the parameter yarn.app.mapreduce.am.resource.mb in the file yarn-site.xml, under the directory /etc/hadoop/conf. After making the change, I needed to restart the cluster to pick up the change.

The next error occurred when I tried to run Talend from a Windows 7 host and tried to connect to a Centos 6 Linux-based CDH5 cluster:

```
83_0004 failed 2 times due to AM Container for appattempt_1413095146783_0004_000002 exited with
exitCode:
1 due to: Exception from container-launch: org.apache.hadoop.util.Shell$ExitCodeException:
/bin/bash: line 0: fg: no job control
org.apache.hadoop.util.Shell$ExitCodeException: /bin/bash: line 0: fg: no job control
```

This was not a problem with Talend, but a known fix appears in Horton Works HDP 2. I assume that it will soon be fixed in other cluster stacks like CDH, but at the time of this writing, I used the Talend application only on Linux.

Finally, the following error occurred because I used insufficient configuration settings in the Talend tPigLoad step.

```
2014-10-14 17:56:13,123 INFO [main] org.apache.hadoop.yarn.client.RMProxy: Connecting to
ResourceManager at /0.0.0.0:8030
2014-10-14 17:56:14,241 INFO [main] org.apache.hadoop.ipc.Client: Retrying connect to server:
0.0.0.0/0.0.0.0:8030. Already tried 0 time(s); retry policy is RetryUpToMaximumCountWithFixedSleep
(maxRetries=10, sleepTime=1000 MILLISECONDS)
2014-10-14 17:56:15,242 INFO [main] org.apache.hadoop.ipc.Client: Retrying connect to server:
0.0.0.0/0.0.0.0:8030. Already tried 1 time(s); retry policy is RetryUpToMaximumCountWithFixedSleep
(maxRetries=10, sleepTime=1000 MILLISECONDS)
```

Because the Resource Manager scheduler address was not being set on the tPigLoad step, the address on YARN defaulted to 0.0.0.0:8030, and so the job hung and timed out.

Summary

You can use visual, drag-and-drop Map Reduce enabled ETL tools, such as Pentaho Data Integrator and Talend Open Studio, for big data processing. These tools offer the ability to tackle the creation of ETL chains for big data by logically connecting the functional elements that the tools provide. This chapter covered only a fraction of the functionality that they offer. Both include an abundance of Map Reduce components that you can combine to create more permutations of functionality than I could possibly examine in these pages.

I created the examples in this chapter using a combination of a Hadoop cluster, which I built using Cloudera's CDH5 cluster manager, and the visual ETL big data enabled tools Pentaho and Talend. I think that the errors that I encountered are either configuration based or will be solved by later cluster stack releases. Remember to check the company websites for application updates and the supplier forums for problem solutions. If you don't see a solution to your ETL problem, don't be afraid to ask questions; also, consider simplifying your algorithms as a way to zero in on the cause of a problem.

Just as I believe that cluster managers reduce problems and ongoing costs when creating and managing Hadoop clusters, so I think tools like Pentaho and Talend will save you money. They provide a quick entry point to the world of Hadoop-based Map Reduce. I am not suggesting that they can replace low-level Map Reduce programming, because I'm sure that eventually you will find complex problems that require you to delve down into API code. Rather, these tools provide a good starting point, an easier path into the complex domain of Map Reduce.

CHAPTER 11

Reporting with Hadoop

Because the potential storage capability of a Hadoop cluster is so very large, you need some means to track both the data contained on the cluster and the data feeds moving data into and out of it. In addition, you need to consider the locations where data might reside on the cluster—that is, in HDFS, Hive, HBase, or Impala. Knowing you should track your data only spawns more questions, however: What type of reporting might be required and in what format? Is a dashboard needed to post the status of data at any given moment? Are graphs or tables helpful to show the state of a data source for a given time period, such as the days in a week?

Building on the ETL work carried out in Chapter 10, this chapter will help you sort out the answers to those questions by demonstrating how to build a range of simple reports using HDFS- and Hive-based data. Although you may end up using completely different tools, reporting methods, and data content to construct the reports for your real-world data, the building blocks presented here will provide insight into the tasks on Hadoop that apply to many other scenarios as well.

This chapter will give you a basic overview of Hunk (the Hadoop version of Splunk) and Talend from a report-generation point of view. It will show you how to source the software, how to install it, how to use it, and how to create reports. Some basic errors and their solutions will be presented along with some simple dashboards to monitor the data. The chapter begins with an introduction to the Hadoop version of Splunk, which is called Hunk.

Note *Reports* show the state of given data sources in a variety of forms (tables, pie charts, etc.) and might also aggregate data to show totals or use colors to represent data from different sources. *Dashboards* provide a single-page view or overview of a system's status and might also contain charts with key indicators to show the overall state of its data.

Hunk

Hunk is the Hadoop version of Splunk (www.splunk.com), and it can be used to create reports and dashboards to examine the state of the data on a Hadoop cluster. The tool offers search, reporting, alerts, and dashboards from a web-based user interface. Let's look at the installation and uses of Hunk, as well as some simple reports and dashboards.

Installing Hunk

By way of example, I install Hunk onto the Centos 6 Linux host hc2nn and connect it to the Cloudera CDH5 Hadoop cluster on the same node. Before downloading the Splunk software, though, I must first create an account and register my details. I source Hunk from www.splunk.com/goto/downloadhunk.

Version 6 is about 100 MB. I install the software by using the Centos-based Linux hadoop account. Given that I am logged into the hadoop account, the download file is saved to the Downloads directory, as follows:

```
[hadoop@hc2nn ~]$ pwd
/home/hadoop/Downloads
```

```
[hadoop@hc2nn Downloads]$ ls -l hunk-6.1.3-228780-Linux-x86_64.tar.gz
-rw-r--r-- 1 hadoop hadoop 105332713 Oct 28 18:19 hunk-6.1.3-228780-Linux-x86_64.tar.gz
```

This is a gzip compressed tar file, so it needs to be unpacked by using the Linux-based gunzip and tar commands. I use the Linux gunzip command to decompress the .tar.gz file and create a tar archive file. The Linux tar command then extracts the contents of the tar file to create the Hunk installation directory. In the tar option, x means extract, v means verbose, and f allows me to specify the tar file to use:

```
[hadoop@hc2nn Downloads]$ gunzip hunk-6.1.3-228780-Linux-x86_64.tar.gz
[hadoop@hc2nn Downloads]$ tar xvf hunk-6.1.3-228780-Linux-x86_64.tar
```

```
[hadoop@hc2nn Downloads]$ ls -ld *hunk*
drwxr-xr-x 9 hadoop hadoop 4096 Nov  1 13:35 hunk
```

The ls -ld Linux command provides a long list of the Hunk installation directory that has just been created. The l option provides the list while the d option lists the directory details, rather than its contents.

Having created the installation directory, I now move it to a good location, which will be under /usr/local. I need to use the root account to do this because the hadoop account will not have the required access:

```
[hadoop@hc2nn Downloads]# su -

[root@ hc2nn Downloads]# mv hunk /usr/local
[root@ hc2nn Downloads]# cd /usr/local
[root@ hc2nn local]# chown -R hadoop:hadoop hunk
[root@ hc2nn local]# exit

[hadoop@ hc2nn Downloads]$ cd /usr/local/hunk
```

The Linux su command switches the current user to the root account. The Linux mv command moves the Hunk directory from the hadoop account Downloads directory to the /usr/local/ directory as root. The cd command then switches to the /usr/local/ directory, and the chmod command changes the ownership and group membership of the installation to hadoop. The -R switch just means to change ownership recursively so all underlying files and directories are affected. The exit command then returns the command to the hadoop login, and the final line changes the directory to the new installation under /usr/local/hunk.

Now that Hunk is installed and in the correct location, I need to configure it so that it will be able to access the Hadoop cluster and the data that the cluster contains. This involves creating three files—indexes.conf, props.conf, and transforms.conf—under the following Hunk installation directory:

```
[hadoop@hc2nn local]$ cd  /usr/local/hunk/etc/system/local
```

Of these three files, the indexes.conf file provides Hunk with the means to connect to the Hadoop cluster. For example, to create a provider entry, I use a sequence similar to the following:

```
[hadoop@hc2nn local]$ cat indexes.conf

[provider:cdh5]
vix.family = hadoop
vix.command.arg.3 = $SPLUNK_HOME/bin/jars/SplunkMR-s6.0-hy2.0.jar
vix.env.HADOOP_HOME = /usr/lib/hadoop
vix.env.JAVA_HOME = /usr/lib/jvm/jre-1.6.0-openjdk.x86_64
vix.fs.default.name = hdfs://hc2nn:8020
vix.splunk.home.hdfs = /user/hadoop/hunk/workdir
vix.mapreduce.framework.name = yarn
vix.yarn.resourcemanager.address = hc2nn:8032
vix.yarn.resourcemanager.scheduler.address = hc2nn:8030
vix.mapred.job.map.memory.mb = 1024
vix.yarn.app.mapreduce.am.staging-dir = /user
vix.splunk.search.recordreader.csv.regex = \.txt$
```

This entry creates a provider entry called cdh5, which describes the means by which Hunk can connect to HDFS, the Resource Manager, and the Scheduler. The entry describes where Hadoop is installed (via HADOOP_HOME) and the source of Java (via JAVA_HOME). It specifies HDFS access via the local host name and name node port of 8020. Resource Manager access will be at port 8032, and Scheduler access is at port 8030. The framework is described as YARN, and the location on HDFS that Hunk can use as a working directory is described via the property vix.splunk.home.hdfs.

The second file, props.conf, describes the location on HDFS of a data source that is stored under /data/hunk/rdbms/. The first cat command dumps the contents of the file, and the extractcsv value refers to an entry in the file tranforms.conf that describes the contents of the data file:

```
[hadoop@hc2nn local]$ cat props.conf

[source::/data/hunk/rdbms/...]
REPORT-csvreport = extractcsv
```

The third file, transforms.conf, contains an entry called extractcsv, which is referenced in the props.conf file above. It has two properties: the DELIMS value describes how the data line fields are delimited (in this case, by commas); and the FIELDS property describes 14 fields of vehicle fuel-consumption data. This is the same fuel-consumption data that was sourced in Chapter 4, where it was used to create an Oozie workflow.

```
[hadoop@hc2nn local]$ cat transforms.conf

[extractcsv]
DELIMS="\,"
FIELDS="year","manufacturer","model","class","engine size","cyclinders","transmission","Fuel
Type","fuel_city_l_100km","fuel_hwy_l_100km","fuel_city_mpg","fuel_hwy_mpg","fuel_l_yr","co2_g_km"
```

Here's a sampling of the CSV file contents via an HDFS file system cat command, which dumps the contents of the file /data/hunk/rdbms/rawdata.txt. The Linux head command limits the output to five lines:

```
[hadoop@hc2nn local]$ hdfs dfs -cat /data/hunk/rdbms/rawdata.txt | head -5

1995,ACURA,INTEGRA,SUBCOMPACT,1.8,4,A4,X,10.2,7,28,40,1760,202
1995,ACURA,INTEGRA,SUBCOMPACT,1.8,4,M5,X,9.6,7,29,40,1680,193
1995,ACURA,INTEGRA GS-R,SUBCOMPACT,1.8,4,M5,Z,9.4,7,30,40,1660,191
1995,ACURA,LEGEND,COMPACT,3.2,6,A4,Z,12.6,8.9,22,32,2180,251
1995,ACURA,LEGEND COUPE,COMPACT,3.2,6,A4,Z,13,9.3,22,30,2260,260
```

Now that some basic configuration files are set up, I can start Hunk.

Running Hunk

Hunk is started from the bin directory within the installation as the Linux hadoop account user.

■ **Note** When you first start Hunk, you must use the `--accept-license` option; after that, it may be omitted.

In either case, you start Hunk by using the splunk command:

```
[hadoop@hc2nn local]$  cd /usr/local/hunk/bin
[hadoop@hc2nn bin]$  ./splunk start --accept-license
```

When starting, Hunk reads its configuration files, so you need to monitor the output for errors in the configuration files' error messages, such as:

```
Checking conf files for problems...
        Invalid key in stanza [source::/data/hunk/rdbms/...] in /usr/local/hunk/etc/system/local/
        props.conf, line 3: DELIMS  (value:  ", ")
```

If any errors occur, you can fix the configuration files and restart Hunk, as follows:

```
[hadoop@hc2nn bin]$  ./splunk restart
```

If all is well, you are presented with a message containing the URL at which to access Hunk's web-based user interface:

```
The Splunk web interface is at http://hc2nn:8000
```

You will need to login with the account name "admin" and the initial password of "changeme," which you will immediately be prompted to change. Once logged in, you will see the Virtual Indexes page, which, as shown in Figure 11-1, displays the provider cdh5 in the family hadoop that was created in the indexes.conf file. If you don't see the Virtual Indexes page, then select Settings and go to Virtual Indexes from the top menu bar.

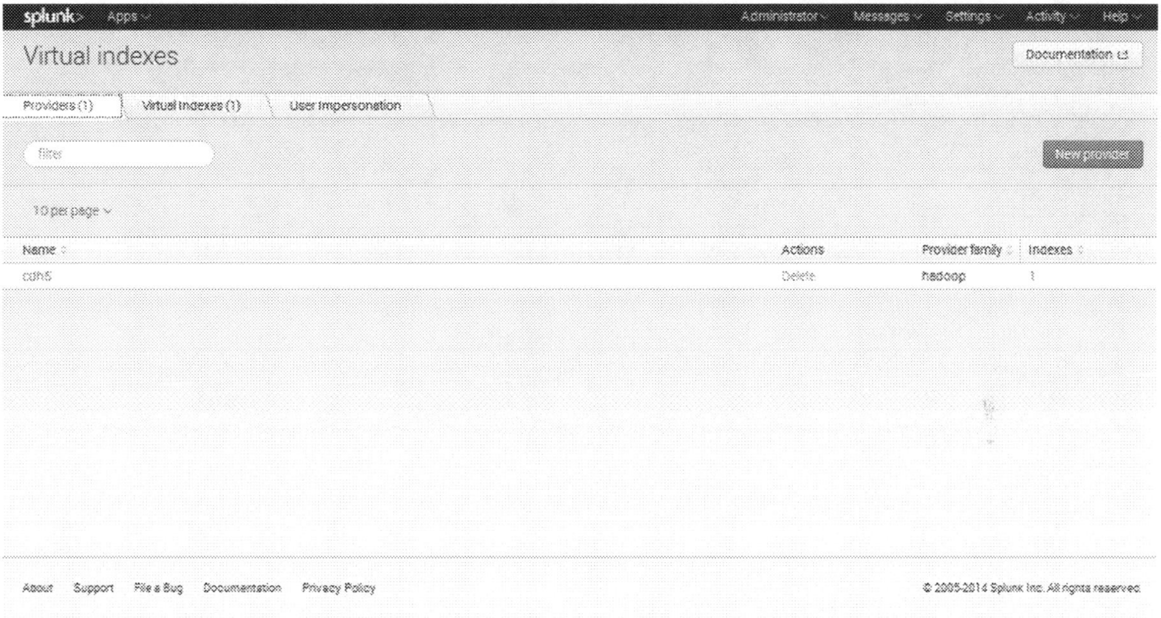

Figure 11-1. *Hunk provider cdh5*

You can click the cdh5 entry to examine the provider's details. The entire list of provider properties is too large to display here, but know that Hunk automatically adds extra entries like `vix.splunk.search.recordreader`, which defines how CSV files will be read. To represent most of the details in Figure 11-2, I arranged the list in two columns.

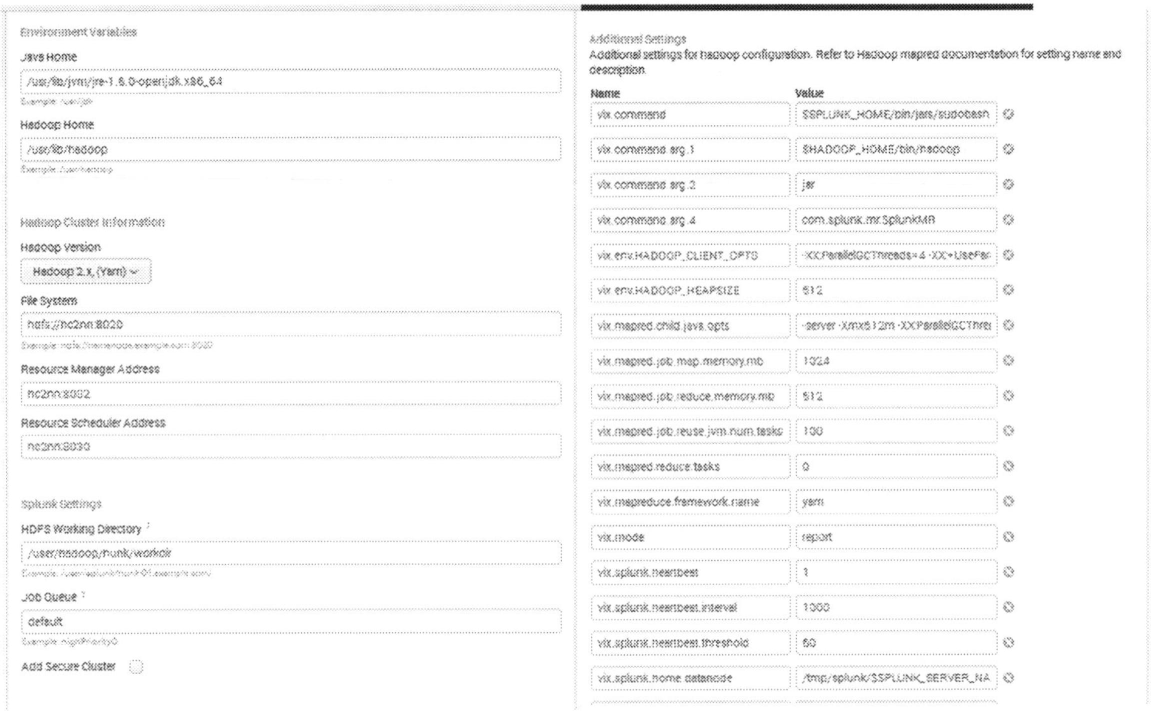

Figure 11-2. *Properties for Hunk provider cdh5*

Note that the Hadoop version in Figure 11-2 is set to YARN to reflect the CDH5 YARN version. It has not been necessary to specify the Hadoop supplier name.

Now, click Cancel to leave the cdh5 properties view, and click on the Virtual Indexes tab. For the chapter example, this tab shows that a single virtual index has been created in Hunk called cdh5_vindex, as shown in Figure 11-3.

Figure 11-3. *Hunk virtual index cdh5_vindex*

Virtual indexes are the means by which hunk accesses the Hadoop cluster-based data. They enable Hunk to use Map Reduce against the data and present the results within Hunk reports. By selecting the cdh5_vindex entry, you can examine the attributes of this virtual index (Figure 11-4). Currently, the entry, which was defined in the props.conf file, doesn't have much detail.It just defines the directory on HDFS where the CSV data is located. Click the Cancel button to exit this property details screen.

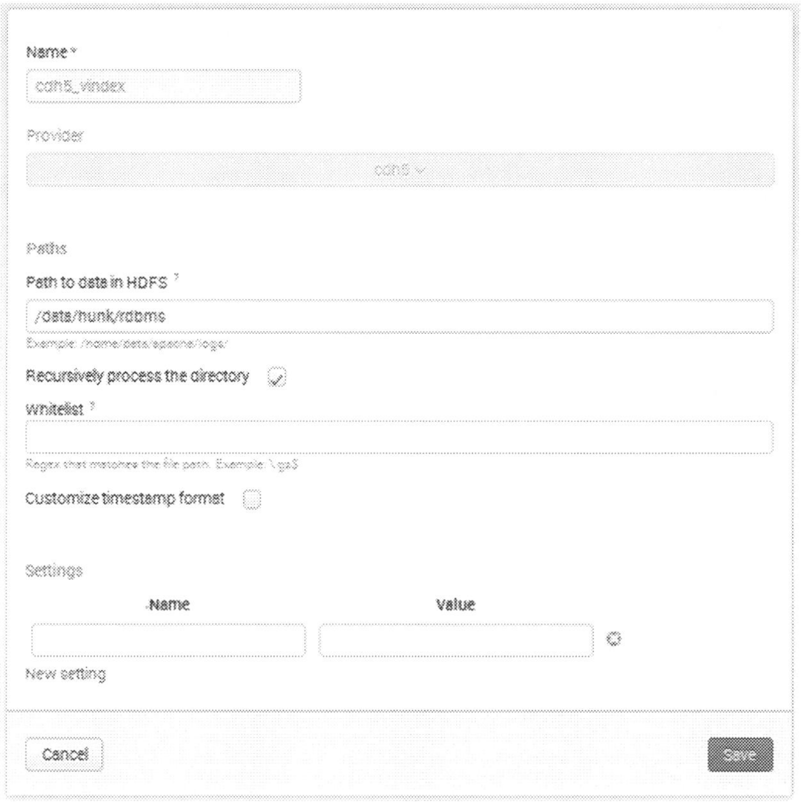

Name *

cdh5_vindex

Provider

cdh5 ⌄

Paths

Path to data in HDFS ?

/data/hunk/rdbms

Example: /home/data/apache/logs/

Recursively process the directory ☑

Whitelist ?

Regex that matches the file path. Example: \.gz$

Customize timestamp format ☐

Settings

Name	Value
		↻

New setting

Cancel   Save

Figure 11-4. *Property details of Hunk virtual index cdh5_vindex*

Creating Reports and Dashboards

Clicking the green menu bar's Search option opens the Search window, as shown in Figure 11-5. It should be noted here that report generation is asynchronous. If the report is complex or the data large, then Hunk may take time to deliver the results. So please be patient, and don't assume an error has occurred if your results do not return imediately.

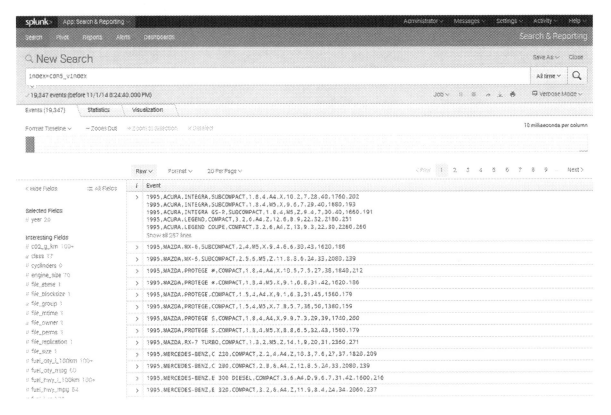

Figure 11-5. *Hunk Search window*

Figure 11-5 shows the raw-data view of the HDFS-based CSV file data I've been working with. I can scroll through the data by using the Previous and Next options on the top right of the screen. The Raw drop-down menu offers three choices for the format of the data displayed: raw, list, or table. In my example, I will build reports from this tabular form of the vehicle data.

The Format drop-down menu's options depend on the display type and on desired affects like data wrapping, line number, and drill down. Notice on the left that fields have been split into two categories: selected fields and interesting fields. The *selected* fields are those that will be displayed in table mode. The *interesting* fields are a combination of those fields defined in the data and those pre-defined by Hunk.

I click any field to view a Hunk Field Summary window, then set the Selected option to Yes to place the field I have chosen into the set of selected fields, as shown in Figure 11-6. When the mode is changed to Table, that field appears in the data in the order in which it appears in the selected list. There are a number of pre-defined reports available, like Top Values, that can be used to create single-column reports. Also, the top 10 values are displayed with counts and percentages.

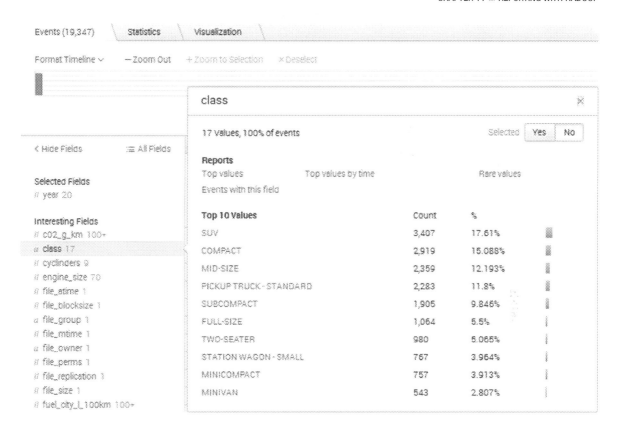

Figure 11-6. *Hunk Field Summary window*

By selecting the Year field, and then selecting the Top Values report option in the menu that pops up, I can begin to build a simple report that will show overall volume in my data by year. The report is created under the Visualization tab. There, I find drop-down menu options for the report's display, as shown in Figure 11-7. For example, I can switch the display from a bar chart, to a single line graph, to a pie chart. There is also an option to change the format for each display type and an option to change the underlying job settings.

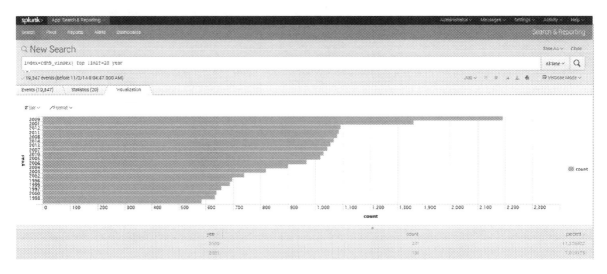

Figure 11-7. *Hunk Field bar report*

The search command that has created this report is displayed at the top of the page as index=cdh5_vindex | top limit=20 year. This means that the original search (index=cdh5_vindex) has been piped to the top command, with a limit of the 20 topmost values for the field year. The report is acting on the year field and displaying a default bar graph with the year value on the Y axis and the volumes on the X axis.

Figure 11-8 is a concatenated view of the report type, format, and job menus to show the options that are available by which you can change the appearance and job details of a report.

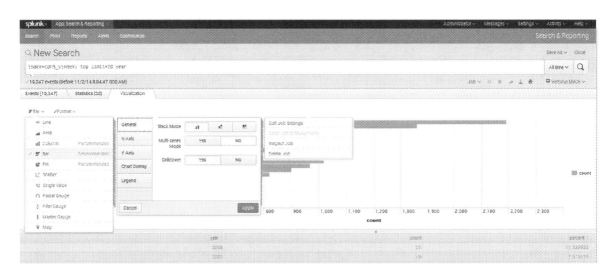

Figure 11-8. *Hunk report menu options*

Figure 11-9 shows all variations on the available options in a single Format menu to change the appearance of the report. For instance, I could define the attributes of the X and Y axis, general drill down, titles, data intervals, and position of the legend. I can modify these attributes until I am satisfied with the appearance of my report, then click the Apply button to put those changes into effect.

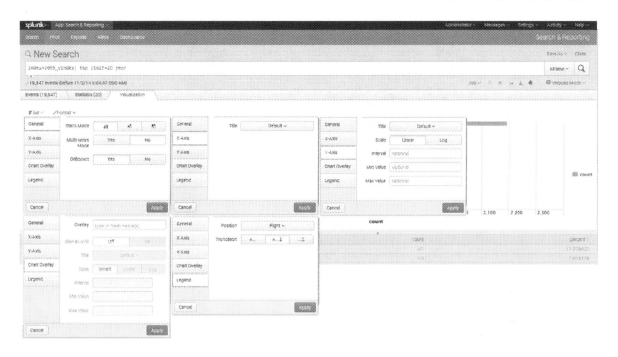

Figure 11-9. *Hunk options for format changes*

I click the Save As option to save the report so that it can be used later.

I have saved a number of reports in this manner within Hunk; they are single-column reports that cover areas like sales volume per year, vehicle manufacturers, and vehicle fuel types. I can view the saved Hunk reports by selecting the Reports option in the green menu bar. The Reports page, as shown in Figure 11-10, displays details like the report's title, ownership, apps, and sharing status. It also shows whether the report has been disabled, and it offers the option to edit or open the report again.

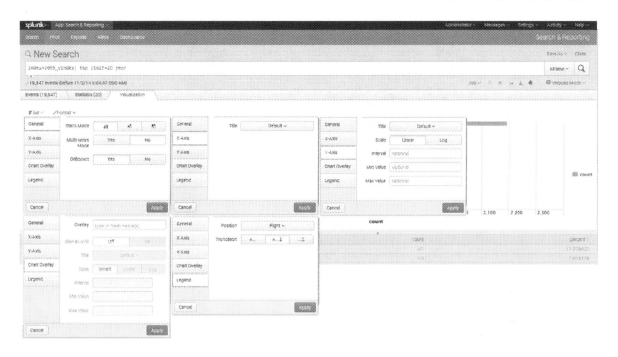

Figure 11-10. *Hunk report status*

Many of the reports shown in Figure 11-10 are built into Hunk, but I created the fuel types, sales volumes, and manufacturers reports. When I save a report, I can choose to add it to a dashboard that I can create at the same time. For example, in Figure 11-11, I create a dashboard called "Vehicle" to contain the three reports I've just mentioned.

Figure 11-11. *Hunk dashboard for reports created*

To open the dashboard, I choose the Dashboards menu option, then select the Vehicle dashboard. The display is populated from the current data, giving me, on a single page, an up-to-date view of the state of the data. A simple example of what can quickly be created to represent HDFS-based data, Figure 11-11 shows two pie charts that display top manufacturers and fuel types. The bottom bar graph displays sales volume by year. Once the report entries are on the dashboard, I can edit them to add titles, or drag and drop them into the best positions to give an at-a-glance impression of the data status.

I can also create complex *report expressions* to construct multi-column reports. For example, the following report expression creates a table that contains the minimum CO_2 emissions by manufacturer and model and was supplied by Ledion Bitincka, an architect at Splunk, in response to a question I submitted to the Splunk forum): `index=cdh5_vindex | stats min(cO2_g_km) AS minco2 BY  manufacturer, model | sort 20 minco2`. Thus, the original search expression `index=cdh5_vindex` is passed to a `stats` function that finds the minimum CO_2 emissions value from the column `cO2_g_km` in grams per kilometer and saves it to a new column name called `minco2`. The results placed in this column are grouped by manufacturer and model, just as you would group non-aggregated columns in SQL. Finally, the data is sorted and the top 20 rows are displayed, creating the report shown in Figure 11-12.

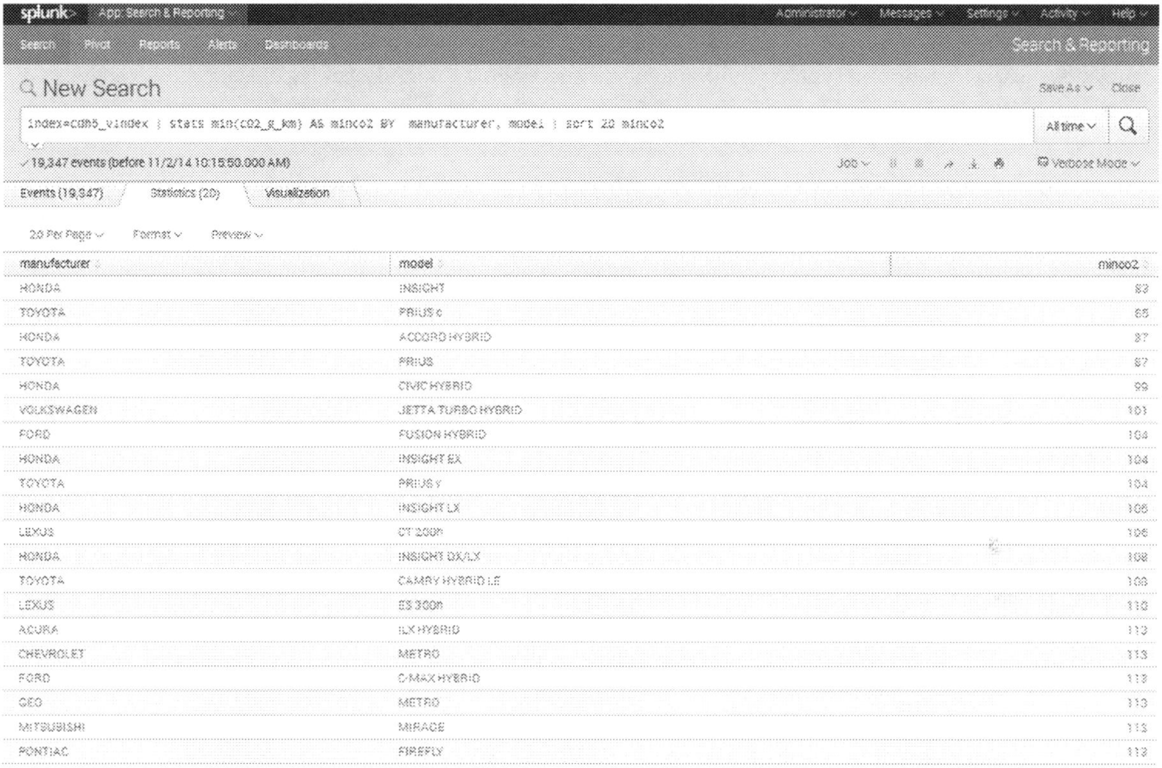

Figure 11-12. *Hunk multi-column report*

Note that because a stats function was used to create an aggregated column in the Figure 11-12 table, the resulting report was displayed in the Statistics tab. Also, the tabs at the top of the page contain a count of records used in creating the table. The example uses 19,347 records from the raw data for the Events tab and 20 records in the Statistics tab.

For additional ideas on what can be accomplished with Hunk, take a look at the splunk.com website, particularly the answers forum. You'll find ideas about and examples of generating complex reports. Remember that reports can be created from multiple Hadoop-based sources in Hunk, and that lookup tables can be used to enrich your data.

Potential Errors

I encountered some problems during the Hunk installation and use, probably because of configuration or installation mistakes. This section shows you what happened and how I fixed the errors.

For instance, I found the following error in a search log when a search failed:

```
[hadoop@hc2nn hunk]$ pwd
/usr/local/hunk
[hadoop@hc2nn hunk]$ find . -name search.log
./var/run/splunk/dispatch/1414868687.5/search.log

[cdh5] Error while running external process, return_code=255. See search.log for more info
[cdh5] RuntimeException - Failed to create a virtual index filesystem connection: java.net.
UnknownHostException: hc2nn. Advice: Verify that your vix.fs.default.name is correct and available.
```

I found the report search logs by running the Linux `find` command from the Hunk install directory /usr/local/ hunk. I knew that the search log file was named search.log, based on forum answers on `splunk.com`. The error was caused by the fact that the user running Hunk did not have access to the Splunk working directory /user/hadoop/ hunk/workdir.

A search log generated this second error, as well:

```
10-29-2014 19:14:07.385 ERROR ERP.cdh5 -  SplunkMR - Failed to create a virtual index filesystem
connection:  java.net.UnknownHostException: hc2nn.semtech-solutions.co.nz. Advice: Verify that your
vix.fs.default.name is correct and available.
```

The error was caused because HDFS was not running. The Hunk report had been run before the Hadoop servers completed their startup sequence.

This error, again from the search log, seems to be an indication of another underlying problem:

```
10-30-2014 18:22:55.398 ERROR SearchOperator:stdin - Cannot consume data with unset stream_type
10-30-2014 18:22:55.453 ERROR ExternalResultProvider - Error in 'SearchOperator:stdin': Cannot
consume data with unset stream_type
```

In this case, I had incorrectly created the configuration for my CSV file processing in the configuration files. Although from Splunk forum entries I understand that it should not be necessary, adding entries to the files props. conf and transforms.conf solved this problem.

I initially tried running Hunk from a remote Linux server—remote from the Hadoop cluster that it was trying to connect to. The following test uses the HDFS file system `ls` command to examine the / directory on the cluster via the HDFS-based URI `hdfs://hc2nn:8020/`:

```
[hadoop@hc1r1m1 ~]$ hdfs dfs -ls hdfs://hc2nn:8020/

ls: Call From hc1r1m1/192.168.1.104 to hc2nn:8020 failed on connection exception: java.net.
ConnectException: Connection refused; For more details see:  http://wiki.apache.org/hadoop/
ConnectionRefused
```

The access failed, so before attempting to use Hunk, it is important to check that the Hunk Linux user on the server where it is installed has access to Hadoop.

The following errors were displayed when Hunk was started in the Linux command window:

```
Checking conf files for problems...
                Invalid key in stanza [source::/data/hunk/rdbms/...] in /usr/local/hunk/etc/system/
local/props.conf, line 3: DELIMS
(value:  ", ")
                Invalid key in stanza [source::/data/hunk/rdbms/...] in /usr/local/hunk/etc/system/
local/props.conf, line 4: FIELDS

(value:  f1,f2,f3,f4,f5,f6,f7,f8,f9,f10,f11,f12,f13,f14)
                Your indexes and inputs configurations are not internally consistent. For more
information, run 'splunk btool check --

debug'   Done

All preliminary checks passed.
```

These errors were caused by incorrect configuration file entries while I was learning to use Hunk. I changed the configuration file entries and restarted Hunk to solve this problem.

If you encounter any of these or additional errors, another good resource to consult is the answers section of the splunk.com website, including the support menu options. Try to be a good Splunk citizen by adding as much detail as possible to any questions or answers that you might post on the forum. If you find a solution to your problem, then post that solution to help future users.

Talend Reports

Expanding on the big data ETL work discussed in Chapter 10, it's time to examine the reporting capabilities of the Talend Enterprise big data product—specifically, the profiling functionality. With Talend, you can check the quality of Hive-based data and build reports from table-based data.

Installing Talend

To download the Talend Enterprise big data application, I go to the URL www.talend.com/download, then select the Big Data tab and click the Download Free Trial button. I need to enter my details, so I can't simply execute the Linux wget command from the Linux command line, as in previous download examples in this book. The package download is 2 GB, so it will take some time.

The license that was automatically emailed to me does not allow me to access the profiling function. I need to request a different license from Talend to "unlock" the profiling function on the user interface. (You can contact Talend via www.talend.com/contact to request similar access.) The Sales Solutions Group Manager at Talend kindly supplied the license, while others at Talend offered help and documentation so I could develop the example I will present here.

I install the Talend software on the Centos Linux host hc1nn, using that machine as a client to access the Centos 6 Cloudera CDH5 Hadoop cluster whose name node resides on the server hc2nn. To unpack the software, I use the root account, accessing it via the Linux su (switch user) command:

```
[hadoop@hc1nn ~]$ su -
```

I move to the Linux hadoop account Downloads directory, where the package was downloaded, and I examine the downloaded file using the Linux ls command to create a long listing:

```
[root@hc1nn ~]# cd /home/hadoop/Downloads
[root@hc1nn Downloads]$ ls -lh
-rw-r--r-- 1 hadoop hadoop 2.0G Nov  3 18:01 Talend-Tools-Installer-r118616-V5.5.1-installer.zip
```

The Talend release file is a zipped archive; I unpack it with the Linux unzip command:

```
[root@hc1nn Downloads]# unzip Talend-Tools-Installer-r118616-V5.5.1-installer.zip
```

The following commands display the unpacked directory via a Linux ls command. The Linux cd command then moves into the unpacked Talend software directory TalendTools-5.5.1-cdrom. After that, the Linux ls command again provides a long list of the contents of the unpacked software:

```
[hadoop@hc1nn Downloads]$ ls -ld TalendTools-5.5.1-cdrom
drwxr-xr-x 2 root root 4096 Jun 18 11:25 TalendTools-5.5.1-cdrom

[root@hc1nn Downloads]# cd TalendTools-5.5.1-cdrom

[hadoop@hc1nn TalendTools-5.5.1-cdrom]$ ls -l
total 2081688
-rwxr-xr-x 1 root root 2095344252 Jun 18 11:25 dist
-rwxr-xr-x 1 root root    6171835 Jun 18 11:25 Talend-Tools-Installer-r118616-V5.5.1-linux64-
installer.run
-rwxr-xr-x 1 root root    6003334 Jun 18 11:25 Talend-Tools-Installer-r118616-V5.5.1-linux-
installer.run
-rw-r--r-- 1 root root   18288640 Jun 18 11:25 Talend-Tools-Installer-r118616-V5.5.1-osx-
installer.app.tar
-rwxr-xr-x 1 root root    5829599 Jun 18 11:24 Talend-Tools-Installer-r118616-V5.5.1-windows-
installer.exe
```

The two files with "linux" in their names are used to install Talend on Linux-based hosts. The other files are used to install Talend on the Windows and Mac OSX operating systems. The file named "dist" is the largest file in the release and contains the actual distributed software for the installation. Given that I am installing onto a 32-bit Linux host, I use the Linux-based file that does not have 64 in its name.

Before I install the Talend software, however, I need to install the Oracle Sun version of the JavaSDK. The Talend installation will fail if this JavaSDK isn't available. Because I generally use the Java OpenJDK, I install this Java release under /usr/local. I download the latest available JavaJDK from the URL http://www.oracle.com/technetwork/java/javase/downloads/jdk8-downloads-2133151.html to Centos Linux host hc1nn as the Linux hadoop user.

The following commands show the Downloads directory within the Linux hadoop account's home directory via the Linux pwd command. They also show a long file listing of the downloaded JavaJDK file using the Linux ls command:

```
[hadoop@hc1nn Downloads]$ pwd
/home/hadoop/Downloads

[hadoop@hc1nn Downloads]$ ls -l jdk-8u25-linux*
-rw-rw-r-- 1 hadoop hadoop 162406890 Nov  3 19:23 jdk-8u25-linux-i586.tar.gz
```

This is a compressed tar archive file (it has a file type of .tar.gz), so I uncompress it using the Linux gunzip command. This produces a tar archive with a file type of .tar. I then use the Linux tar command to unpack it, using the option xvf, where x means extract, v means verbose, and f allows the tar file name to be specified. Then, a long file list shows that the unpacked software resides in a directory called jdk1.8.0_25:

```
[hadoop@hc1nn Downloads]$ gunzip jdk-8u25-linux-i586.tar.gz

[hadoop@hc1nn Downloads]$ tar xvf  jdk-8u25-linux-i586.tar

[hadoop@hc1nn Downloads]$ ls -ld jdk1.8.0_25

drwxr-xr-x 8 hadoop hadoop 4096 Sep 18 11:33 jdk1.8.0_25
```

Having extracted the JavaJDK, I use the Linux su (switch user) command to change to the root account and move the software to /usr/local/. The Linux mv (move) command moves the jdk1.8.0_25 directory. The Linux cd (change directory) command moves to the /usr/local/ directory. The Linux chown (change owner) command recursively changes the ownership of the Java release to the Linux hadoop account. Finally, the export command sets the JAVA_HOME variable to the path of this new Java release:

```
[hadoop@hc1nn Downloads]$ su -
Password:
 [root@hc1nn ~]# cd /home/hadoop/Downloads

[root@hc1nn Downloads]# mv jdk1.8.0_25 /usr/local
[root@hc1nn Downloads]# cd /usr/local
[root@hc1nn local]# chown -R hadoop:hadoop jdk1.8.0_25
[root@hc1nn local]# export JAVA_HOME=/usr/local/jdk1.8.0_25
```

It is this Oracle Sun JavaJDK that the Talend release requires, so with the JAVA_HOME variable set I can attempt to install the software. As the root user, I change the directory to the Downloads directory and then change the directory to the directory containing the unpacked Talend software. After that, I run the non-64 bit linux.run file. The ./ prepended to the name specifies that the installation file will be sourced from the current directory:

```
[root@hc1nn local]# cd /home/hadoop/Downloads
[root@hc1nn Downloads]# cd TalendTools-5.5.1-cdrom

[root@hc1nn TalendTools-5.5.1-cdrom]#
./Talend-Tools-Installer-r118616-V5.5.1-linux-installer.run
```

The installation is simple. I just accept all defaults and install only the client application, not the server, as the server is not required for this example. As mentioned earlier, I also need a platform-enabled license file supplied by Talend to enable the profiling function. This file is specified during the installation. I install the software to the default path /opt.

The Talend application also needs an rpm build packaging component; the yum installation command, run as root, installs the necessary software, as follows:

```
[root@hc1nn TalendTools-5.5.1-cdrom]# yum install rpm-build
```

Finally, the Talend client software can be started from the studio subdirectory of the Talend software installation directory /opt/TalendTools-5.5.1. The 5.5.1 string shows the version of the Talend Enterprise software that has been installed. The list obtained using the Linux ls command shows that there are numerous Linux .sh and windows .ini files available for starting Talend. Given that I have installed Talend on a 32-bit Linux host, however, the .sh file has "linux" in its name but lacks the "64," which denotes a 64-bit architecture. The ./ in the final command indicates that the file should be run from the current directory. The "and" character (&) denotes the command should be run in the background:

```
[root@hc1nn TalendTools-5.5.1-cdrom]# cd /opt/TalendTools-5.5.1/studio/

[root@hc1nn studio]# ls *.sh *.ini
```

```
commandline-linux.sh                     Talend-Studio-linux-gtk-x86.sh
commandline-linux_x86_64.sh              Talend-Studio-solaris-gtk.ini
commandline-mac.sh                       Talend-Studio-solaris-gtk-x86.ini
Talend-Studio-linux-gtk-ppc.ini          Talend-Studio-win32-wpf.ini
Talend-Studio-linux-gtk-x86_64.ini       Talend-Studio-win32-x86.ini
Talend-Studio-linux-gtk-x86.ini          Talend-Studio-win-x86_64.ini

[root@hc1nn studio]#  ./Talend-Studio-linux-gtk-x86.sh &
```

Some basic connection details need to be specified, then Talend will be available for use. The next section will take care of these details.

Running Talend

If multiple Talend clients were being installed, it would have made sense to install the Talend server and have the Talend clients connect to that server. The server would then provide access to a storage repository, which would be based on SVN. In this way, work created via one client could be shared among multiple users. But because only a single client was installed and no server, the connection specified is to the local host file system. Figures 11-13 and 11-14 show that the connection details are specified as local and a workspace directory is specified under the studio client directory under /opt.

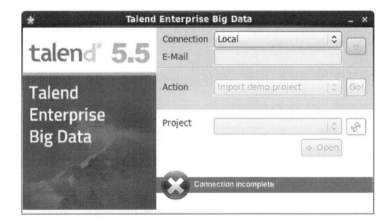

Figure 11-13. *Talend local connection*

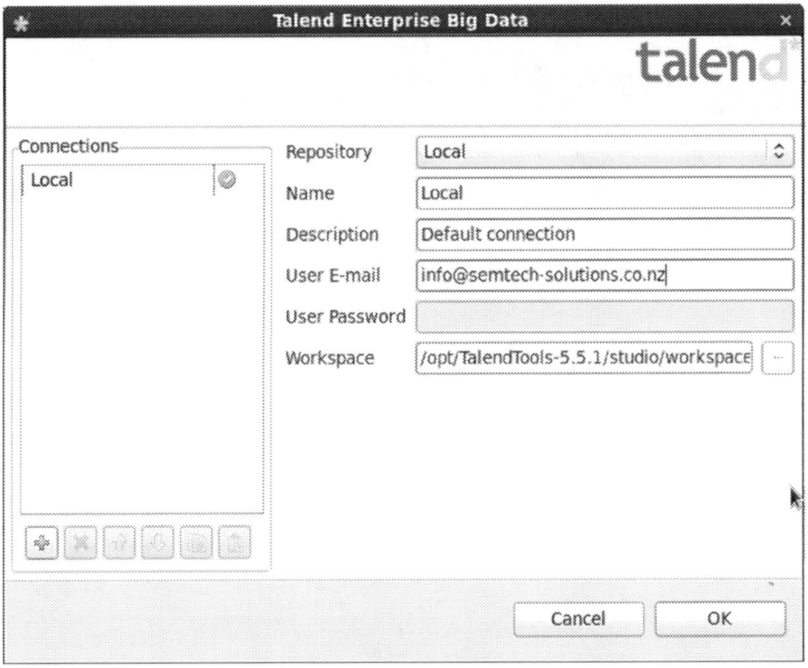

Figure 11-14. *Details of Talend local connection*

Although it's not shown as currently active in Figure 11-13, the Import Demo Project option is enabled when the connection details are specified. I then click the Go! button to install the big data code I use for this example. Figure 11-15 shows the installed Talend client running; note that I have opened one of the job examples and have attempted to connect to Cloudera CDH5. At this point, I am prompted to install any missing Talend or third-party libraries that are required. Indeed, the yellow banner in Figure 11-15 is prompting a library installation. I install what is required now to avoid possibility of errors later. (See also the "Potential Errors" section later.) Also, note that I have placed a red box around the top-right menu option that will be used to change the interface perspective.

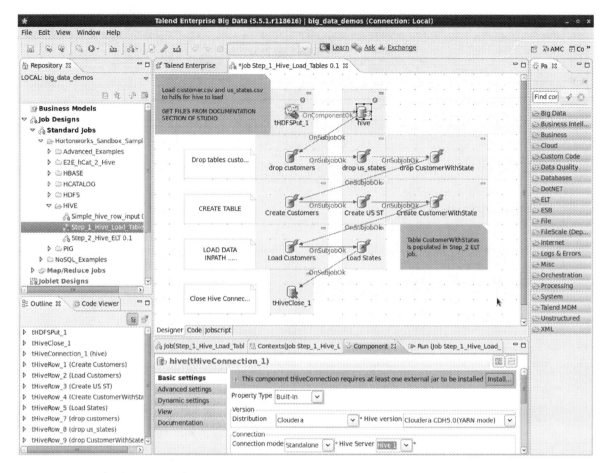

Figure 11-15. *Talend Enterprise client*

Figure 11-16 shows an example of the necessary libraries being installed. This process may be slow, but I am patient until it is 100 percent completed. I won't describe the integration perspective shown in the Figure 11-16, as it was described in Chapter 10.

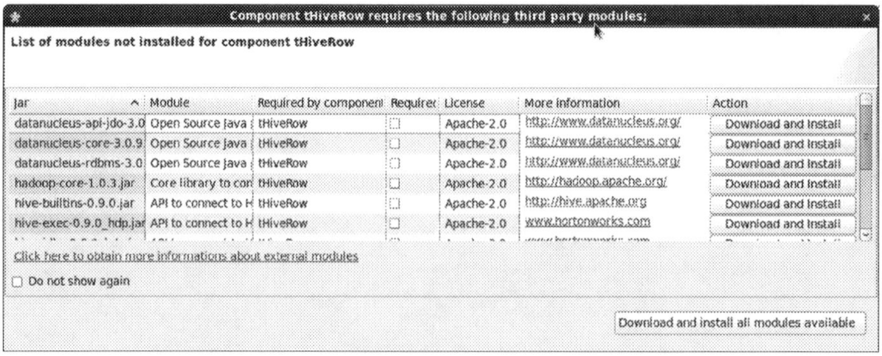

Figure 11-16. *Talend libraries installation*

It is important at this point that I check to see that the correct license is installed so that data quality reports can be generated. From the client user interface, I select the Help option, then select the About License option to bring up the window that is depicted in Figure 11-17. Notice also that the term "Talend Platform" is displayed in the figure. This phrase is important, as it denotes that the correct type of license is being used. (To obtain the necessary access and licenses, contact Talend at www.talend.com/contact.)

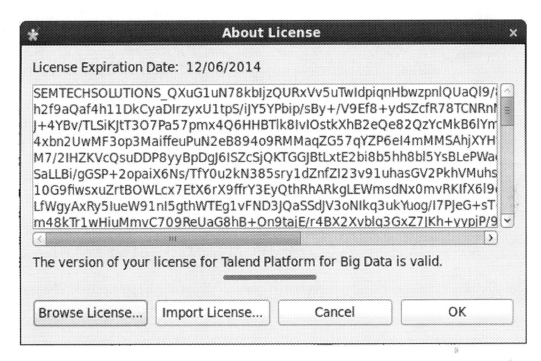

Figure 11-17. Details on required Talend license

The profiling perspective can now be accessed by clicking the red-outlined drop-down menu in the top right of Figure 11-15. I select Other. Next, I choose the "profiling perspective" option, as shown in Figure 11-18.

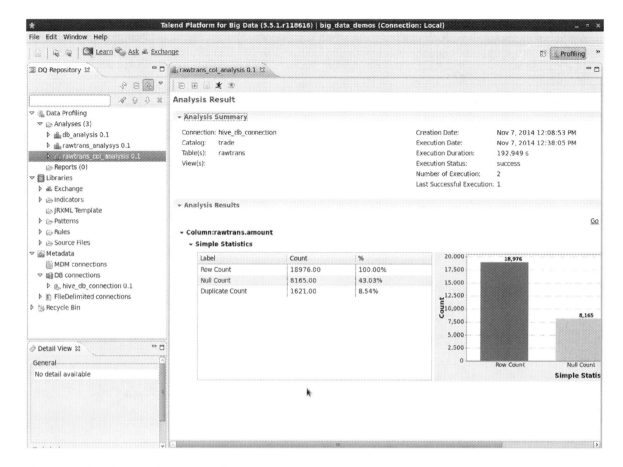

Figure 11-18. *Talend's profiling perspective*

The Repository pane on the left of the screen shown in Figure 11-18 has a list of previously created analysis reports and metadata database connections. Earlier, I had created a Hive-based Cloudera CDH5 connection named hive_db_connection 0.1, as you can see in the list under DB Connections. I also opened the Hive rawtrans table column analysis report named rawtrans_col_analysis 0.1. The items are expanded in Figure 11-18 to familiarize you with the display.

To run Talend profiling reports against a CDH5-based Hive data warehouse, I need to know a number of properties about the Hive installation: which host it is running on, what port number to use to connect to it, the Linux-based user name of the account to use, the password for that account, and the version of Hive in use. I know that Hive is installed on my cluster on the server hc2nn, and that the account used will be called hadoop. I also know the password for that account, plus I know that the Hive version that I am using for Talend is version 2. That means that the only property I need to determine is the metastore port number. Given that I know all logs will be stored under /var/log for Cloudera CDH5 servers, I obtain that information as follows:

```
[hadoop@hc2nn hive]$ pwd
/var/log/hive

[hadoop@hc2nn hive]$ ls -l
total 3828
drwx------ 2 hive hive    4096 Aug 31 12:14 audit
-rw-r--r-- 1 hive hive 2116446 Nov  8 09:58 hadoop-cmf-hive-HIVEMETASTORE-hc2nn.semtech-solutions.
co.nz.log.out
-rw-r--r-- 1 hive hive 1788700 Nov  8 09:58 hadoop-cmf-hive-HIVESERVER2-hc2nn.semtech-solutions.
co.nz.log.out

[hadoop@hc2nn hive]$ grep ThriftCLIService hadoop-cmf-hive-HIVESERVER2-*.log.out | grep listen |
tail -2

2014-11-08 09:49:47,269 INFO org.apache.hive.service.cli.thrift.ThriftCLIService:
ThriftBinaryCLIService listening on 0.0.0.0/0.0.0.0:10000
2014-11-08 09:58:58,608 INFO org.apache.hive.service.cli.thrift.ThriftCLIService:
ThriftBinaryCLIService listening on 0.0.0.0/0.0.0.0:10000
```

The first command shows, via a Linux pwd (print working directory) command, that I am in the directory /var/log/hive. (Note: use the cd command to move to that directory, if necessary.) Then, using the Linux ls command with the -l option to provide a long listing, I check to see which log files exist in this Hive log directory. Finally, I use the Linux grep command to search the HIVESERVER2-based log file for the string ThriftCLIService. I pipe (|) the output of this search to another grep command, which searches the ouput further for lines that also contain the text "listen." Finally, I limit the output to the last two lines via the Linux tail command with a parameter of -2. The output contains the port number that I need at the end of the line. Then, 10000 is the default port number that will be used in the Talend Hive connection for this section.

So, now I am ready to create a Hive database connection. I can do this by right-clicking the DB Connections option in the Repository pane. Then, I select Create DB Connection to open a form that offers a two-step process for creating the connection.

The first section requests the name, purpose, description, and status of the connection. Take care to make the name meaningful. The second step (shown in Figure 11-19) gives the actual connection details. That is, the database type is set to Hive and the server/port are defined as hc2nn/10000, as previously determined. The Linux account login for the CentOS host hc2nn is set to hadoop, along with its password. The Hive version is set to Hive2, while the Hadoop version and instance are set to match the Hadoop cluster being used, Cloudera/CDH5. Finally, the jdbc string, the Java-based method that Talend will use to connect to Hive, is set to a connection string that uses the hostname, port, and Hive version.

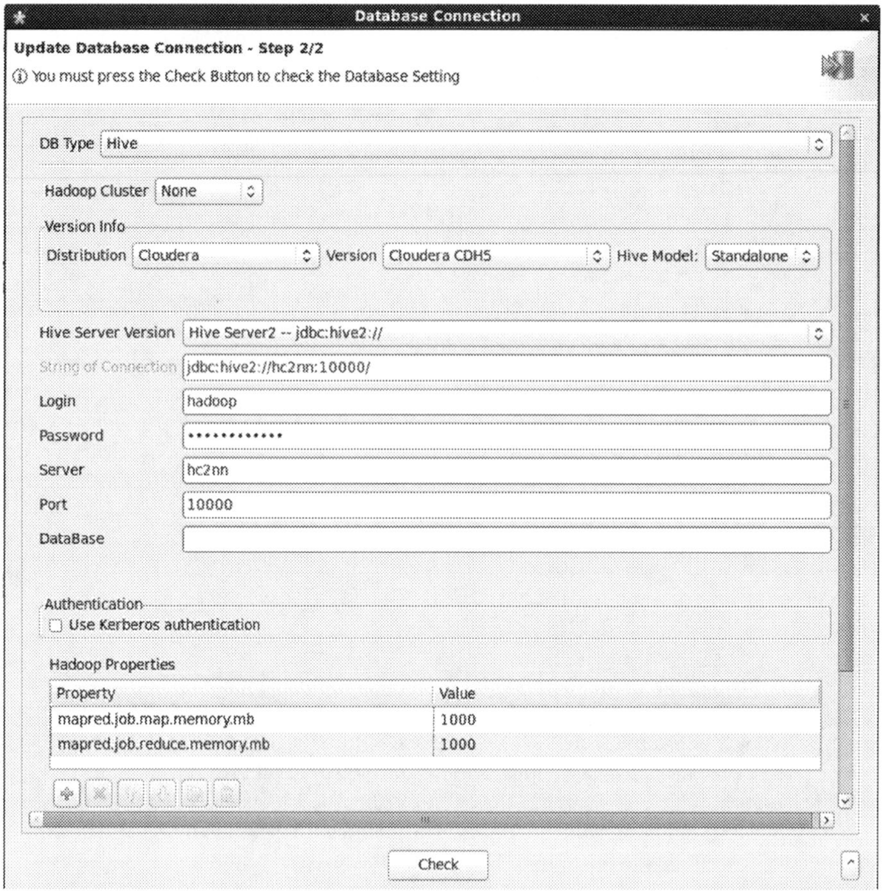

Figure 11-19. *Talend-Hive database connection*

Some variables have been added to set the memory used by the Talend Map Reduce jobs in terms of map and reduce functions. A value of 1000 MB is set for each function via the variables mapred.job.map.memory.mb and mapred.job.reduce.memory.mb. To check this connection before saving it, I click the Check button. In case of problems, I first ensure that the cluster is functioning without error, that HDFS is accessible, and that Hive can be accessed and have Hive QL scripts run against it. (You also might use the Hue interface for this.) The connection works, so I save it. It subsequently will appear in the Repository pane.

By expanding the objects under the Hive database connection in the Repository pane, I can view a table to the column level, as shown in Figure 11-20.

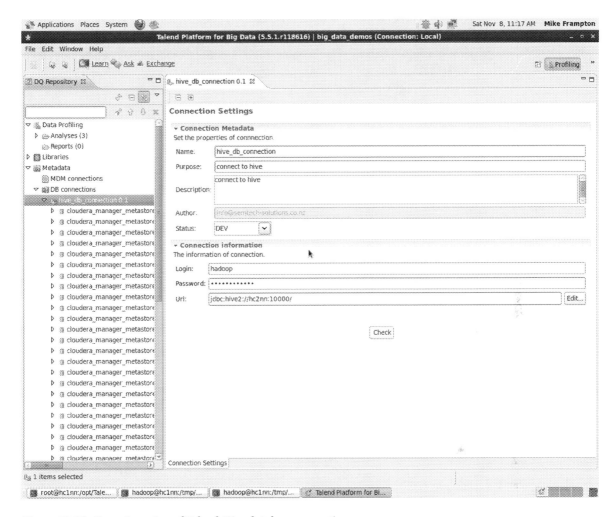

Figure 11-20. *Repository view of Talend-Hive database connection*

At this point, I can create a range of reports based on the underlying Hive table data.

Generating Reports

By using the Splunk/Hunk product at the start of this chapter, I was able to quickly create some reports and develop a dashboard based on HDFS data. When I create Talend reports based on Hive table data, I can start to think about the quality of the data that's residing on HDFS and Hive.

As you remember from Chapter 9, you can create Hive external tables to represent HDFS data. In this section, I create reports that represent the column data in the Hive rawtrans table of the trade information database. The content of the data in that table is not relevant; it is the functionality of the Talend data-quality reports that I concentrate on here.

To create the reports that this section will use, I first need to create two rules for data quality under Libraries, then Rules, then SQL in the Repository pane, and one regular expression pattern by going to Libraries, then Patterns, then Regex, then Date. The regular expression rule for date is copied from a similar pre-existing rule in the same location, called date MM DD YYYY. I simply right-click it and select duplicate.

Shown in Figure 11-21, the new Hive-based rule is now called "Hive Date DD MM YYYY" and will be used to check Hive table-based dates. Note that the connection type has been set to Hive. The actual regular expression rule, which appears at the bottom right, basically checks a date column to confirm that it contains a date of the form that's either 01/12/2014 or 01-12-2014.

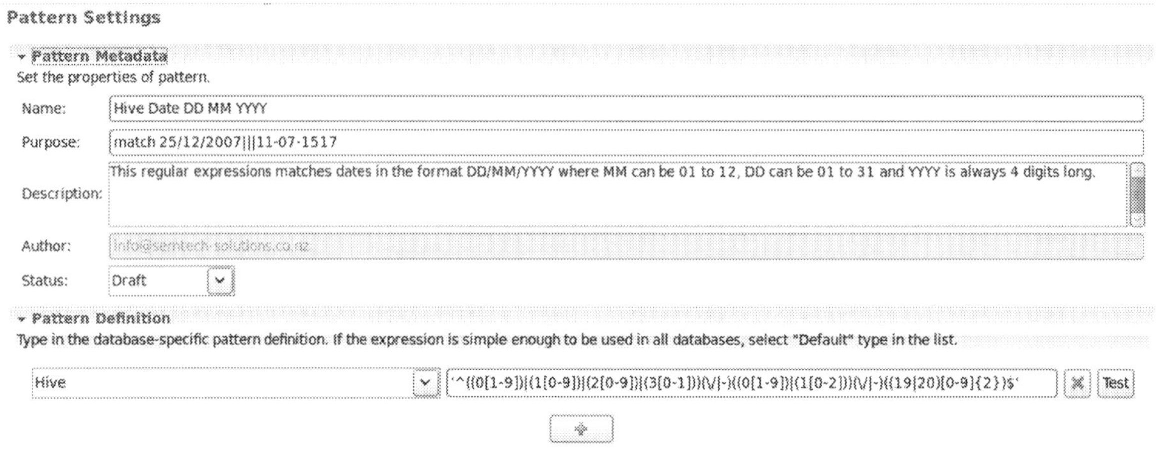

Figure 11-21. *Regular expression rule for dates*

You can create SQL-based data-quality rules as well. For example, to do so, I right-click the SQL folder under Libraries, then choose Rules within the Repository pane and select the option New Business Rule. Figures 11-22 and 11-23 show two rules I created to test column values.

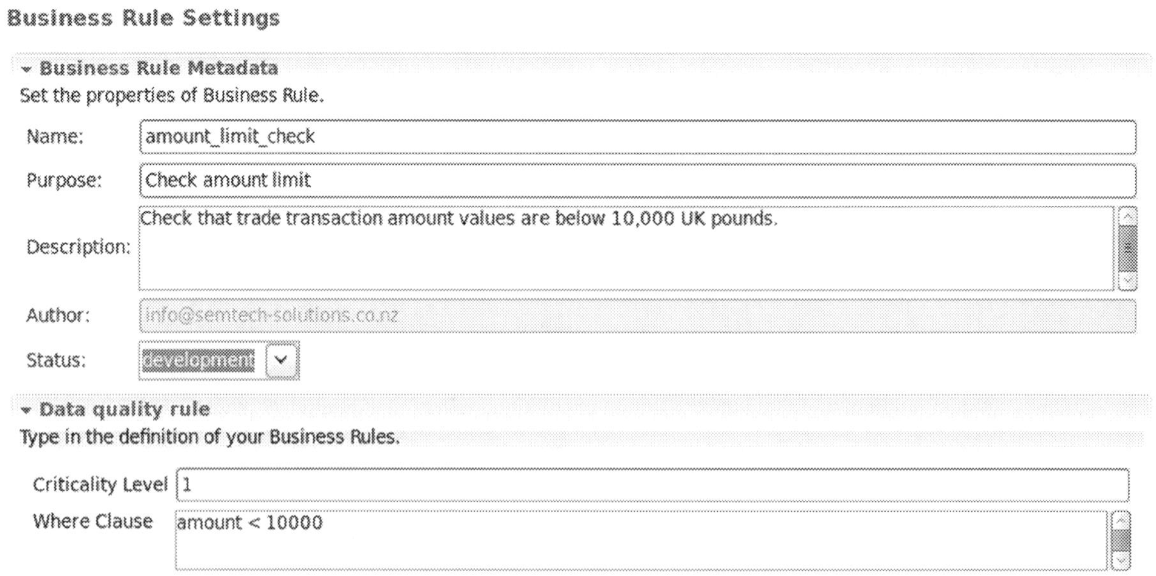

Figure 11-22. *SQL data-quality rule for "amount" column*

For instance, Figure 11-22 tests the size of the trade.rawtrans table "amount" column by checking that no transaction values exceed 10,000 English pounds sterling. The text in the Where clause field is added to the Hive QL that is generated by Talend and is run against Hive to create the report's content. In this example, I have not amended other fields except to add the rule's name, purpose, and description.

Figure 11-23 is a similar SQL-based data-quality rule, except that in its Where clause field it uses the Hive QL length function to check the length of the rawtrans table's Supplier field data in the trade database. Of course, this field is a string, so this check ensures that a maximum length for the supplier name is adhered to. Checks like this are useful during data-migration exercises.

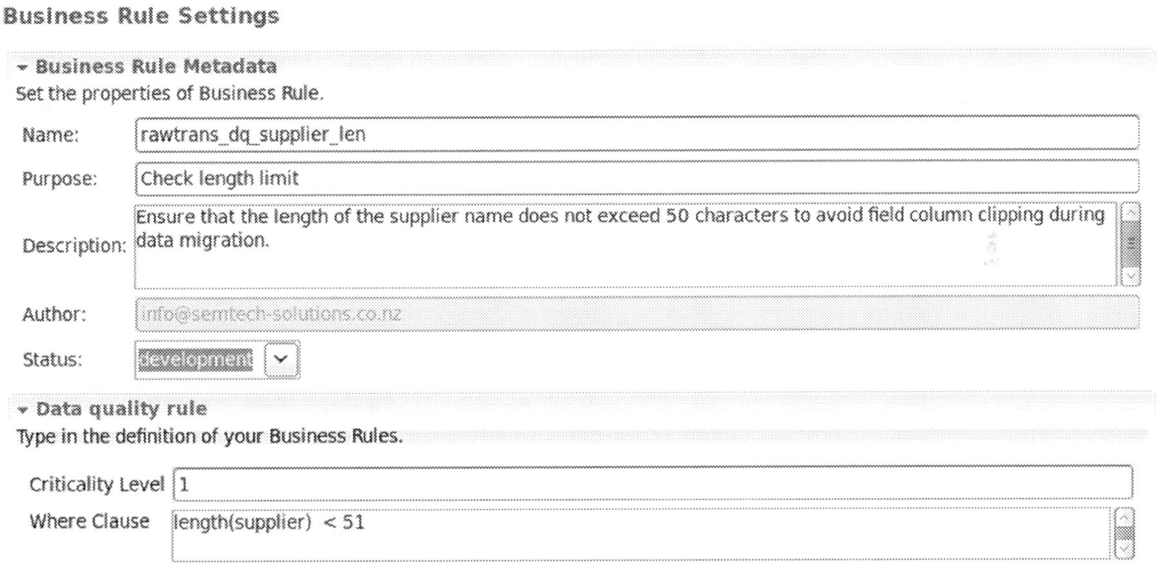

Business Rule Settings

▾ Business Rule Metadata
Set the properties of Business Rule.

Name:	rawtrans_dq_supplier_len
Purpose:	Check length limit
Description:	Ensure that the length of the supplier name does not exceed 50 characters to avoid field column clipping during data migration.
Author:	info@semtech-solutions.co.nz
Status:	development ⌄

▾ Data quality rule
Type in the definition of your Business Rules.

Criticality Level	1
Where Clause	length(supplier) < 51

Figure 11-23. *SQL data-quality rule for "Supplier" column*

Now that the rules have been created, it is possible to create some Talend reports on data quality in the trade. rawtrans Hive table. I concentrate on column-based reports and create single- and multi-column reports, as well as reports based on SQL and regular expressions. (Note: you can create new reports by right-clicking the Analysis folder in the Repository pane and selecting the New Analysis option.) Each type of report has the set of control buttons shown in Figure 11-24.

Figure 11-24. *Control buttons for reports*

The first two buttons shown in Figure 11-24 contract and expand the report display to either hide or show charts and/or details. The third button saves any changes, while the fourth is used to run the report. The final "eye" icon refreshes the report charts.

Single-Column Reports

I create a single-column analysis report by right-clicking the Analysis folder under Data Profiling in the Repository pane. Next, I choose the report folder and type of column analysis in the pop-up window that is shown in Figure 11-25.

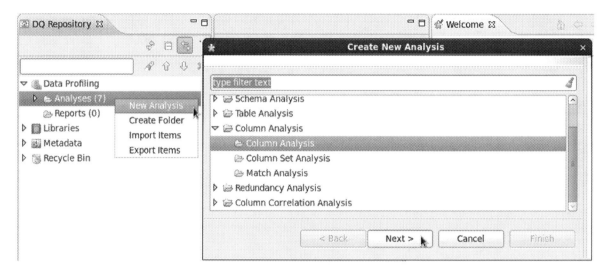

Figure 11-25. *Choosing the type of report to be created*

I then click Next and enter a report name and a report description, as shown in Figure 11-26.

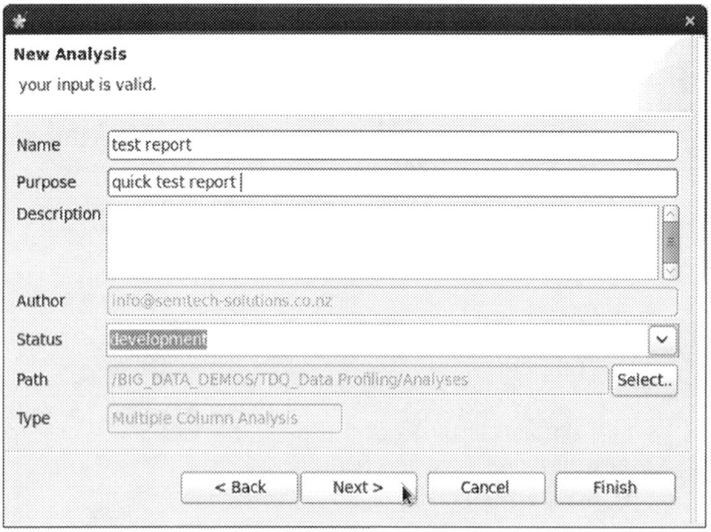

Figure 11-26. *Describing and naming the report*

I click Next again to choose the database table column for the report. I expand the Hive DB connection until the appropriate database table column can be selected, and then I select Finish, as shown in Figure 11-27.

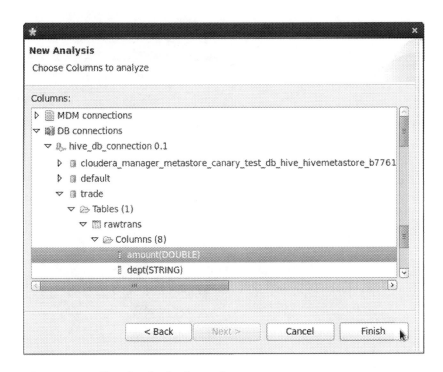

Figure 11-27. *Choosing the database column*

Figure 11-28 shows the Analysis Results tab of a single-column report drawn from the rawtrans "amount" column. It shows both table and bar graph, with the total and null counts for the column. It also shows the spread of values as the duplicate, distinct, and unique counts for this column.

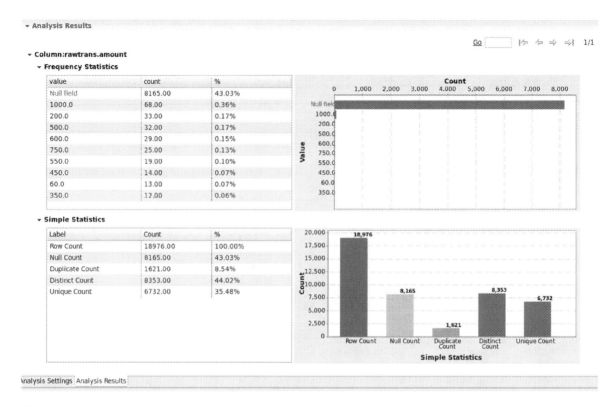

Figure 11-28. *Results for a single-column report*

Multi-Column Reports

In the same way that a single-column report is created, multi-column reports can be generated. When I select the columns on which the report will be drawn, I press the Control key and select "multiple columns."

The report shown in Figure 11-29 was created from four columns in the trade.rawtrans table: "amount", "department" (dept), "supplier", and "export area" (exparea). The Data mining Type parameter tells Talend what type of data it is examining. For the "amount" column, I set it to Interval because that column contains numeric values; I set it to Unstructured Text for the other columns, as they are strings. The simple statistics displayed in Figure 11-29 include the column row count and counts for distinct, duplicate, and unique values. Clicking the Run icon then runs the report and produces the bar graph shown on the right.

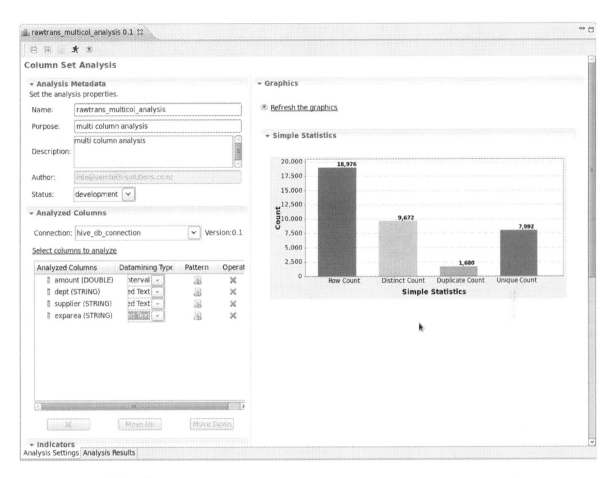

Figure 11-29. *Muliple-column report*

These reports are fine for a simple check of the table column data, but what about generating and using more complex rules for data quality that check data values and attributes? Well, that is where SQL and regular expression-based rules can help.

Reports Based on SQL Rules

Reports on data quality are created when the options Table Analysis and then Business Rule Analysis are chosen when creating a report. For example, I click Next and enter a report name in the form. I click Next again and expand the database Hive connection until I can select the appropriate table. I click Next, and I select the SQL-based rule that was created earlier. Finally, I click Finish to generate the report. The resulting report, shown in Figure 11-30, uses the SQL-based rule that states amount < 10000 and shows that more that 93 percent of the data values in the "amount" column fail this data-quality rule.

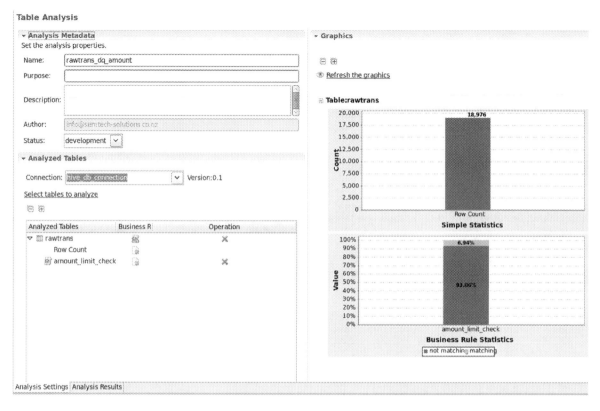

Figure 11-30. *SQL-rule results for data-quality report on "amount" column*

The SQL-rule statistics in Figure 11-30 show failures in red and passes in green. The report shown in Figure 11-31 checks the length of the supplier string column and displays a failure rate of around 29 percent. Thus, these simple reports demonstrate that there are improvements needed in the quality of data in this table.

Figure 11-31. *SQL-rule results for data-quality report on "supplier" column*

These reports check the content and data ranges of tabular data, but what about the actual structure or format of the data? Regular expressions are the tool for this job.

Reports Based on Regular Expressions

This section uses the date-based regular expression rule that was created previously. I begin the example by creating a new report in the same manner as I did for the single-column report. In the report's Analyzed Columns section (Figure 11-32), I click the Pattern icon next to the column to be checked—in this case, it is the "paydate" column, which is a date-based string column. This invokes the Pattern Selector window. There, I navigate to Regex, then to date, and to "Hive Date DD MM YYYY", which is the regular expression-based rule that I created earlier. I select that rule, and click OK. Now, I click the Run icon to populate the report's graphs, as shown in Figure 11-32.

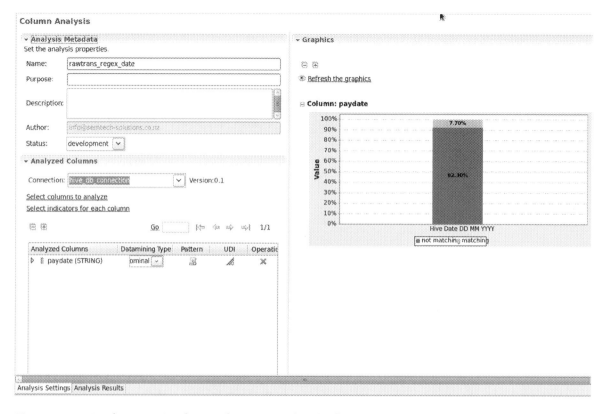

Figure 11-32. *Regular-expression data-quality report on "paydate" column*

The Figure 11-32 report shows the output and the fact that over 92 percent of the data in the "paydate" column has failed the basic date-format test, probably because it is null. The content here is not important, but the usefulness of these reports is. If you take care when selecting the rules for generating a data-quality report, you can produce a set of reports like this that have great importance, especially when you are attempting to ensure the quality of HDFS- and Hive-based data.

Potential Errors

I did encounter a few errors while using Talend's profiling. Generally these were not problems with Talend itself but, rather, were configuration issues. By examining my solutions, you will be able to either avoid these errors or use them to devise your own solutions.

I received the following error, followed by a message stating that I needed to install the library Zql.tar:

```
zql/ParseException
```

This issue was caused by my not installing all third-party libraries when being prompted to do so. The solution is simple: just click Finish and accept the licensing, then patiently wait for the libraries to install.

```
Two errors were caused by my setting up the Hive connection incorrectly. Specifically, I received
the following error:Failed to run analysis: rawtrans_analysys
Error message:
Error while processing statement: Failed: Execution Error, return code 1 from
org.apache.hadoop.hive.ql.exec.mr.MapRedTask
```

and the following error was in the Hive log file /var/log/hive/hadoop-cmf-hive-HIVEMETASTORE-hc2nn.semtech-solutions.co.nz.log.out:

```
assuming we are not on mysql: ERROR: syntax error at or near "@@"
```

The port number should have been set to 10000 for the hiveserver2 address. I used the value 9083, which was the port value defined in the `property hive.metatstore.uris` in the file hive-site.xml under the directory /etc/hive/conf.cloudera.hive.

```
There was the following error regarding an RPM component:There was an error creating the RPM file:
Could not find valid RPM application:
RPM-building tools are not available on the system
```

The error occurred because an RPM build component was missing from the Centos Linux host on which Talend was installed. The solution was to install the component using the `yum` command `install`.

Finally, this short error occurred while I was installing the Talend client software and it implied that the Talend install file called "dist" was corrupted:

```
Unable to execute validation program
```

I don't know how it happened, but I solved the problem by removing the Talend software release directory and extracting the tar archive a second time.

Summary

Relational database systems encounter data-quality problems, and they use data-quality rules to solve those problems. Hadoop Hive has the potential to hold an extremely large amount of data—a great deal larger than traditional relational database systems and at a lower unit cost. As the data volume rises, however, so does the potential for encountering data-quality issues.

Tools like Talend and the reports that it can produce offer the ability to connect to Hive and, via external tables, to HDFS-based data. Talend can run user-defined data quality checks against that Hive data. The examples presented here offer a small taste of the functionality that is available. Likewise, Splunk/Hunk has the potential for generating reports and creating dashboards to monitor data. After working through the Splunk/Hunk and Talend application examples provided in this chapter, you might consider investigating the Tableau and Pentaho applications for big data as well.

You now have the tools to begin creating your own Hadoop-based systems. As you go forward, remember to check the Apache and tool supplier websites. Consult their forums and ask questions if you encounter problems. As you find your own solutions, post them as well, so as to help other members of the Hadoop community.

Index

▓ C

▓ D, E

■ W, X

workflow.txt file, 141

■ Y

Yet another resource
        negotiator (YARN)
    component package installation, 40
    configuration process, 40
    core-site.xml file, 41
    Data Node machines, 45
    file mapred-site.xml file, 42

file yarn-site.xml file, 42
hdfs-site.xml file, 41
logging and history data, 45
ls command, 41, 45
mapred-site.xml file, 44
recursive switch (-R), 45
yarn-site.xml file, 43–44
yum commands, 40

■ Z

ZeroMQ, 178
ZooKeeper server, 182

Get the eBook for only $10!

Now you can take the weightless companion with you anywhere, anytime. Your purchase of this book entitles you to 3 electronic versions for only $10.

This Apress title will prove so indispensible that you'll want to carry it with you everywhere, which is why we are offering the eBook in 3 formats for only $10 if you have already purchased the print book.

Convenient and fully searchable, the PDF version enables you to easily find and copy code—or perform examples by quickly toggling between instructions and applications. The MOBI format is ideal for your Kindle, while the ePUB can be utilized on a variety of mobile devices.

Go to www.apress.com/promo/tendollars to purchase your companion eBook.